The Handbook of Environmental Chemistry

Volume 3 Part A

Edited by O. Hutzinger

Anthropogenic Compounds

With Contributions by
R. Anliker, G.C. Butler, E.A. Clarke, U. Förstner,
W. Funke, C. Hyslop, G. Kaiser,
C. Rappe, J. Russow, G. Tölg, M. Zander, V. Zitko

With 61 Figures

Springer-Verlag Berlin Heidelberg GmbH 1980

Professor Dr. Otto Hutzinger

Laboratory of Environmental and Toxicological Chemistry
University of Amsterdam, Nieuwe Achtergracht 166
Amsterdam, The Netherlands

ISBN 978-3-662-15998-9

Library of Congress Cataloging in Publication Data
Main entry under title: Anthropogenic compounds.
(The Handbook of environmental chemistry; v. 3, pt. A–).
Includes bibliographies and index.
1. Pollution – Environmental aspects. 2. Pollution – Toxicology. 3. Environmental chemistry.
I. Butler, Gordon Cecil, 1913–. II. Series: Handbook of environmental chemistry; v. 3, pt. A–.
QD31.H335 vol. 3, pt. A, etc. [QH545.A1] 80-16609
ISBN 978-3-662-15998-9 ISBN 978-3-540-38522-6 (eBook)
DOI 10.1007/978-3-540-38522-6

Originally published by Springer-Verlag Berlin Heidelberg New York in 1980
Softcover reprint of the hardcover 1st edition 1980

2152/3140-543210

Preface

Environmental Chemistry is a relatively young science. Interest in this subject, however, is growing very rapidly and, although no agreement has been reached as yet about the exact content and limits of this interdisciplinary discipline, there appears to be increasing interest in seeing environmental topics which are based on chemistry embodied in this subject. One of the first objectives of Environmental Chemistry must be the study of the environment and of natural chemical processes which occur in the environment. A major purpose of this series on Environmental Chemistry, therefore, is to present a reasonably uniform view of various aspects of the chemistry of the environment and chemical reactions occurring in the environment.

The industrial activities of man have given a new dimension to Environmental Chemistry. We have now synthesized and described over five million chemical compounds and chemical industry produces about hundred and fifty million tons of synthetic chemicals annually. We ship billions of tons of oil per year and through mining operations and other geophysical modifications, large quantities of inorganic and organic materials are released from their natural deposits. Cities and metropolitan areas of up to 15 million inhabitants produce large quantities of waste in relatively small and confined areas. Much of the chemical products and waste products of modern society are released into the environment either during production, storage, transport, use or ultimate disposal. These released materials participate in natural cycles and reactions and frequently lead to interference and disturbance of natural systems.

Environmental Chemistry is concerned with reactions in the environment. It is about distribution and equilibria between environmental compartments. It is about reactions, pathways, thermodynamics and kinetics. An important purpose of this Handbook is to aid understanding of the basic distribution and chemical reaction processes which occur in the environment.

Laws regulating toxic substances in various contries are designed to assess and control risk of chemicals to man and his environment. Science can contribute in two areas to this assessment; firstly in the area of toxicology and secondly in the area of chemical exposure. The available concentration ("environmental exposure concentration") depends on the fate of chemical compounds in the environment and thus their distribution and reaction behaviour in the environment. One very important contribution of Environmental

Chemistry to the above mentioned toxic substances laws is to develop laboratory test methods, or mathematical correlations and models, that predict the environmental fate of new chemical compounds. The third purpose of this Handbook is to help in the basic understanding and development of such test methods and models.

The last explicit purpose of the Handbook is to present, in concise form, the most important properties relating to environmental chemistry and hazard assessment for the most important series of chemical compounds.

At the moment three volumes of the Handbook are planned. Volume 1 deals with the natural environment and the biogeochemical cycles therein, including some background information such as energetics and ecology. Volume 2 is concerned with reactions and processess in the environment and deals with physical factors such as transport and adsorption, and chemical, photochemical and biochemical reactions in the environment, as well as some aspects of pharmacokinetics and metabolism within organisms. Volume 3 deals with anthropogenic compounds, their chemical backgrounds, production methods and information about their use, their environmental behaviour, analytical methodology and some important aspects of their toxic effects. The material for volume 1, 2 and 3 was each more than could easily be fitted into a single volume, and for this reason, as well as for the purpose of rapid publication of available manuscripts, all three volumes were divided in the parts A and B. Part A of all three volumes is now being published and the second part of each of these volumes should appear about six months thereafter. Publisher and editor hope to keep materials of the volumes one to three up to date and to extend coverage in the subject areas by publishing further parts in the future. Plans also exist for volumes dealing with different subject matter such as analysis, chemical technology and toxicology, and readers are encouraged to offer suggestions and advice as to future editions of "The Handbook of Environmental Chemistry".

Most chapters in the Handbook are written to a fairly advanced level and should be of interest to the graduate student and practising scientist. I also hope that the subject matter treated will be of interest to people outside chemistry and to scientists in industry as well as government and regulatory bodies. It would be very satisfying for me to see the books used as a basis for developing graduate courses in Environmental Chemistry.

Due to the breadth of the subject matter, it was not easy to edit this Handbook. Specialists had to be found in quite different areas of science who were willing to contribute a chapter within the prescribed schedule. It is with great satisfaction that I thank all 52 authors from 8 contries for their understanding and for devoting their time to this effort. Special thanks are due to Dr. F. Boschke of Springer for his advice and discussions throughout all stages of preparation of the Handbook. Mrs. A. Heinrich of Springer has significantly contributed to the technical development of the book through her conscientious and efficient work. Finally I like to thank my family, students and colleagues for being so patient with me during several critical phases of preparation for the Handbook, and to some colleagues and the secretaries for technical help.

I consider it a privilege to see my chosen subject grow. My interest in Environmental Chemistry dates back to my early college days in Vienna. I received significant impulses during my postdoctoral period at the University of California and my interest slowly developed during my time with the National Research Council of Canada, before I could devote my full time to Environmental Chemistry, here in Amsterdam. I hope this Handbook may help deepen the interest of other scientists in this subject.

Amsterdam, May 1980 O. Hutzinger

Contents

Mercury

G. Kaiser and G. Tölg

Cadmium

U. Förstner

Polycyclic Aromatic and Heteroaromatic Hydrocarbons

M. Zander

Volume 1, Part A: **The Natural Environment and the Biogeochemical Cycles**

Volume 2, Part A: **Reactions and Processes**

List of Contributors

Dr. R. Anliker
ETAD
Clarastr. 4–6
CH - 4005 Basel 5
Switzerland

Dr. G. C. Butler
Div. of Biological Sciences
National Research Council
of Canada
Ottawa, Canada K1A OR6

Dr. E. A. Clarke
ETAD
Clarastr. 4–6
CH - 4005 Basel 5
Switzerland

Prof. U. Förstner
Institut für Sedimentforschung
Universität Heidelberg
D - 6900 Heidelberg
Federal Republic of Germany

Prof. W. Funke
II. Institut für Technische Chemie
Universität Stuttgart
D - 7000 Stuttgart 80
Federal Republic of Germany

Dr. Colleen Hyslop
Div. of Biological Sciences
National Research Council
of Canada
Ottawa, Canada K1A OR6

G. Kaiser
Max-Planck-Institut für Metall-
forschung
D - 7070 Schwäbisch Gmünd
Federal Republic of Germany

Prof. C. Rappe
Dept. of Organic Chemistry
University of Umeå
S - 901 87 Umeå
Sweden

Dr. J. Russow
Hoechst AG
D - 6230 Frankfurt/M. 80
Federal Republic of Germany

Prof. G. Tölg
Max-Planck-Institut für Metall-
forschung
D - 7070 Schwäbisch Gmünd
Federal Republic of Germany

Prof. M. Zander
Rütgerswerke AG
D - 4620 Castrop-Rauxel
Federal Republic of Germany

Dr. V. Zitko
Fisheries and Environmental
Sciences
Fisheries and Oceans
Biological Station
St. Andrews, N. B.
Canada EOG 2XO

Mercury

G. Kaiser, G. Tölg

Max-Planck-Institut für Metallforschung, Institut für Werkstoffwissenschaften, Laboratorium für Reinststoffe
D-7070 Schwäbisch Gmünd, Federal Republic of Germany

Historical Background

The story of mercury can be traced back to prehistoric times. A precise dating is, however, impossible because reliable written records are lacking [1]. The first evidences of the use of mercury originate from the ancient Chinese, who used the metal and its principal ore cinnabar as a medicine to prolong life [2] and cinnabar for the preparation of red ink [3]. Often the Hindus [4], the Egyptians [5, 6], the Hettities [7], and the Assyrians [8] were credited with the use of mercury. Positive proofs for this assumption are, however, still lacking [9, 10]. The metal is said to have been known very early in Persia [9, 11] but a chronological assignment is impossible [12]. The Phoenicians exploited cinnabar in Spain from the 8th century B.C. but there is no direct evidence of their involvement with the metal [14]. In the 5th century B.C. cinnabar was used as a pigment by the Greeks [13, 15] but Aristotle is reputed to be the first in Europe who mentioned the metal itself [16].

By the first century B.C. the preparation of mercury by roasting cinnabar and distilling off the metal was well known [17]. Roman writers describe for the first time the process of amalgamation [18] for the recovery of gold from garments [19]. The first recorded mention of an amalgamation process being worked on a large scale appeared in the 12th century in Egypt [9] and was technologically applied in Mexico and South-America to process silver in the 16th century [20]. About one and a half pounds of mercury were used to produce one pound of silver. The life expectance of the native miners was about 6 months [21]. Already in about 1567 Paracelsus described a therapy for mercurial diseases of miners [22].

Throughout the Middle Ages mercury was used as an intermediate to produce gold and silver from base metals [23–26] and for the treatment of various diseases [5, 10, 27–29]. The endeavour to cure syphilis with Hg and its compounds persisted till the 19th century [30] although the toxic nature of mercury was already reported by ancient authors [31] and the danger of mercury vapour had been demonstrated adequately in 1493 [22].

During the 18th century as chemistry slowly evolved from alchemy into a science, the physical and chemical properties of mercury were investigated [32–35], entraining a growing use of the metal and its compounds [36, 37].

The first anthropogenic release of mercury into the environment began with the industrial revolution. The steam engine was invented in 1705 entailing an increased consumption of fossil fuels, and in 1892 a new technique for the production of chlorine and caustic soda by electrolysis using a mercury cathode. Moreover in 1900 organo-mercury compounds ("chlorophenol-quecksilber") were introduced as fungicides to treat seed and from about 1950 as slimicides [21]. The quantities discharged into the environment remained unnoticed and were disregarded until serious hazards that occurred in the 1950s in Japan and Sweden were brought to light.

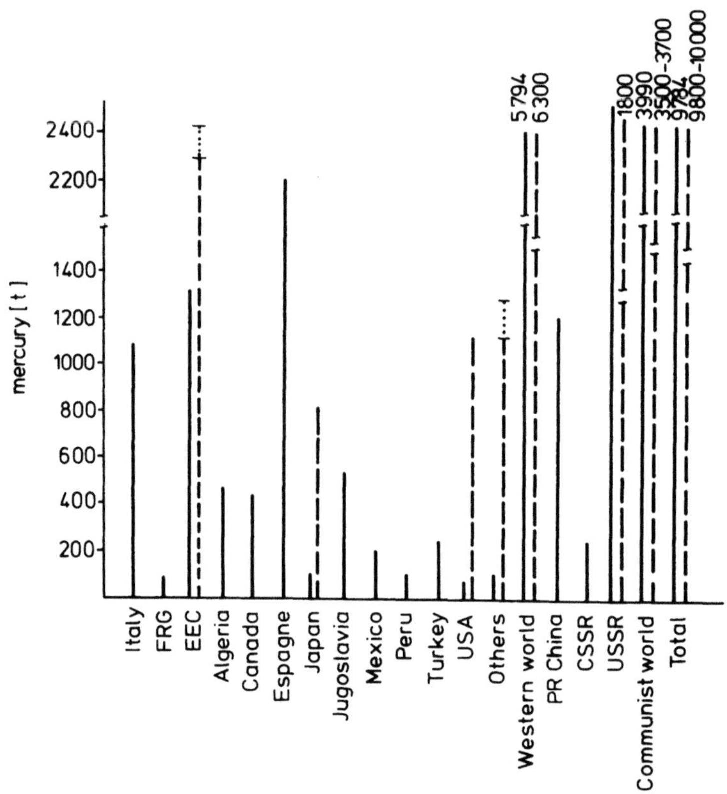

Fig. 1. Production and consumption of mercury in 1973 [68, 54]
—— production; ----- consumption; EEC: European Economic Community

Production-, Use-, Shipment-, and Release Data

Occurrence. Mercury is commonly found in nature as the red sulfide (cinnabar) and in lesser amounts as the black sulfide (metacinnabar), the formula of which may be assumed as (Hg,Zn,Fe)(S,Se) [38–42]. It is also found in a number of minerals in which it is not an essential constituent, and in which it substitutes for other elements [38, 41]. It is often combined with pyrite, quartz, calcite, dolomite, stignite and others. Major deposits of cinnabar are in Spain, Italy, Yugoslavia, USSR, USA, China, Mexico [29, 40–42]. The Hg content in ore ranges from 0.3–2% [39, 43–45]. About 80% of the world supply of the last years came from these countries [46] (Fig. 1).

The deposits of mercury were formed when hydrothermal solutions from hot springs or volcanic activity penetrated unstable geological formations to replace porous sandstone or limestone formation with mineral solution containing mercury [39, 40].

Preparation. Mercury is still prepared, in principle, as described in the 16th century [47–49]. The mined ore is crushed, ground and concentrated by flotation before being roasted in a kiln [50, 51] at 500–600 °C in the presence of air, and sometimes with added iron and calcined lime to remove sulfur. The liberated mercury passes over with the combustion gases and is condensed in water-cooled condensers. The recovery is $\geq 95\%_0$. Mechanical impurities can be removed by passing the metal through a perforated paper or leather. Contaminating heavy metals can be dissolved from mercury by pouring it in a thin jet through diluted nitric acid. The purity of the metal is $\geq 99\%_0$. The metal is further purified by either threefold distillation or by electrolysis [52] and is commercially offered normally in 3 purity categories (I technically pure $99.995\%_0$, II chemically pure $99.999\%_0$, III analytically pure $99.9995\%_0$), but also $99.999995\%_0$ [53] and $99.9999999\%_0$ [110] qualities are available.

Production. Quantities for the global production in 1973 are reported between 8,747 t [46] and 9,784 t [54]. It is assumed that the amount of the produced Hg kept nearly constant at about 10,000 t (Table 1) up to 1978.

Consumption. Data in Table 1 and Fig. 1 indicate declining rather than increasing global use of mercury. Precise data from eastern bloc countries are lacking. Most of the requirements of these countries are supplied by imports [54]. In 1973 the use of mercury in the USSR amounted to 1,800 t [55]. The production in Red China largely supplies its own requirements.

Anthropogenic Discharged Mercury

Specific Uses and Discharges. Table 1 shows the use of mercury by final use. Data up to 1976 are available only from the USA and the FRG [54, 56]. The table does not show the potential capacity to pollute the environment by the manufacture of the corresponding compounds by the chemical industry. An indication of the size of pollution problems posed by the use of mercury is given by the following examples valid for the USA:

Table 1. Consumption of mercury classified by use [54, 56]

Industrial division	Consumption [t]					
	EEC	USA		FRG		
	(1973)[a]	1973	1975	1973	1975 [b]	1976 [b]
Electrolysis						
net cons.	730	451	520	372	137	226
invest.	300	35				
Electrotechnique and instruments	280	868	690	60.1	61.1	81.5
Paints	70	262	250	18.7	5.6	12.4
Catalysis	60	23	25	42	12	16
Agriculture	95	63	30	50.3	31.9	31.4
Dental use	135	92	60	31	25	26
Pharmaceutical products	40	21	20	0.5	0.5	0.5
Laboratory products	170	23	15	40	61.3	78.1
Others and stock	690	35	160	190	44.2	46.7
Total	2,570	1,873	1,770	808	377.6	514.6

[a] In 1969 the total amount was 2,830 [84]
[b] If exportation and increase in stock are considered the consumption amounts to about 330 t [56]
EEC: European Economic Community

1. Mercury losses in industry are assumed to amount to 10,850 t over a period 1944–1959 [57].
2. Up to 1974 recycled mercury accounts for less than 20% of the total consumption [58]. (About 500 t from 2,900 t total consumption.)
3. It is estimated that in 1968 the chloralkali industry used 590 t only to maintain inventory [58].

Electrolysis. The chloralkali industry is usually the biggest consumer (Table 1) and has been one of the biggest polluters. In recent years this industry made every effort to reduce the emissions [59] as can be seen from the consumed and the emitted mercury quantities (Table 2), which have been evaluated in the FRG [56].

Although the mercury cell capacity might be replaced by the diaphragm process by which less mercury is released to the environment, there is a trend to the former on economical grounds.

Electrical Apparatus and Control Instruments. Mercury finds widespread use in fluorescent, and discharge lamps, in industrial power rectifiers, and to a great extent in mercury-cell batteries. The most part thereof is assumed to be lost, e.g., by breakage of thermometers [60], waste of fluorescent lamps

Table 2. Decrease of discharged mercury [g Hg/t Cl_2] in the chloralkali electrolysis from 1972–1976[a] [56]

Pathway of emission	Year					
	1971	1972	1973	1974	1975	1976
	(1.7)	(1.7)	(2.2)	(2.3)	(1.8)	(2.05)
Waste water	25	25	10	9	9	5
Outgoing air	15	15	13	13	13	10
Different products	29	24	15	13	8	5
Dump	20	20	20	20	20	10

[a] Figures in parentheses indicate production of chlorine in 10^6 [t/yr]

[61–63] and batteries. In control instruments mercury metal is used in barometers, gauges, thermometers, lamp seals, electrical switches, etc.

Recent developments aim at substituting mercury in dry batteries [64, 65]. Thermometers with an infra-red indication are in development [65].

Catalysis. Mercury chloride and sulfate are used for converting acetylene into vinyl chloride and acetate, (PVC, PVA production). The catalyst is regenerated and recycled [66]. In addition to this, mercury compounds are used for the conversion of acetylene to acetaldehyde and for the preparation of dye raw materials. Mercury in effluents from factories converting acetylene to a variety of products has received particular attention in the Minamata incidence in Japan [67].

Paints. The fungicidal effect of some mercury compounds is taken advantage of in the production of protective paints. Mostly organic mercury compounds such as phenyl mercury acetate, oleate and dodecenylsuccinate are used. In 1969 about 3×10^6 t paints were produced in Europe [68], which is 40% of the world production [69]. This corresponds to about 5,000 t of mercury which were painted onto surfaces [68]. The stability of the mercury compounds in the paints is quite low. Photochemical breakdown and vaporization of both, mercury compounds and of the metal reduce the mercury content in paints quickly [70, 71].

Recently fungicidal compounds of zinc, copper and phenyl and sulphur derivatives have been tested as substitutes for mercury [72, 73].

Agriculture. Inorganic and organic (alkyl, alkoxy, aryl) mercury compounds have been used as seed dressing (potatoes, grains, flower bulbs, cotton, etc.) and as foliage sprays against plant diseases. These uses are dangerous because mercury compounds are brought in direct contact with the ambient environment and thus contaminate plants and birds [74]. Administrative laws have promoted replacement of mercurials by substitutes [68], whose efficacy is, however, smaller than that of mercurials.

Amalgamation. In electrolytic processes mercury is used for the recovery of metals (Zn), furthermore as a reducing agent and for dental fillings. Today amalgam residues are collected and recycled. They amount in FRG to 5 t annually [56]. Dental amalgams are today partly replaced by artificial products (acrylic-, epoxy resins) [75].

Pharmaceuticals. Mercury compounds are used for their antiseptic and preservative properties in soaps, cosmetics, antiseptic preparations. Some cases of mercury intoxication by absorption of mercury into human skin are known [76]. Most of the mercury thus used is lost to the environment via sewage and drain waters.

Pulp and Paper. Organic mercury compounds (especially phenyl mercury acetate) have been used to prevent microorganisms (bacteria, fungi, algae etc.) from growing in pulp. In recent years official regulations have been issued to eliminate mercury from those papers which come into contact with food. None the less mercury is found in paper and board products because cellulose seems to concentrate mercury from contaminated caustic soda [77].

The pulp and paper industry has recently improved with respect to water pollution, but the air pollution via the incineration of the products remains. Artificial papers derived from polyolefines, may bring further improvement since substantially less slimicides are required in their production. In many countries the use of mercury in slimicides has been banned by governmental action. 14 t are estimated to be released into the environment on a world-wide basis [78].

Other Uses. Smaller amounts of Hg are used in the production of plastics, in the tanning industry, and as heat transfer agent [68].

Table 3. Global mined fossil fuels and ores and released mercury during burning and smelting processes in 1970

Raw material	Mined quantity [t]	Mercury content [μg/g]	Assumed average concentration [μg/g]	Released mercury [t]	Ref.
Crude oil	$2\text{–}3 \cdot 10^9$	0.005–2	0.04	10^2	[80, 81, 89, 95]
Bituminous, anthracitic coal	$2.18 \cdot 10^9$	0.012–33			[81, 84, 95, 96]
Lignite	$0.77 \cdot 10^9$	0.036–0.056			[87]
Coal (all types)	$3 \cdot 10^9$		1	$3 \cdot 10^3$	[80]
Natural gas	$0.6\text{–}1.35 \cdot 10^9$	0.18	0.04	20	[80, 81]
Sulfide ores (Cu, Pb, Zn)	$1 \cdot 10^9$	0.3–1,000		$1.5\text{–}20 \cdot 10^3$ [a]	[80, 81]
Phosphorites Bauxite Minerals for cement preparation				$2.5\text{–}3 \cdot 10^2$	[80]

[a] Mercury produced in the smelting process is estimated to amount to 30,000 t [81]

Table 4. Comparison of global, natural and anthropogenic mercury emission

Pollution	Discharged mercury [t]	Ref.
Nature:		
Volcanoes, geysers, weathering	$0.5-5\cdot10^3$	[81, 97]
Degassing of crustal materials	$25-150\cdot10^3$	[98]
Evaporation from ocean	$23\cdot10^3$	[99]
River, glacial ice runoff	$3.8\cdot10^3$	[98]
Man:		
Mercury industry [a]	$6-10\cdot10^3$ [b]	[54, 95]
Processing of minerals [c] and ores	$1.5-20\cdot10^3$ [b]	[80, 81]
Burning of fuels	$0.1-8\cdot10^3$	[81, 88, 93–95]

[a] Evaluated up to 1974
[b] Depending on the efficacy of recycling
[c] Based on data from 1970

Processing of Ores. The quantity of mercury discharged through stacks in the smelting process of sulfide ores (Cu, Pb, Zn) [79] is reported between 1,500 t/yr [80] and 30,000 t/yr [81] depending on the assumed level of mercury. Thus with zinc roaster gases of a Finish company 20 t were reported to be discharged annually [55]. In addition, emissions in the processing of phosphorites, bauxite, minerals of iron and manganese (oxides) have to be considered (Table 3).

In the production of sulfuric acid from zinc ores mercury can be removed from roasting gases with a newly developed technique [82].

Fossil Fuel Combustion. Although the mercury content of fossil fuels is small the burning of large quantities constitute an enormous pollution hazard [83–88]. The mercury levels in coal depend strongly on its origin [89]. In the USA for instance levels between 0.012 μg/g [84] and 33 μg/g [88, 90] were found. For lignite mined in the FRG levels lie between 0.036–0.056 μg/g [87]. Upon incineration of the fuels about 90% of the mercury is released into the atmosphere via the flue gas [84, 86, 91]. About 10% distribute in furnace bottom ash, precipitator ash and drainwater [87, 91]. In 1970 the global coal consumption was about 3×10^9 t [92]. This corresponds to a release of mercury of 3000 t assuming an average content of 1 μg/g [88] (Table 4).
According to figures on record, the concentration of mercury in fossil fuel ranges between 0.005–33 μg/g [81, 93, 94], and discharged mercury between 80 [81] and 1,800 t/yr [94] based on an annual consumption of 2×10^9 t.

Naturally Released Mercury

Natural polluters are volcanoes [100–103], geysers [104, 105], thermal fluids [100, 103, 106] and the earth crust itself by weathering and erosion of rocks. The latter process is held responsible for emissions from 500 [97] up to 150,000 t/yr [81, 98]. These quantities can be calculated both from the mercury concentration in the air and its precipitation by rainfall [98, 107]. The quantities that enter the oceans by river and ice cap runoff, 3.8×10^3 t [98], are small in comparison with the mercury stock of $5–20 \times 10^7$ t in the ocean [108] (Table 4). The immense reported range of discharged mercury quantities can be traced back partly to the different average level of mercury in the respective matrix used for the calculation. Additionally, it is not known how much mercury is recycled by industrial processes today. It is assumed that about 50–80% of the global consumed mercury (about 10^4 t) is lost to the environment. The amount released by burning of fossil fuel and in smelting of metals and ores can not be estimated. At first view one ought to assume that there is no impact on the atmospheric and hydrospheric mercury burden by man, since the quantities released by natural processes [98, 107] are larger (Table 4).

Analysis of glacial samples from Greenland indicate, however, a significant increase in mercury depostiton during the course of time [98]. The reasons for this increase are believed to be less due to industrial pollution than to those activities which result in greater exposure of the earth's crust through alteration of terrestrial surfaces thus allowing more mercury to enter the atmosphere [98].

Global reflections by these emissions are not expected but an impact on a local ecosystem can occur if industrially derived mercury is discharged uncontrolled into the environment.

Chemistry

Elemental Mercury

The chemical symbol Hg for mercury was derived both from the latin name Hydrargyrum, i.e., liquid silver, and "argentum vivum" meaning live or quick silver, or from the planet Mercury and the Roman God. Mercury can easily

Table 5. Some physical properties of mercury

Atomic weight	200.59
Melting point	38.9 °C
Boiling point	357.3 °C
Density	13.595 g/cm^3 (0 °C)
Vapour pressure	$0.189 \cdot 10^{-3}$ mm (0 °C); $1.22 \cdot 10^{-3}$ (20 °C); $2.8 \cdot 10^{-3}$ (30 °C)
Solubility in water	$6 \cdot 10^{-6}$ g/100 g (25 °C)
Ohmic resistance	$95.76 \cdot 10^{-8}$ Ω m (20 °C)[a]

[a] Depends on purity

be obtained in a pure state by heating of mercuric oxide [109]. There are seven stable and eleven unstable known isotopes [110, 111]. The most useful of them being ^{203}Hg (half life:47 days, β-, γ-emitter) and ^{197}Hg (half life:65 h, γ-emitter).

Mercury is a glistening silvery metal. Some important properties for elemental mercury are compiled in Table 5 [112] and for some mercury compounds in Table 6.

The reactions of mercury with some common reactants are briefly compiled in Table 7.

Removal from Contaminated Rooms. On account of the high vapour pressure mercury evaporates quickly into the air after being spilled. It can be removed from breathing air by sucking it through a filter consisting of different layers of $CaCl_2$, NaI, and activated carbon, and from laboratory air by gassing the room with H_2S and by covering the floor and the benches with

Table 6. Properties of some inorganic Hg(I) and Hg(II) compounds [112, 113]

	Solubility [g/100 g water]	Boiling (B), Sublimation (S) Decomposition (D) and Melting (M) point [°C]
Hg(I) compounds		
Hg_2F_2	D	570 D
Hg_2Cl_2	$2\cdot10^{-4}$	400 S
Hg_2Br_2	$4\cdot10^{-6}$	345 S
Hg_2I_2	$2\cdot10^{-8}$	140 S, 290 D
$Hg_2(NO_3)_2\cdot2\,H_2O$	D	70 D
Hg_2O	i.	100 D
$(Hg)_2SO_4$	0.06 (25 °C)	D
Hg_2S	i.	D
Hg(II) compounds		
HgF_2	-	645 D
$HgCl_2$	6.6 (20 °C)	277 M
$HgBr_2$	0.62 (25 °C)	241 M
HgI_2	$6\cdot10^{-3}$ (25 °C)	257 M
HgO (yellow, red)	$5.3\cdot10^{-3}$ (25 °C)	350 D
HgS (α)	$1\cdot10^{-6}$ (18 °C)	583.5 S
(β)	i.	583.5 S
$HgSO_4$	D	D
HgSe	i.	vcc.S
$Hg(NO_3)_2\cdot H_2O$	S	79 M
$Hg(NO_3)_2\cdot1/2\,H_2O$	v.s.	79 M
$Hg\,(Me)_2$		92.5 B
$Hg\,(Et)_2$	i.	159 B
$Hg\,(Ph)_2$		121.8 S
MeHgCl		170 M (S > 100)
EtHgCl	i.	193 M (S > 100)
PhHgCl		271 M
PhHgAc	sl.s.	149 M

Key: i: insoluble; sl.s.: slightly soluble; v.s. very soluble; vcc.: vaccuum

H_2S-water. Iodine carbon has proved especially useful. Splashed or spilled mercury can easily be collected by taking it up with a capillary connected with a glass container and a pump.

Table 7. Reaction of mercury with some common reactants [110]

Reactant	Conditions	Reaction products
Noble gases	In discharge tubes	HgAr, HgKr [114]
Halogens	At room temperature on excess of hal.	Mercurous halide [115] Mercuric halide [116]
Oxygen, air	At about 350 °C, room temp. (u.v., electron bombardment	HgO (Hg, O_2 at temp. >350 °C
Ozone	Room temperature	HgO
S, Se, Te	On heating	
Dry hydrides HX (X = F, Cl)	≥ 200 °C	HgX_2
H_2S, NH_3, PH_3, AsH_3 etc.	> 200 °C	
ICl	Room temperature	$HgCl_2$, HgI_2
NO_2	Room temperature	$Hg_2(NO_2)_2/Hg_2(NO_3)_2$
Conc. H_2SO_4	Room temperature	Hg I, Hg II, sulfates
HNO_3	Room temperature	Hg I, Hg II, nitrites, nitrates
Ammonia solution	In air	Millons base

Mercury Compounds

A detailed description of mercury chemistry has been given [117, 118]. Here only some common inorganic and organic mercury compounds are cited.

Inorganic Hg (I) Compounds. Studies of various equilibria support that in Hg(I) compounds two Hg atoms are associated to give Hg^{2+}ions.

$$Hg_2^{2+} \rightleftharpoons Hg^0 + Hg^{++}.$$

From the equilibrium constant one can infer [119, 120] that Hg(I) ions are moderately stable towards disproportionation in solution. In spite of this any reagent that reduces the activity of Hg(II) ions compared with that of Hg(I) ions will force the equilibrium to the right. Since many Hg(II) compounds are very insoluble, are slightly dissociated in solution or form stable complexes the number of Hg(I) compounds is limited. For instance, the addition of OH^-, S^{--}, or alkylsulfides to a solution of Hg(I) salts gives Hg and HgO, HgS or Hg(II) complexes of the organic ligands.

Apart from a few soluble salts such as nitrate, chlorate, and perchlorate most known Hg(I) compounds are sparingly soluble (Table 6). A detailed review on Hg complex formation of Hg(I) is available [121].

Mercury (II) Compounds. Hg(II) compounds with highly electronegative anions F^-, NO_3^-, ClO_4^-) have ionic structures; they are dissociated and hydrolysed in aqueous solution. Other halides, the oxide, and sulfide are covalent in nature. They are largely undissociated in water. Mercury forms a host of strong complexes with linear or tetrahedral coordination arrangements. Complexes where mercury is five-, or six-coordinate are less common [121].

Organomercury Compounds. A number of up-to-date reviews consider organomercury compounds in general [122–124], mercury alkyls [125], organometallic reaction mechanisms [126] and complex formation of the methyl mercury cation [127]. This contribution covers only some environmentally relevant compounds.

The formation of monoalkyl mercurials from mercury and methyl iodide in the presence of sunlight was discovered in 1851 [128] and the dialkyls accidentally in 1858 [129] in an attempt to form methyl mercury cyanide by double decomposition. The extraordinary toxicity of these compounds caused fatal poisonings at that time [130]. More comprehensive studies on organomercurials have been resumed by 1900, when the important mercuration reaction which yielded relatively inoffensive aryl compounds was discovered by Dimroth [131]. Mercury acetate and benzene and its derivatives were found to react to give phenyl derivates. These compounds had then already been tested for their fungicidal effect.

Mercury for Metal Substitution (Transmetallation). It is the most universally applied method in organometallic synthesis. General methods include: a) HgX_2 and LiR or AlR_3, b) HgX_2 and RMgX, c) Hg or sodium amalgam and RX (X = halides, sulfates) giving rise to mono-, and dialkyl mercury compounds as in

$$RM + HgX_2 \longrightarrow RHgX + MX$$

$$2\,RM + HgX_2 \longrightarrow R_2Hg + 2\,MX$$

$$RHgX + R'M \longrightarrow RHgR' + MX.$$

Another possibility is a disproportionation reaction as in

$$2\,EtHgI + 2\,NaI \longrightarrow HgEt_2 + Na_2HgI_4.$$

The Grignard route is suitable for the synthesis of primary secondary and tertiary alkyl halides [132, 133].

Mercury for Hydrogen Substitution (Mercuration). In the mercuration reaction compounds such as Hg(II) acetate react readily with aliphatic [132 to 135] and aromatic [126, 136–138] compounds with replacement of –H by –HgOAc, e.g.

$$CH_2(COR)_2 + Hg(OAc)_2 \rightleftharpoons (RCO)_2CH\,HgOAc + HOAc$$

$$PhNH_2 + Hg(OAc)_2 \rightleftharpoons p\text{-}NH_2C_6H_4\,HgOAc + HOAc$$

$$C_6H_6 + Hg^{+}OAc \rightleftharpoons C_6H_5HgOAc + [H^{+}].$$

Phenyl mercury acetate is widely used, and was introduced in the 1920's for seed dressing, and as a fungicide in the pulp industry.

Addition Reaction (Oxymercuration and Related Reactions). Aliphatic organomercury compounds with selected groups containing oxygen or nitrogen can readily be prepared by the reaction of alkanes and to a lesser extent of alkenes with mercury II salts in the presence of appropriate nucleophiles [132, 133].

$$R_1R_2C = C\,R_3R_4 + HgX_2 + R_2YH \longrightarrow R_1R_2C(YR_n)\,C(HgX)\,R_3R_4 + HX.$$

Methoxymethyl mercury acetate is a representative which is still allowed for use as a seed dressing.

Analytical Methods

Many comprehensive reviews on the determination of total mercury and individual compounds exist [43, 139–148]. Therefore, only a brief summary is given here, considering also sources of systematic errors. The analysis can roughly be classified into two groups:

1. Procedures for total mercury (elemental mercury, inorganic, organic mercury compounds)
2. Procedures which discriminate between the respective forms.

Total Mercury Analysis

In the determination of total mercury levels, mercury and its compounds are normally converted into mercuric ion. This presupposes a range of analytical operations which are associated with methodical and systematical errors causing the analytical result to be incorrect by orders of magnitude if environmentally relevant levels as low as pg/g are to be determined. The sources of such errors are complex and are inherent in each step of the analytical procedure, but lie preferentially in the taking, preparation and decomposition of the sample and the separation of mercury from the matrix, unless appropriate precautions are met, e.g., cleaning and purifying all necessary tools and reagents and storing them under cleanroom conditions [149–151].

Taking and Preparation of the Sample. Systematic errors can be encountered in the following steps of the procedures: a) sampling, if the matrix is heterogeneous [152, 153], b) during storage by interactions of the traces of mercury with interfaces (adsorption, desorption) which is dependent on the working material and the matrix to be analysed, by volatilization from solids [163–166], and from liquids [167–172], e.g. caused by bacteria, by diffusion processes which may cause introduction of blanks and losses in mercury if the sample is stored in plastic containers [153, 173, 174], c) with disintegration and pulverizing [153], d) drying [153, 163, 175], and e) lyophilization [163, 176, 179] of the sample. In addition to this there is always the risk of contaminating the sample seriously by insufficiently cleaned tools and devices [153]. With gaseous samples mercury normally has to be preconcentrated directly or after passing a combustion unit. The absorbers have to ensure quantitative trapping (see section preconcentration).

Decomposition of the Sample. In wet oxidation the most commonly applied reagents, are HF, HCl, $HClO_3$, $HClO_4$, HBr, $HBrO_3$, H_2SO_4, H_2O_2, $KMnO_4$, $K_2Cr_2O_7$, $K_2S_2O_8$, V_2O_5 and mixtures thereof. Normally open systems are used [180–187] some of which are partly automated with respect to routine work [188, 189]. Some open systems may involve the risk of introducing blanks from outside and losing considerable amounts of mercury by volatilization. Closed vessels with reflux systems and special receivers [190] evade this disadvantage but so-called pressure decompositions in single, [191–194] and multiple arrangements [195] using vessels with small surfaces and acids which can be obtained extremely pure, e.g. by subboiling point or isothermal distillation [196], are preferable if ng/g and lower levels are to be determined. Volatilization and combustion techniques in open systems [93, 197, 199] may be associated with losses in mercury by incomplete trapping or incomplete decomposition of mercury compounds [198, 200]. Organic matrices can be ashed in closed systems under static conditions [201] or under dynamic conditions [202], using also HF- [203] or UHF- [200, 204] excited oxygen or an oxygen-hydrogen flame [205, 206].

In aqueous solutions organomercurials can be degraded, e.g. with u.v. rays [189, 207], thus avoiding introduction of blanks by the decomposition agents.

Separation of Mercury from the Matrix

Volatilization. From inorganic solids e.g. metals, rocks and some soils mercury can be volatilized by heating the sample (900 °C) in a stream of nitrogen, air or argon [93, 208, 209], from coal by combustion in oxygen [199], from nearly all organic solids by combustion in a HF- or UHF-induced oxygen plasma [200, 203], and from aqueous solutions and decomposition solutions after reduction of the ionic mercury to the elemental form by aeration (cold vapour technique) [210, 211]. Some papers point at interferences encountered in this technique resulting in incomplete release of mercury [153, 212, 215].

Miscellaneous Techniques. From solids a separation is possible by solvent extraction, e.g. with benzene in case of organomercurials [216–218], from aqueous solutions by precipitation [219], co-precipitation [220], precipitation exchange on thin layers, e.g. on ZnS which reacts with ionic mercury to form HgS [221], liquid-liquid extraction [222], e.g. with dithizone [223] and other complex forming agents [224–226], by chromatographic methods, such as ion exchange [227–229], e.g., on Wofatite 1–150 [230, 231], thin layer [232, 223], paper [234–236], and gas chromatography [237–240], by electrophoresis on a macro scale [241], and by electrodeposition which allows separation yet in the lower ng/g and pg/g range [200, 242, 243].

Preconcentration. From gaseous samples mercury can be pre-concentrated by passing the gas stream through impinger flasks containing absorbing solutions, e.g. permanganate sulfuric acid [244, 245], iodine potassium iodide, iodine monochloride [246] and others [247, 248]. Solid adsorbers, e.g. Cu, Ag, Au, Pt [249, 258], Chromosorb W [253], activated charcoal [252, 253, 255, 259], with KI- [260], CdS- [255], Au-impregnated or prepared filters [234, 261], glass fibers [262], glass wool [153] or glass beads [263] enable separation from a gaseous sample or from combustion gases and specific preconcentration.

One should always keep in mind that in gaseous samples different forms of mercury may occur which additionally may be particulate bound [254, 264]. For total mercury analysis the gaseous sample best is passed over catalysts, e.g. CuO at 900 °C [254] or Ag at 600 °C [200, 205] in combustion units to ensure degradation of mercury compounds and complete trapping. Tandem arrangements of the mentioned absorbers enable specific separation and preconcentration of individual mercury forms [261, 263, 265].

Methods of Determination

Determination as a Metal. Stock [266, 267] liberated mercury by heating the sample and trapped it in a cooled capillary. The diameter of the mercury droplets was then measured under a microscope. This way he could determine the mercury in the µg/g range with good accuracy.

Spectrometric Methods

Spectrophotometry (Colorimetry). Dithizone is the most widely used reagent and covers a wide concentration range down to about 0.01 µg/g of mercury [268]. Possibilities of an elimination of interferences from other metals which are complexed by dithizone and the application of a multitude of other reagents are reviewed [43, 141, 143]. For instance, with Brillant Green as low as 1.7 ng of mercury can be determined [269].

Atomic Absorption (AAS). By far the most popular and widely used method of determining mercury is AAS. Flame [270, 271] and graphite tube atomisation can be applied. But better sensitivities are achieved if metallic mercury is carried by a gas stream into an absorption cell in the light path of the spectrometer where the absorption is measured at 253.7 nm or more sensitively at 184.9 nm if the system is purged with N_2 or a noble gas, or where the absorption line is splitted by a magnetic field (Zeeman effect) [272]. Numerous publications, which are extensively reviewed [273–276], describe combinations of cold vapour techniques – involving a subsequent pre-concentration –, with AA spectrometry [211, 277–285] (see also section separation). Some procedures are partly [288–299] and some are totally automated [289, 286, 287] with respect to routine work. Arrangements where decomposition, separation, pre-concentration, and determination are connected closely together, reduce the risk of introducing blanks and losing mercury, e.g. by interchanges of the traces of mercury with large surfaces [153, 300]. An example of a so called multi-stage procedure is shown in Fig. 2 where mercury can be

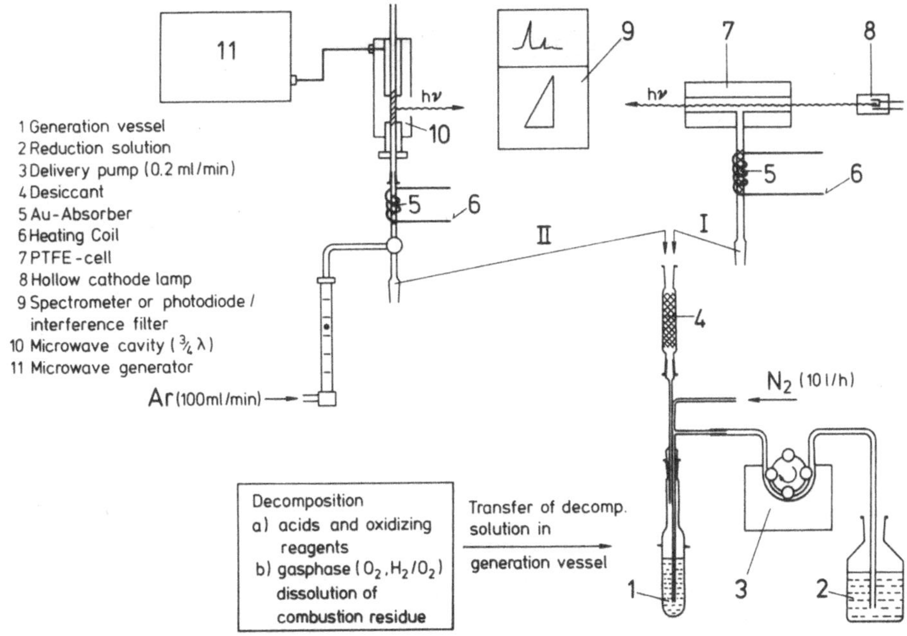

Fig. 2. Determination of total mercury in organic matrices by flameless AAS (I) and OES-MIP (II) after decomposition and cold vapour technique

monitored at 253.7 nm both by atomic absorption and emission with detection limits of 0.5 and 0.05 ng respectively. In this practically all decomposition methods can be applied. For routine analysis of biological samples a semi-automated device (using a mixture of $HClO_3/HClO_4/HNO_3$ for decomposition) has proved especially useful [188].

Atomic Emission (OES). The use of microwave induced gas plasmas (MIP) mainly helium, and argon plasmas as excitation sources in connection with different separation techniques (volatilization, electrodeposition) enable mercury to be determined down to the pg/g range [153, 200, 301–306]. Poorer sensitivities are achieved with inductively coupled high frequency plasma- (ICP) [307], radio frequency plasma- [308], and arc- [309] excitation.

Atomic Fluorescence. Similarly to AAS, AFS techniques have greatly improved, for instance by using electrodeless discharge tubes [310] or a separation step, e.g. cold vapour technique or amalgamation. This way detection limits of 3 ng (0.06 ng/g) can be obtained [311, 312].

X-ray Fluorescence (XRF). Detection limits in the µg-range can be achieved if the sample is directly applied [313, 314]. In connection with separation steps, e.g. cold vapour technique [315] and concentration steps, e.g. precipitation exchange [221] or ion exchange [316], even ng/g levels can be detected.

Electroanalytical Methods. These include mainly potentiometry [317, 318] coulometry [319, 320] dc- and ac-polarography [321–324], amperometry

[325–329] anodic stripping- [330–335] and differential pulse anodic stripping voltammetry [336, 336 b] and anodic stripping chronopotentiometry [337]. With these methods mercury can be determined in the µg/g to the ng/g range.

Neutron Activation Analysis (NAA). NAA enables sensitive determination of mercury and excludes the risk of introducing blanks provided that the sealing of the sample and of standards happen under blank controlled conditions in the comparative NAA [338, 339]. Detection limits down to 1 ng are reported if interferences are excluded by chemical separation of the activated mercury from the matrix [340–343].

Chromatographic Methods. Thin layer chromatography (TLC) [344], and paper chromatography [345] are used to determine mercurials directly in liquids, and in solids after decomposition, and more sensitively after extraction with, e.g. dithizone [268] but are mostly used as separation methods in combination with other more sensitive detection systems (see below). Gas chromatography (GC) allows more sensitive determination of both inorganic mercurials after transfer into suitable derivatives [346, 364] and organic mercurials directly in solutions [347] or after solvent extraction [216–218, 237, 239, 348, 349, 366]. A detailed review is given elsewhere [146].

Miscellaneous Methods. These include radio-release [350], catalytic methods which take advantage of the capability of mercury to catalyse or inhibit reactions [351–353], and mass spectrometry to analyse for mercury in small natural samples [354].

Distinction Between Individual Mercury Compounds

The occurrence of various forms of mercury in the environment (elemental mercury vapour, inorganic and organic mercurials), and the high toxicity of mercury vapour and some organomercurials necessitate distinction and determination of the individual species. From air the individual forms can specifically be absorbed or adsorbed (see section separation). For liquids and decomposition solutions TLC, paper chromatography, and electrophoresis with use of various complexes, papers, coated slides, and developers have been successfully applied down to the upper ng/g range [237, 355–357]. GC is an efficient technique for separation of individual inorganic and organic mercurials and in connection with sensitive detectors, e.g., flame ionization (FID) and electron capture (ECD) a very sensitive determination method. Discrimination and determination of mercurials by chromatographic methods are exhaustively studied and reviewed [146, 218, 349, 367]. These methods are often connected with other sensitive detection systems, e.g., high performance liquid chromatography (HPLC) with voltammetry [643] or GC with flameless AAS [358–360] or OES-MIP [361–363] or with mass spectrometry [254, 368]. Thus combining selectivity and high sensitivity to enable detection of individual mercury compounds down to the pg/g range.

With the aid of differential pulse anodic stripping voltammetry mercury species complexed by organic ligands can be discriminated and sensitively determined [368 a].

Transport Behaviour in the Environment

Transport into the Environment

Mercury is released into the environment mainly as particulate matter, elemental vapour, $HgCl_2$-vapour, inorganic mercurous and mercuric compounds, methyl mercury (II) compounds, dimethyl mercury, and phenyl mercury compounds [264]. Natural discharges occur almost always in relatively low concentrations and widely distributed. In contrast to this, man-made mercury enters the environment at only a few locations but in relatively large quantities, which are assumed to amount from a fraction [88, 94, 98, 370] to an equal order of magnitude of the natural burden [99, 371].

The following sources must be considered:

Natural sources

- volcanic activity, geysers and thermal fluids
- weathering of rocks
- degassing of the earth mantle
- transpiration and decay of vegetation
- emanation from the ocean

Anthropogenic sources

Natural Input

Atmosphere. There are no data on emission from volcanoes, and geysers [100, 372, 373]. It is assumed however, that the amounts greatly exceed those released from deposits [94]. Weathering of rocks does not significantly contribute to the atmospheric burden. Mercury sulfide impounded in rocks [374] is resistant to solubilization through weathering, and enters the geocycle mostly in form of mechanically degraded particulate matter. In this form it may, however, undergo chemical and microbial transformation to the elemental form [143, 375]. When passing through the soil further transformation e.g. into organo mercurial with aid of bacteria [164] enable mercury to reach the atmosphere.

Decay and transpiration of land plants are another source for the atmospheric mercury burden [376]. An estimation of the global quantity released by transpiration of soil and vegetation amounts to 44,000 t/yr [108] without giving details about the exact origin of this quantity. A considerable but not yet determined amount of mercury is released from the surface of the oceans [95, 644]. One of the mechanisms which effect transition from the hydrosphere to the atmosphere is the bursting of gas bubbles [377–379]. The aerosols thus formed as well as those lifted from the land surface, can be transported great distances and are distributed on a global scale if the particle size is small (< 10 μm) [95, 380].

Hydrosphere. The flux of mercury from the continents to the oceans by river and ice cap runoff (3.8×10^3 t/yr) is much less than that from the continents to the atmosphere ($2.5–15 \times 10^4$ t/yr). For the most part weathering of rocks contributes to the river runoff in form of fine particles [94, 99, 381].

Oxidation of sulfide ores may result in mercurous and mercuric ions which are readily leached by rainfall and reach the oceans by runoff under flow, and groundwater [382]. In certain areas mercury bearing deposits, thermal springs and mine drainage contribute significant amounts to streams. All these processes are affected by physical, chemical and microbial processes and geological conditions when water passes individual strata. An elucidation of the pathways, and an assignment of the released mercury to a definite source, is very difficult on account of their complex mechanisms.

The transition of man-made mercury into the environment follows a similar pattern. The greater part is discharged directly via stacks and flues (industry, space heating, transport facilities, generation of energy) or with effluents into the aqueous environment.

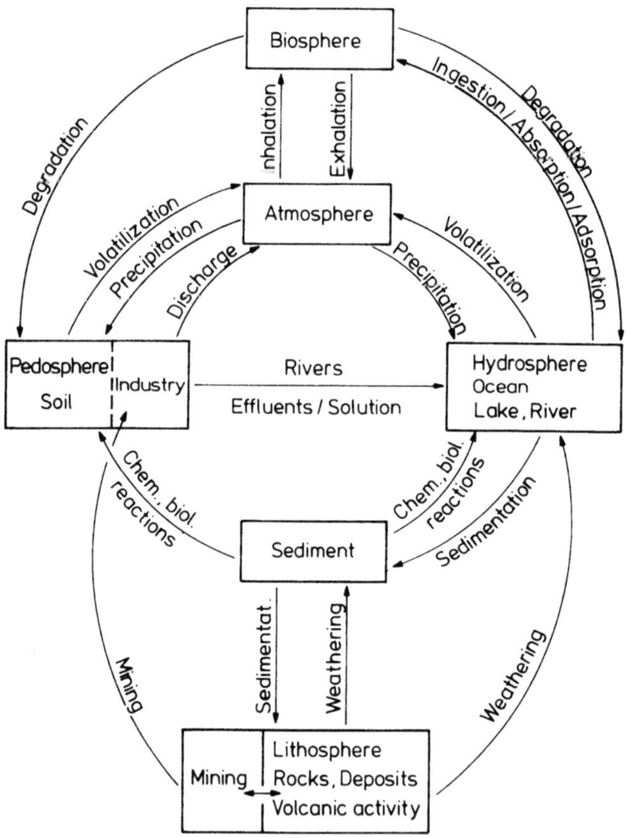

Fig. 3. Global mercury cycle

Figure 3 shows a simplified model of the exchange of mercury between the different compartments of the environment. A comparison of pre-man with present day cycles of mercury [99, 108] show a global impact by man's activity upon the environmental mercury burden. Such balances are, however, to be considered reservedly as a series of differing data exist depending on the assumed background and average concentrations of Hg in the corresponding matrices used for the calculations as e.g., weathering of rocks [94, 381–384] Hg stock in ocean [385, 386] and in the earth crust [41], atmospheric burden, natural output from the earth [88, 108].

Such calculations can only give approximate values with an uncertainty of about half an order of magnitude.

Transport in the Environment

Atmosphere. The individual forms of mercury in the atmosphere contribute to the overall mercury burden in the following way: mercury vapour 4‰, Hg(II) halide 25‰, monomethyl mercury 21‰, dimethyl mercury 1‰, particulate form 4‰ [264]. The existence and proportion of individual forms depend on many factors [375].

Statements on the percentage of mercury vapour and particulate bound mercury are in conflict on occasion. They range from about 4‰ [263, 387, 388] up to 50‰ [376] for particulate bound mercury, which is susceptible to transportation or removal by impactation, and dry or wet precipitation [389, 390]. Varying proportions in the air near the ground reflect the irregular transport by winds. In meterological terms the horizontal dispersion is several orders of magnitude greater than the vertical [391] and therefore most fallout will occur near the place of emission [392]. This contrasts, however, with high mercury levels in Greenland ice [98], and studies which did not observe any noticeable reduction of gaseous mercury levels during rain storms. The decrease is due to an increase in ventilation [393]. Jet streams carry pollutants from the industrial areas of the northern hemisphere in a concentrated band around the globe. The pollutants are precipitated beneath them. Higher mercury levels of samples from below northerly jet stream paths and adjacent latitudes confirm this assumption [372]. The regional and global circulation depends on meterological factors, e.g. wind speed and direction, rainfall intensity, and atmospheric stability. Reviews comment on numerous contradictory views on the effectiveness of atmospheric processes in the removal of mercury from the air [264], and give a model for the calculation of evaporation and recycling rates [371].

Some authors doubt the theory of a global circulation of mercury. There is no correlation between the long distance transport of SO_2, NO_2 and mercury in air [394].

Hydrosphere. The transport of a trace element in an aqueous medium is determined by several factors [395]. For mercury, the following factors are important: dissolution of ionic species and inorganic compounds [94, 396, 397], adsorption on and coprecipitation with solids, e.g. Fe_2O_3 [398], limonite [399] or clay [400, 401], incorporation in a crystalline structure [402], cationic exchange [399], formation of complexes with organic molecules, e.g. sulfur containing proteins and humic material etc., sorption and ingestion by viable biota [108]. Pelargic organisms agglomerate mercury bearing particles promoting sedimentation. Thus mercury is removed by stream sediments and related fine grained materials within a distance of a few km after being introduced into streams [382, 403] depending on the composition of the aquatic medium, redox potential, pH, temperature, amount of the suspended sediment, mineralogical-chemical nature of the sediment, presence of complexing agents and existence of aquatic biota. In rivers, sediments are transported with a speed between 4 and 80 km/yr [404, 405]. Mercury in lakes will be deposited and covered with layers of other deposited materials at a rate between 5 mm/yr and less than 1 mm/yr depending on whether the lake is eutrophic or oligotrophic [95]. When the mercury pollution source is eliminated, mercury will be slowly released from the bed sediment until a steady state condition is reached.

The most important physical influences on the distribution of mercury are motions of any kind such as currents, waves, turbulent mixing processes which are extensively commented upon [380]. Some mathematical models were introduced which allow general prediction of the behaviour of pollutants [406, 407] if chemical and physical behaviour in addition to runoff, topography, locally induced currents for coastal areas [408] as well as action of bacteria, yeast and other microflora and the composition of the bottom sediment are known. Thus inorganic mercury which is highly preconcentrated in bottom sediments [383, 409, 410] reaches the aquatic food chain [412, 413] via the methylating activity of bacteria [164,411] and it is widely distributed in the biota (Fig. 4).

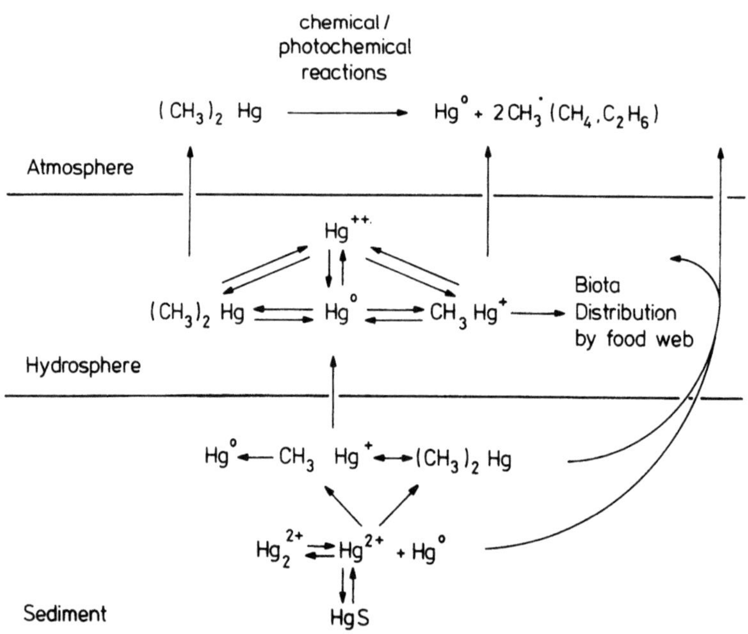

Fig. 4. Interconversion of mercurials and their mobility in the aquatic environment [413, 646]

Soil. Mercury distribution in soils has a characteristic profile [414, 415] (see: accumulation of Hg in soil) and its mobility appears to be due to redox potential, pH [416], drainage, type of soil [417], and other factors [418–420]. Thus sulfur containing amino acids and proteins form very strong soluble complexes [441, 422]. Humic acids form strong complexes of relatively low solubility [418, 423]. Investigations on selective extraction suggest that both metallic and ionic mercury are adsorbed in the form of a humate [424] since none of the common and stable mercury compounds including HgS were found [425, 426]. Thus, the leaching into deeper layers is small. The mobility of mercury in soil depends on many factors, e.g. reduction by chemical processes, microbes, plants, and other living organisms or biotransformation into volatile mercury compounds [166, 428–431]. Models describe the behaviour of mercury in soil and evaluate time constants between 36 and 3,600 yr [371, 427].

Aquatic Food Chain. Mercury, which arrives at the aquatic environment for the most part as inorganic and phenyl mercury [432], is quickly adsorbed by organic and inorganic particulates. This particulate matter is deposited in sediments, where, in turn, inorganic mercury may be transformed into methyl mercury. Phytoplankton (the main primary producers in the aquatic environment cf. [433]) as well as Zooplankton concentrate both inorganic and alkylated mercury [409, 434–436] and they thus enter the food chain efficiently (Fig. 5).

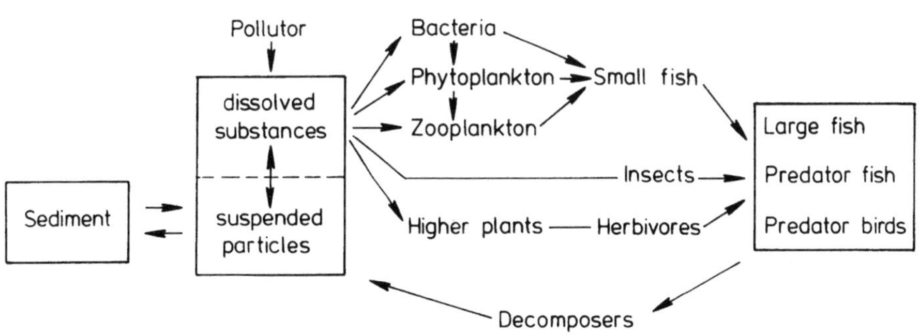

Fig. 5. Mercury cycle in the aquatic environment [645, 646]

A significant transfer from coastal pollution sources into the open ocean marine biota, possibly occurs through the food web connecting inshore plankton – where the mercury concentration is relatively high – to higher trophic levels rather than by direct transport through water [436] (see also accumulation in marine biota).

Terrestrial Food Chain. Mercury may enter the terrestrial food chain by way of seed eating species. Comparative analyses on feathers of museum birds showed an increase in mercury concentration approximately at that time when seed dressing with methyl mercury compounds started [437, 438]. In vivo studies on terrestrial fauna, e.g. predatory birds [439], game, singing birds, and rodents [440] with contaminated feeding confirmed the origin of increased levels in tissue and eggs. After the use of alkylmercurials for seed dressing was banned, the mercury levels in wild-life decreased substantially [441].

Plants take up small amounts of mercury in ionic complexed [442] and gaseous form through leaves [443], also from dry fallout [435], and via roots [442] (see also accumulation in plants). A transport from the leaves to the roots and into the fruits is more likely than the converse [444–450]. A repercussion on man from the uptake of mercury by plants from soil and the atmosphere is as yet unknown. Figure 6 displays a flow chart of the movements of mercury in the ecosystems.

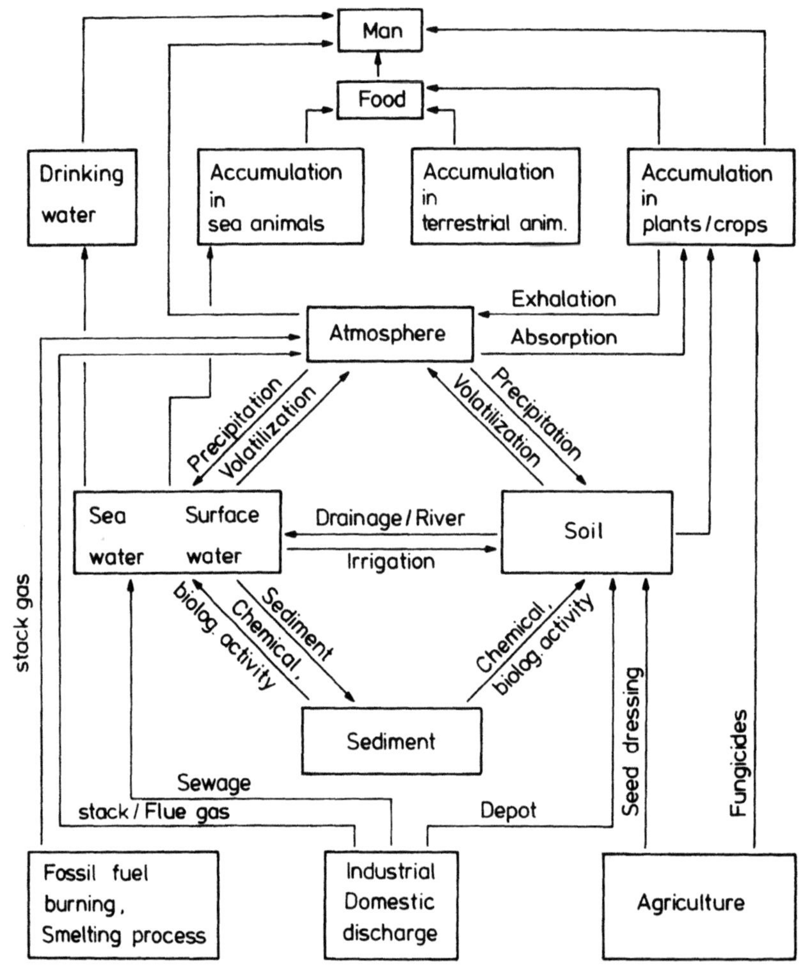

Fig. 6. Flow chart of industrially derived mercury (modified from [609])

Chemical, Biochemical, and Photochemical Reactions

At all times the disproportionation reaction

$$Hg_2^{2+} \rightleftharpoons Hg^{2+} + Hg^0$$

is prominent in the consideration of reactions in respective media [413]. Which compound prevails, depends on its solubility and to which extent metallic mercury enters or leaves the system, and on external factors that affect the biosphere [451].

Conversion Between Inorganic Forms

$Hg^{2+} \rightarrow HgS/HgSe$: Wherever sulfide and selenide ions are present, mercury sulfide or selenide form owing to the great affinity of mercury for sulfide sulfur ($K_S = 10^{53}$) and selenide. The conditions under which HgS is stable in aqueous solutions can be evaluated by the Eh-pH diagram [94, 396]. HgS seems also to be stable under anaerobic conditions. In excess of sulfide ions the complex HgS_2^{2-} is formed [452] depending on the pH [453]. A reaction which is believed to occur in soils but on which no exact information is available.

$HgS \rightarrow Hg^{2+}$: Humic compounds (fulvic-, humic acid, humin) increase the solubility of HgS by complex formation [424, 426]. It seems likely that an enzymatic reaction [413, 454] oxidizes the sulfide to sulfite and sulfate releasing bivalent mercury ions, which then undergo further conversion.

$Hg^{2+} \rightarrow Hg$: The transformation from the cationic to the elemental state can occur chemically under suitable reducing conditions, e.g. in the presence of humic acid [397] or by bacterial cultures (Pseudomonas), yeasts, and other microflora [455–457]. As a method for detoxification under strictly anaerobic conditions the reduction of Hg^{2+} to Hg^0 becomes an important consideration [458].

$Hg^0 \rightarrow Hg^{2+}$: The oxidation depends on the redox potential in a medium which can be calculated from the formula

$$E = 850 + 30 \log \frac{[Hg^{2+}]}{\alpha}$$

where α is an estimation of the strength of the binding between bivalent mercury and the available complex forming substance [459]. For mercury complexes with organic soil α has been calculated to be $>10^{21}$ [460]. This means that oxidation of metallic mercury to inorganic bivalent mercury takes place in an aquatic environment if an organic substance and oxygen are present [452], e.g.

$$\begin{array}{ccc} & C_2H_5Hg^+ & \\ & \downarrow & \\ ArHg^+ \longrightarrow & Hg^{2+} & \rightleftharpoons CH_3Hg^+ \rightleftharpoons Hg(CH_3)_2. \\ & \downarrow & \\ & CH_3O\text{-}(CH_2)_n\,Hg^+ & \end{array}$$

Conversion Between Organic and Inorganic Forms

$R{-}O{-}(CH_2)_nHg^+ \rightarrow Hg^{2+}$: Alkyloxyalkyl mercury compounds are unstable in acid media. At pH = O the half-life of the reaction is about 10 min, in humic soil (pH = 5), three days [452].

$ArHg^+$, $R{-}Hg^+ \rightarrow Hg^{2+}$: Organomercury compounds can be degraded chemically and biochemically and by the effect of u.v. radiation. In general, the stability of the compounds decreases as the carbon chain increases.

Different papers describe the breakdown of organomercurials into inorganic mercury in aqueous solution by u.v. radiation from low pressure lamps [461] and with a special inversion radiator in the presence of oxidizing agents [207].

$Hg^{2+} \rightarrow CH_3Hg^+/(CH_3)_2Hg$: Mercuric ion can be abiotically methylated by, e.g. methylcobalamin ($B_{12}{-}CH_3$) [462, 463] or trimethylsilyl salts [464], and biotically by enzyme systems [357].

Two different pathways for methylation are reported. One in the presence of cell free extracts of a bacterium strain [465], $B_{12}CH_3$ and ATP in anaerobic conditions where the methyl group is transferred in a nonenzymatic reaction to the mercury ion and $B_{12}CH_3$ is regenerated enzymatically. The other reaction is an enzymatic methylation of mercury bound to homocystein as observed, e.g. in cells of Neurospora Crasia [357, 358].

Biological methylation has been observed in river, lake and sea water [164, 466] in soils [467, 468], and in sediments where HgS also is methylated under aerobic conditions mostly in the top layers by the reaction of various strains [412]. The reaction begins with a chemical oxidation of the sulfide followed by biological methylation. CH_3Hg^+ and $(CH_3)_2Hg$ are formed according to aerobic or anaerobic conditions respectively [459]. From soils a methylating substance can be extracted whose efficacy is dependent on temperature, mercury concentration, pH, and type of soil [430, 431]. Humic acid seems to affect the methylation [431].

Some microorganisms in the intestine of yellow sea tuna and in the slime of the fish have the capability to methylate mercury [469].

$CH_3Hg^+ \rightarrow Hg^0$: Microbial degradation occurs in river, sea, and lake sediments [470]. Thus the environmental methyl mercury concentration is maintained at a minimum by the continuous cycle of breakdown and formation [471].

$(CH_3)_2Hg \rightarrow Hg^0$: In the atmosphere dimethyl mercury is photolysed by u.v. radiation whereby radicals might form [472]:

$$(CH_3)_2\,Hg \xrightarrow{\text{u. v.}} CH_3^0 + CH_3\,Hg^0 \xrightarrow{\text{u. v.}} Hg^0 + C_2H_6.$$

Further decomposition of the monomethyl mercury radical leads to metallic mercury and another methyl radical which can abstract hydrogen or recombine, to give rise to methane or ethane respectively [457]. Both organic and inorganic compounds in the atmosphere yield elemental mercury in presence of sunlight [418]. Dust particles onto which mercury is adsorbed may act as activation sites for photochemical processes [376]. A direct photolytic cleavage of dimethyl mercury in the troposphere is unlikely, however, due to the influence of OH, O(^{1}D), and O_3 [473].

Conversion Between Organic Forms

$CH_3Hg^+ \rightarrow (CH_3)_2Hg$: The formation of dimethyl mercury from bivalent mercury in the presence of vitamin B_{12} [457], e.g. in decomposing fish or in sediments runs over monomethyl mercury as an intermediate.

$(CH_3)_2Hg \rightarrow CH_3Hg^+$: Dimethyl mercury is unstable at low pH values. During its breakdown monomethyl mercury is assumed to form as an end product or as an intermediate metabolite.

Transalkylation Reaction

$CH_3\,Hg^+ \rightarrow CH_3Se^{3+}/(CH_3)_2Se^{2+}$: A transfer of methyl groups from methyl mercury to selenium has been observed in in vivo studies. Dimethylselenide is released [474]. Selenium salts can also attack the Co–C bond in presence of thiols [475] resulting in a transfer of the methylgroups of the Hg–Co cycle to the selenium cycle. Moreover, anhydrous selenium salts react with methyl mercury to give dimethylselenide as a major product [476].

Metabolism

Metabolism of mercurials has been investigated with labelled compounds and was found to correlate with a series of factors, e.g., type of compound (inorganic mercury differs from organomercurials, which themselves differ greatly from each other. Arylmercurials are rapidly metabolized whereas the metabolically stable alkylmercurials resist degradation into the inorganic form [477]), the species to which mercurials are administered, dose, mercury body burden and the toxic effects of the individual compounds. A detailed, excellent review has been given [478].

Mercury and its compounds can enter the organism *via* the lungs by inhalation, the gastrointestinal tract by ingestion, the skin and via the placenta into the fetus.

Uptake of Inorganic Mercury

Elemental Mercury Vapour. The elemental form penetrates the skin of humans [479], and animals [480], and on account of its sparing solubility in water, it penetrates on inhaling far down the bronchial tree to the alveoli [481]. In animals between 25 and 100% are retained [482–484], whereof a part was found to be in the lung – with a half-life of 5–10 h [485], and in the blood [481]. A similar deposition mechanism is assumed to occur in humans as can be inferred from autopsies [486]. But less than 0.1% is absorbed by blood and organs if mercury is administered to the gastrointestinal tract [487].

Mercury Compounds. In general, aerosols of inorganic compounds are absorbed to a lesser extent than mercury vapour [488]. Deposition in the air ways and the lungs depends on particle size and density. Half-lives in the peripheral lung tissue lie between a day and 1 yr [489]. Gastrointestial absorption of salts is governed by their solubility and may amount to 20% for mercuric acetate [490], but less than 2% and 8% for mercuric chloride for mice [491] and humans [492], respectively. Penetration of skin has been observed with mercuric oxide, ammoniated mercuric chloride [493, 494] with potassium mercuric iodide [495], mercuric chloride [496] and with various other compounds in pigs.

Organic Mercury Compounds

Alkylmercurials. Respiratory uptake of methyl mercury iodide, chloride and dicyandiamide were found on various animals [497, 498]. Within 45 s of exposure 50–80% of the offered dimethyl mercury was absorbed by mice [499].

Gastrointestial absorption has been studied with methyl mercury chloride on humans [500], and on mice [491] with ethyl mercury on cats [501] and with various alkyl mercury salts on rats [502]. Methyl mercury dicyandiamide is absorbed from water solution through the skin of guinea pigs [503]. Placental transfer was observed with methyl mercury salts in mice [504], and guinea pigs [505], and humans [506, 507] (see also toxicology).

Aryl Mercury Compounds. Phenyl mercury acetate (aerosol/particle size 0.6–1.2 μm) is absorbed by animals upon inhalation within 1 h [498], penetrates the skin of rats (25% within 24 h) [508] and of humans [509] and is better absorbed from the gastrointestinal tract than inorganic mercury as was found in experiments on various animals. Measurements of the excretion in the faeces indicate absorption rates between 10% [510] and 40% [490]. Only limited mercury levels were found in the foetus indicating a limited placental transfer [511, 512].

Biotransformation

Inorganic Mercury. In contrast to in vitro studies on blood where elemental mercury is quickly oxidized – no differences in distribution and toxicity between inhaled mercury vapour and absorbed mercuric salts, and a binding by haemoglobin solution rather than by plasma were observed [513] – in vivo studies on various animals show a higher uptake by the brain which allows the conclusion that mercury in blood passes the lungs in elemental form [514, 515], and is only slowly converted into ionic form by enzymes [516]. The reverse process can also occur [483, 514]. 67%–84% of the total blood mercury is found to be in blood cells immediately after exposure to mercury vapour as opposed to 25–31% if mercury ions are injected intravenously to animals [515]. A large part of elemental mercury is taken up by the erythrocytes where it may be dissolved in the lipid structure [477].

Organic Mercury Compounds. Investigations with different methyl mercury salts on various animals showed no definite difference in metabolism [501, 517, 518]. After monomethyl mercury administration mercury is mainly found in the blood cells. The extent depends on the species of the animal and on the dose administered [499].

Dimethyl mercury administered to mice by inhalation or intravenous injection was found to be in fat deposits [499]. Two kinds of transformation of monomethyl mercury can be assumed. A metabolic transformation of the methyl groups in situ, or a breakage of the covalent bond between carbon and mercury. On the one hand the slow and even elimination of mercury after administration of monomethyl mercury to various animals indicates a rather high stability of the covalent bond. More than 90% of injected methyl mercury dicyandiamide was still found as organomercury after 6 weeks in liver, spleen, and blood, 75% in plasma and brain and 55% in kidney [519]. On the other hand there is evidence of a small breakage of the covalent bond in liver [499], and in the intestinal lumen [422, 520]. 20–90% of dimethylmercury administered to mice are rapidly exhaled, the remainder was metabolized within 20 minutes after administration into methyl mercury ion and was detected mainly in liver and bronchi [521].

There is no difference in metabolism between different salts of methyl mercury in rats [501, 522]. The compounds are almost exclusively firmly bound to the haemoglobin in the red cells [522]. Eight days after administration in the organic form more than 94% were still detected in liver [523] and brain. Metabolism into inorganic mercury takes place in the organs mentioned with time but mainly in the kidney (34% after 8 d) [524].

Aryl Mercury Compounds. Investigations are mostly restricted to phenyl mercury. No measured differences in metabolism of the different salts have been established in animals [501, 504, 522]. High levels are found to be in the blood – for the most part attached to blood cells – in liver and in kidney, not more than 20 and 10% respectively in the form of organic mercury [525]. In another study 85% of a subcutaneous administration dose appeared in the urine and about 5% in the breath within 4 days [526] which indicates relatively quick breakage of the mercury carbon bond probably after ortho-hydroxylation [527].

Alkoxyalkyl Mercury Compounds. The metabolism of this type of compound has mainly been investigated with methyloxyethyl mercury salts in animals, indicating a fairly rapid breakage of the carbon mercury bond [528]. Within 24 h about 50% of a single dose was exhaled together with ethylene and carbon dioxide. The percentage of the organic mercury in the kidney decreased from 50% after a few hours to nearly zero after one day. About 10% of the mercury was excreted in the urine first in the organic then exclusively in the inorganic form [528, 529].

In conclusion we can summarize that alkyl mercury compounds (mainly methyl mercury) have the highest stability in the body. The highest levels are found to be in the blood according to the declining order.

Alkylmercury > phenylmercury > inorganic mercury

They are found also in tissue of kidney, liver and brain. The distribution to the brain is very slow but mercury which is present there as methylmercury has a long half-life time.

The excretion occurs mainly via faeces, via kidney into the urine, the hair and to a very small extent via the milk [500, 520, 530, 531]. The normal excretion with urine is about 10 μg/24h. Levels over 40 μg are assumed to be due to an intoxication [532].

Biodegradation – Decontamination of Polluted Areas

Biological degradation is a natural process of decontamination and detoxication of polluted systems. In addition to this, measures have been proposed to restore areas locally polluted by man [43, 451, 533].

Biological Degradation. Microbes have the capability to degrade inorganic [534] and organic mercury compounds, as has been observed to occur in lake sediments [535], soil [536], sludge [537], and in model tests to study biodegradation of methyl mercury compounds [470].

A series of factors, e.g., type of microorganism, mineral salt-composition in the medium, supplementary nutriants, pH, temperature, and light have been shown to affect this process [451]. The bacterial strain *Pseudomonas aeruginosa* obtained from aquatic media was found to convert mercury ion to elemental mercury [166, 538]. The strain K 62 from the genus *Pseudomonas* isolated from soil, which is capable of mercury uptake and conversion, was used to remove mercurials that were present in industrial waste waters. This strain shows a high resistance to both inorganic and organic mercurials which are loosely adsorbed onto the cell surface [534, 540]. The cell wall is then biologically stimulated to induce vaporization of mercury, a process which might be prompted by a gaseous substance secreted from the bacterial surface [541, 542] or the mercurial might be chemically transformed into a form, e.g., elemental mercury, which is more volatile [543]. Furthermore, selected strains of bacteria even show a high degree of tolerance of mercury. For instance, *Pseudomonas* and *Pseudomonas* like bacteria exhibit growth inhibition at mercury concentrations in the percent range [541, 542]. The bacteriostatic activity of mercurials towards bacteria may be a result of a different type of chemical or biological binding. In a pertinent study it has been established that the mercurial is not deposited in the cell wall of the bacterium but is attached to the cytoplasma [544].

Removal of Contaminated Sediments by Dredging. This kind of decontamination of an aquatic ecosystem has been investigated by laboratory experiments [533], and practically exercised by dredging lakes [545, 546]. The dredged sediments can be deposited in settling ponds or they may be buried.

This should, however, happen together with sand, silicates, or inert clays in order to bind mercury, thus avoiding recontamination by drainage water [533].

Conversion of Mercury to Mercuric Sulfide. Techniques which have been proposed are

a) covering a mercury sediment with FeS or FeS_2, enabling formation of mercuric sulfide by exchange of the sulfide ion [533],

b) change of the redox potential, which is to a very large extent determined by the concentration of dissolved oxygen in an aquatic medium [547].

The rate of biological conversion of mercury depends on it. Thus conversion of aerobic to anaerobic conditions, e.g., by adding oxygen consuming easily degraded organic substances such as glucose [533] or plants [547] favours the reduction of sulfate to sulfide and with it also the formation of mercuric sulfide. This is sparingly soluble and undergoes methylation only at very high concentrations. Simultaneously the redox potential in anaerobic environments can become so low that the oxidation of elemental mercury which is necessary for biological methylation hardly occurs.

Conversion of Mercury into Dimethyl Mercury by Raising the pH. The process of biological methylation of mercury is determined by the pH value [548]. The composition of microorganisms changes with pH levels. Higher pH favours those producing dimethyl mercury, which can evaporate while lower pH those forming monomethyl mercury [410, 549] which is more likely to accumulate in aquatic biota. Lower pH values adjusted with $CaCO_3$ yielded higher methyl mercury levels in fish [550].

Other Techniques. Proposed methods are: coverage of the bottom of mercury contaminated lakes with a plastic coating, amalgamation of mercury with metals and the use of shrimps, crabs and clams to biologically extract mercury in aquatic media [551].

Most of these methods are only of theoretical value because they are either too costly, as large areas have normally to be restored, or are associated with ecological damage.

Accumulation

The extent to which mercury has been accumulated in the different ecosystems can be ascertained if the respective background concentrations are known.

Atmosphere. Background concentrations vary between 0.001 and 50ng/m^3 [264] depending on the extent of urbanization. An average value of 1–2 ng/m^3 is assumed [107]. Much higher values were measured over industrialized areas and mercury deposits (Table 8). In air various forms of mercury occur, which can be partly particulate bound [254, 264, 387]. Thus, e.g. in a speciation measurement, elemental mercury (1–15 ng/m^3) monomethyl, dimethyl, diethyl mercury (150–250 ng/m^3), and particulate bound mercury (1–10 ng/m^3) was found [254].

Mostly total mercury concentrations are given which depend on numerous factors, e.g., site and altitude [254, 375, 552] – lower levels are found at higher

Table 8. Mercury levels in air

Description	Concentration range [ng/m^3]	Ref.
Global average	1–10	[89, 533, 267, 627]
Atlantic (1977)	0.4–20	[627]
FRG (1977)	2–37	[627]
Russia	< 10	[674]
USA (San Francisco Bay)		
Summer	2–50	[376]
Winter	1–25	
Chicago	2–10	[387]
Industrial areas	150–550	[254, 533]
Urbanization	150–400	[254, 388]
Japan (non industr.)	< 14	[675]
Air over deposits mines, geysers	30–10^6	[89, 533, 676]
Air over agricult. area (fungicides)	10^4	[675]
Volcanic exhalations		
Russia	100–9,600	[89]
Hawai	730–40,000	[100]

elevations; temperature and barometric pressure [552], sunlight, wind speed and direction [389, 553]. Daily but also diurnal differences have been established [256, 388].

Analyses of permanent ice sheet indicate an increase of the total mercury burden in course of time. The average concentration in ice for the period 800 B.C. up to 1952 was found to be 60 ± 17 ng/kg as opposed to 125 ± 52 ng/kg for the period 1952–1965 [98].

Hydrosphere. Part of the atmospheric mercury is washed out by rain. Levels in rain water lie between 0.005 and 0.48 µg/l [267, 417, 554]. The content of mercury in an aquatic environment (Table 9) depends on many factors. Mostly total mercury concentrations are given [555, 556]. A discrimination between inorganic and organic forms has been made for river and coastal sea water wherein organic mercury makes up about half of the total portion [557].

In ocean water a vertical distribution of mercury from about 0.1 µg/l at the surface to 0.15–0.27 µg/l at greater depths [558, 559] appears to be due to the uptake of mercury by plankton and the subsequent conveyance to depths by marine biota [409].

For the applicability of surface waters for the drinking water supply national and international guidelines exist [560, 561] which are compared with mercury level of some German waters (Fig. 7).

Sediments. Mercury which enters rivers, lakes and oceans for the most part ends up in the sediments [108, 380, 563, 564] (Table 10). There it is accumulated with a distinct increase towards the surface [563], which might be

Table 9. Mercury levels in aquatic media

Description	Concentration range [μg/l]	Ref.
Rain water		
unpolluted	0.02 -0.48	[264[a], 267, 417, 554]
near volcano	0.25	[100]
Surface water		
unpolluted	0.1	[555]
Drainage water		
unpolluted	0.05	[684]
Ground water		
unpolluted	0.01 -0.46	[403, 685]
near Hg-deposits	1-1000	[43[a]]
River water		
unpolluted	0.01 -0.2	[267, 403, 686]
near Hg-deposits	0.5 -100	[43[a]]
Rhine (Wiesbaden)	0.03 -8.4	[556[a]]
Lake water		
Ontario (Canada)	0.048	
Lake Constance (FRG)	0.03 -0.38	[556[a]]
Sea water		
North sea (1934)	0.03	[267]
(Belgium 1972)	0.03 -0.76	[556[a]]
Atlantic	0.001-1.6	[688, 687, 689]
Greenland sea	0.016-0.364	[690]
Pacific near shore	0.012-0.15	[559]
Hot springs and mineral waters	0.01 -20	[43[a], 555]
Oil field brines and saline waters	0.1 -230	[43[a], 555]

[a] Reviews

explained by the relatively high mobility of mercury in the interior of the anaerobic sediment and its continuous concentration in new deposits [380]. Profile analyses (Fig. 8) [565, 566] suggest that the preconcentration ensue from man's activity. Detailed compilations of mercury in sediments are available [556, 564, 567].

Marine Biota. Plankton and zooplankton the first links in the aquatic food chain take up and concentrate both inorganic and alkylated mercury com-

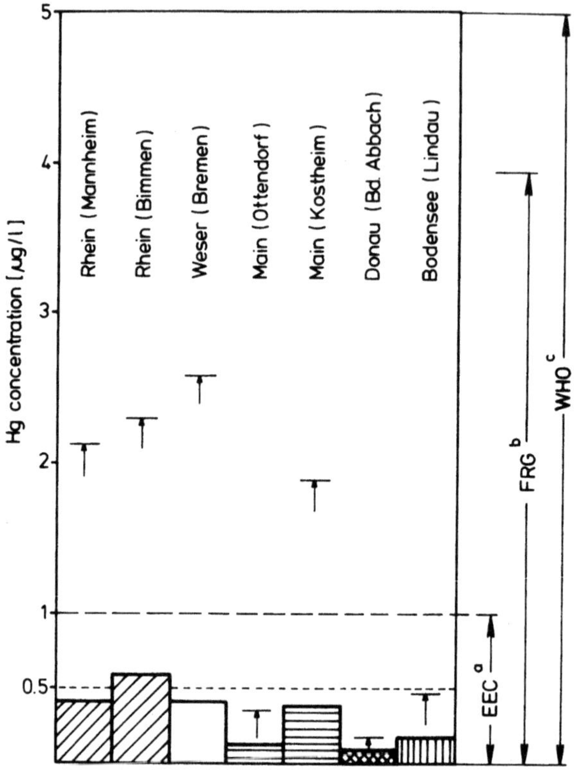

Fig. 7. Average levels of dissolved mercury in river and lake water (FRG) and threshold limit values (TLV) for the applicability of waters as drinking water supply [561]. –: Maximum levels, **a:** TLV of the EEC [560], **b:** TLV of the FRG (1975 [562]), **c:** Internat. Standard (WHO)

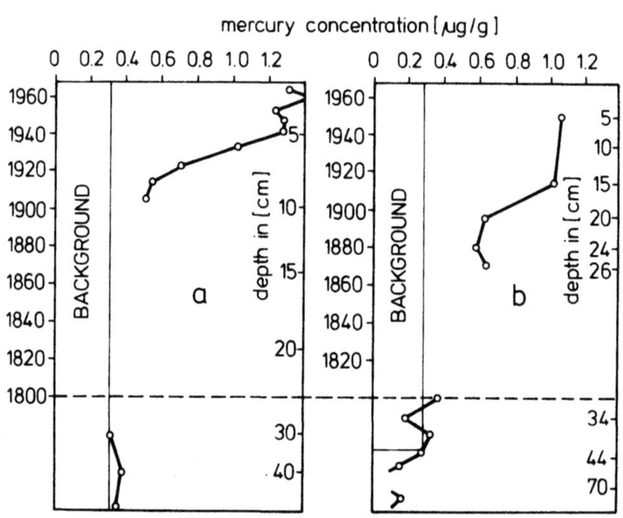

Fig. 8. Mercury in sediments of lake Ontario *a* [566] and lake Windermere *b* [565]

Table 10. Mercury levels in river-, lake-, and sea-sediments

Description	Concentration range [μg/g]	Remarks	Ref.
Background level	~ 0.06	19th century	[380]
Unpolluted waters	~ 0.05		[267, 95]
Ocean	0.1 -1	Uplitted sedim.	[385]
North Sea	0.01-5.7		[556]
Lake of Ontario	0.35-1	Fraction < 63 μm	[556]
Wisconsin river and lakes	0.4 -2.7		[680]
Wisconsin river and lakes	684	Vicinity of chloralkali industry	[680]
River sediment (Rhine, Koblenz)	4.5		
Swiss lakes	0.01-2.23	Fraction < 63 μm	[681]
Lake Sangchris (Illinois, USA)	0.037	Before coal-fired power plant operation (1965)	[54]
Lake Sangchris (Illinois, USA)	0.049	Power plant in operation (mean 1968-1973)	[54]
Minamata Bay (Japan)	2,010	Wet weight	[682]

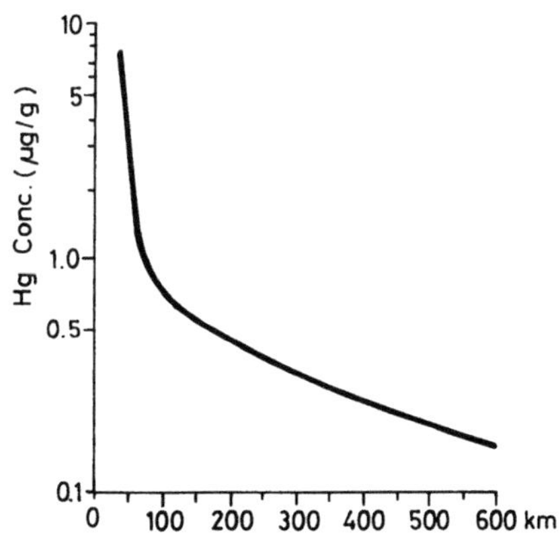

Fig. 9. Concentration of mercury in plankton in relation to the distance from North American Coast [436]

pounds by direct assimilation from the adjacent medium (Fig. 9) [409, 434 to 436]. Concentration factors up to 100,000 are reported [568]. Higher trophic levels feed upon these organisms thus forming a biological magnification from algae feeders (mercury concentration of 0.001–0.18 μg/g) to predators such as pike, tune and shark (mercury concentration 0.01–5.82 μg/g) [569].

In fish concentration factors of 5,000 up to 100,000 are reported [69, 568, 570] because they take up mercury by ingestion and from the adjacent water.

In pike caught, e.g., at various distances down stream from a paper mill up to 8 μg/g [571], and in rainbow trout exposed to methyl mercury (60 ng/g, 1 h a day) 17.4 μg/g were found [568]. A comparison of museum specimen and fish caught in a river with a chloralkali plant clearly shows up an exponential increase of mercury. Similar results yielded comparative studies on osprey and grebe [571] (Fig. 10).

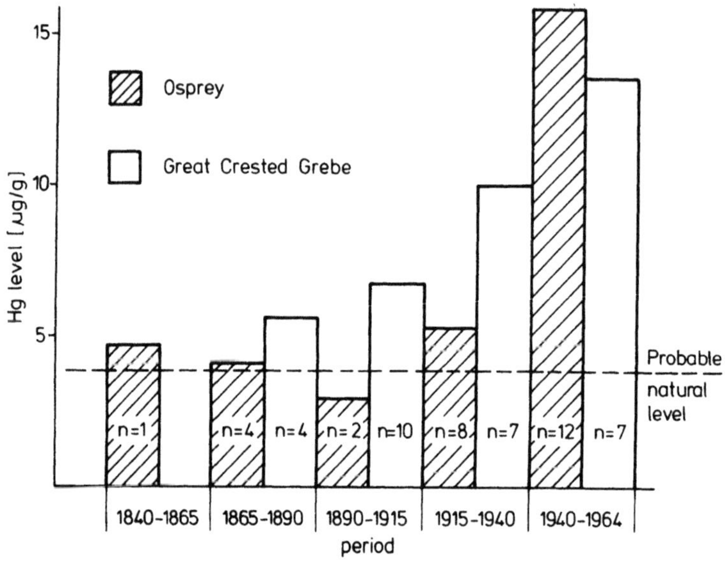

Fig. 10. Mercury levels in feathers of osprey and great crested grebe [571]

The highest concentrations in fish were found in the liver, kidney and muscle [568, 572, 573] (Table 11) but also in gills and skin depending on the water being contaminated with inorganic or organic mercurials [574]. Dimethyl mercury and some monomethyl mercury compounds can directly be taken up by diffusion across the gills [575], while pike and trout are able to concentrate orally-ingested protein bound with methyl mercury in muscle tissue [576].

Mercury was not found to accumulate in tissues of water plants [572].

The mechanism of accumulation is not clear but seems to be a function of metabolic rate in individual fish, differences in selection of food objects as the fish matures, or the fish's epithelial surface area [572, 577]. Detailed studies review the accumulation of the mercury in water organisms regarding species, mercurial [556, 572, 578], exposure time, and distribution to different organs [556, 579].

Soil. The global average concentration of mercury in soil is estimated to be somewhere between 50 [580] and 100 ng/g [95]. Figures range from 0.1 to 5 μg/g, depending on numerous parameters (see transport through soil).

Table 11. Total mercury concentration in some aquatic organisms

Species	Concentration range [µg/g]	$\bar{x}$ [µg/g]	Remarks	Ref.
Plants	0.03–0.64	0.2	Ruhr (1970–1972)	[556]
Plankton	0.1 -5		Depending on distance from shore	[436]
Fish:				
pike	0.19–0.59	0.44	The Netherlands (1970)	[556]
pike	1.2 -8		Vicinity of paper mill	[571]
rainbow trout	2.8		FRG (1934)	[267]
	0.02–0.08	0.03	UK (1972)	[556]
perch	0.57–1.9	0.85	The Netherlands (1970)	[556]
			Sweden (1967)	[571]
Organs of pike:				
heart muscle	1			
liver	0.78			
kidney	0.64			
gill	0.3			
scales	0.1			

Locally, close to strong polluters, such as chloralkali plants [581], coal-fired power plants [582] and deposits [583] the mercury levels can build up to as much as 10 µg/g and more. While in rocks mercury is distributed more or less homogeneously with depth, in soils it has its highest concentration in the upper 5 to 20 cm [175] (Fig. 11a, b).

A profile analysis of a high bog (Fig. 11c) suggests this accumulated mercury to be of anthropogenic origin. From that a background level of about 18 ng/g can be derived as opposed to about 250 ng/g at a depth of about 10 cm. On the immediate surface evaporation as a result of chemical and biological processes yield lower levels. The effects of mercury enriched soil on the terrestrial food chain are not yet known.

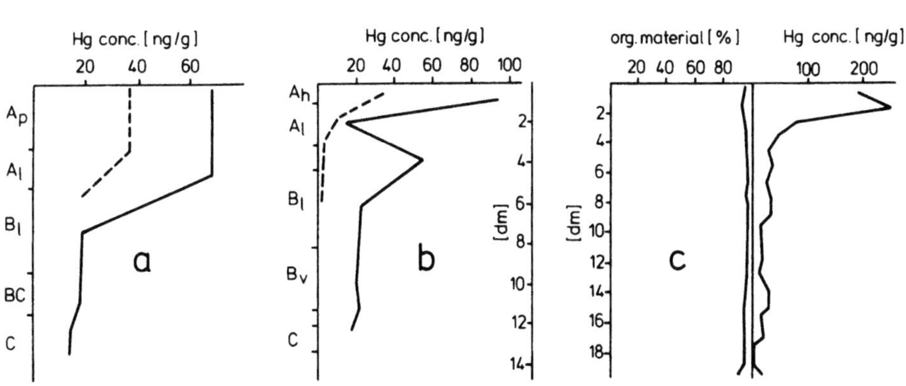

Fig. 11 a–c. Mercury distribution in soil [414], **a**: Arable, **b**: Forest, **c**: High bog,
– – – – – ‰ Humus

Terrestrial Plants and Fruits. Some plants take up mercury from soil depending on type of soil, plant, and form of mercury [100, 585–588]. For instance, mercurous or mercuric mercury chloride is taken up by the root system of lettuce, and carrot plants [435], pines and deciduous tree [589, 175] but there is little translocation into the aerial parts [586]. Mercury levels in soil ≥ 1 mg/g reduce the yield of cultivated plants by 50% [585]. Grain, grown from dressed seed has up to two times the mercury content as crops from untreated seeds [590, 591]. In the application of methyl mercury [592], all parts of the plant contain methyl mercury [593].

These data are contradicted by other investigations which assume mercury not to be taken up by plants from a contaminated soil [444, 446, 448]. A translocation from leaves – after being sprayed or after uptake of airborne mercury [594] – into the root system [447], as well as into fruits [449, 450], e.g., limes [595] potatoe tubers [596, 597] the pulp of tomatoes, and into rice [598] is more likely to occur. A comparison of mercury contents in food stuffs, from 1934 [267] with present-day data (Table 12) [69, 556, 578, 579, 585, 587, 614]

Table 12. Mercury levels of some foodstuffs

Foodstuff	Concentration range [μg/g]	Ref.
Vegetables:		
fresh	0.001–0.05	[586]
canned	n.d. –0.06	[586]
Fruit:		
apples	0.002–0.18	[556]
Eggs:		
egg white		[599]
total Hg	0.023	
methyl Hg	0.023	
Meat:		
pork	0.003–0.5	[556]
ox		[599]
total Hg	0.074	
methyl Hg	0.068	
canned Meat	0.01	[556]
Prepared food		
baby food	n.d. –0.02	[679]
sauces	n.d. –0.02	
Flour	<0.005–0.1	[556]
Mushrooms		
Yellow Bolete	3.17 –8.77	[556, 678, 599]
Field Mushroom	3.21 –6.09	[556, 678]
Drinking water (Lake Constance)	0.01 –0.08	[556]
Beer, Cider	0.01 –0.02	[556]

with consideration of mushrooms [599] and proportion of methyl mercury [600] does not show a significant increase unless mercury containing waste from industrial, agricultural or mining processes are discharged into local water systems or where plant, fruit and seed treatment have contaminated game, other wild life or food [601, 602].

Terrestrial Animals and Man. In general, terrestrial wild-life has lower natural contents of mercury than aquatic organisms [578]. There is a difference in distribution and accumulation in different organs [578] depending on the type of mercurial [499, 511].

For instance, organs of pheasants from an area where seed was treated with methyl mercury [511] or which where continuously fed for 60 days with methyl mercury dressed seed [615] contained up to 10 and 40 times more mercury respectively, than normal birds. High mercury levels were found to occur in the liver, spleen, and kidney of a doe in the vicinity of a chloralkali plant [594]. Generally, in animals high concentrations were found to occur in muscles, kidney and liver but mainly in the kidney and especially in some parts of the brain with acute and prolonged elemental mercury exposure [478, 578, 603].

For an evaluation of the risk of accumulation in different critical organs and different kinds of exposure a mathematical model for the kinetics of mercury exchange has been proposed [604]. An accumulation of mercury in

Table 13. Mercury in human tissues [612]

Tissue	Concentration range [μg/g]
Blood total[a]	0.005 -0.02
plasma[a]	0.002 -0.01
serum[a]	0.012
Bone	0.45
Brain	0.005 -2.94
Gastro-Intestinal Tract	
unspecified	0.075
stomach	0.0083-2.27
Hair	1.25 -7.6
Heart	0.005 -0.15
Kidney	0.0063-2.75
Liver	0.005 -3.7
Lung	0.01 -0.25
Muscle (Skeletal)	0.004 -0.71
Nails	0.07 -7.27
Ovary	0.2 -2.14
Pancreas	0.05 -1.14
Placenta	0.06 -0.12
Skin	0.003 -3.34
Spleen	0.004 -1.5
Urine[a]	4.3 -114

[a] μg/ml

man can easily be established by analyses of e.g., hair and blood of fish eaters [605–608]. Normal blood is reported to contain between 5 and 20 ng/ml, but that of fish consumers up to 100 ng/ml. In hair the concentration factors are yet higher [441, 610]. In exceptional cases (Minamata patients) up to 249 μg/g [611] were found as opposed to about 5 μg/g in hair of persons with no occupational exposure [609]. Detailed compilations of mercury in human tissues of nonexposed [612] (Table 13) and exposed persons [613] are available.

Persistence

Aquatic Environment. Polluted river, lake, and sea sediments are a serious hazard because the mercury being confined may remain active by methylation processes for some 100 yr [43] even if the source of pollution is eliminated [616]. The persistence of methyl mercury is relatively high because it is metabolized very slowly. Retention times of one year up to 3 yr [286] depending on the species are reported (Table 14). The strong binding of

Table 14. Methylmercury half-life times in fish [575]

Species	Half-life (days)
Flounder	400–700
Perch	500
Pike	500–700
Eel	900–1000

methyl mercury with fish can not even be disrupted by boiling or frying [77]. Biological half-lives of inorganic and phenyl mercury compounds generally are shorter than those of methyl mercury in all aquatic species [619].

Terrestrial Environment. Generally the persistence in animals depends on the mercury compound. For instance 90% of phenyl mercury acetate administered intravenously, intramuscularly, and orally to rats, chicks, and dogs is metabolized into inorganic mercury and then excreted within 96 h [525] and 40–60 days after administration of a single dose [603]. Ethyl mercury similarly administered is detectable in kidney and in liver for at least 21 days [621]. Methyl mercury, and elemental mercury vapour taken up by the brain remain there for a long time [622]. In soil the persistence of fungicides is many months [487, 578]. Compounds of this kind penetrate, the pulp of tomatoes, after application and persist for 2–3 weeks [623]. A review summarizes the distribution and retention of mercury in plants and fruits after several treatments with fungicides [43].

Humans. The evaluation of the retention of mercury in humans is mostly restricted to whole body measurements and depends on the type of mercurial and dose administered. Long half-lives occur in the brain if humans are exposed to mercury vapour [77, 622]. For methyl mercury a half-life of 70–76 days is frequently stated [500, 531, 624–626].

Biological Effects and Toxicity

Biological and Toxicological Effects

The biological and toxicological activity of mercury which is reviewed in general [575, 630–633] and with special regard to inorganic [634] and organic mercury compounds [635], epidemiology [43, 68], Minamata disease [636], and genetic effects [637] depends on the form in which it is taken up, the route of entry into the body (skin, inhalation, ingestion), and on the extent to which mercury is absorbed.

There is copious evidence to subdivide the individual forms into the following categories in declining order of biological and toxicological activity:

Alkylmercury salts (methyl, ethyl) > mercury vapour >
inorganic mercury-, phenyl- and methoxyethyl mercury salts.

Aryl mercury is largely converted to the inorganic form and handled as such in the body.

Short chain alkyl mercury compounds are more soluble in lipids than are those of mercury (II) or elemental mercury. They are also about 100 times more soluble in lipids than in water [638] enabling CH_3Hg^+ to penetrate more readily into cells than inorganic forms. Lipotropy, affinity to SH-groups (thiols) [639, 640], and other biological interactions such as inhibition of enzyme systems [457, 633, 641, 642], cause alkyl mercurials to be 10–100 times more toxic than soluble inorganic forms. Inorganic protein-bound mercurials are absorbed to a low degree in the intestinal tract and injuries heal quickly if the exposure ceases. In contrast to this methyl mercury derivatives are almost totally absorbed causing irreversible lesions, implying genetic effects which cause both damage of reproductive cells – inheritable damage to following generations – and of the genetic material in the chromosomes of ordinary cells – disturbances of the nuclear material, which regulates cell function, thus giving rise to carcinogenesis and teratogenic damage [637]. Breakage and abnormal chromosome division have been shown to occur in concentrations as low as 0.05 and 0.6 μg/g for phenyl and methyl mercury and methoxymethyl mercury respectively in experiments on plants [658, 659, 660], animals [658, 661], and on humans [662]. The dominant effect is on the spindle fiber mechanism which is responsible for the distribution of chromosomes into equal sets in the daughter cells. CH_3Hg^+ partially inactivates this mechanism thus producing cells with erroneous distributions of single chromosomes (Mongolism is one of the congenital disorders, which depend on it [637]).

The type of genetic damage actually observed in humans (Japan, Iraq) indicates that the same mechanism is acting as in the animal experiments. Numerous investigations have been conducted to study, e.g., embryotoxic and teratogenic effects [653, 666, 664].

Selenium has been found to protect against the toxicity of organic [677] and inorganic [677, 699, 700] mercurials by liberating dimethylselenide from the methyl mercury cycle [457]. Animals given a high dose of mercuric compounds lethal to controls survived when treated with selenium, and a decreased passage of mercury into foetuses and into milk occurred. A change of distribution and retention within the body has been observed [700].

Symptoms of Intoxication

Inorganic Mercury. Inhaled mercury vapour injures the respiratory tract and the oral cavity, e.g. sore mouth, ulcerated gums, etc. arise [633], manifested as coughing, bronchial inflamation, chest pains, vomiting, excitement, tremors, irritability, diarrhoea and respiratory arrest [617, 620, 633]. Longer exposure may lead to death [486, 620]. Disturbances by dental fillings has as yet not been shown, however, in some cases, allergic reactions of the lips and the oral mucous membranes have been observed [691].

Ingestion of dissociated salts of bivalent mercury causes precipitation of proteins upon contact with the mucous membranes of the gastrointestinal tract and produces local pain, gastric pain, and vomiting. In acute poisoning organic changes arise, such as renal failure with all sequences, and inflamation of the oral cavity, which are both reviewed in detail [630]. Typical chronic poisoning, mostly caused by occupational exposure involves injury of the central nervous system [628], which takes effect in characteristic tremor of the hands and other parts of the body, erethism [628] a peculiar form of psychic disturbance, decreased productivity, increased fatique, loss of memory and self confidence [629], injury to the kidney [647], vascular symptomatology [648], idiosyncracy [649], and effects upon the skin which are particularly marked with mercuric chloride [650].

Organic Mercury Compounds. Organic mercurials are absorbed to the skin, by inhalation, and by ingestion [633]. Methyl mercury chloride discharged into the Minamata river initiated the first disease (Minamata disease) caused by environmental pollution [636, 651, 652, 692]. As a result over 100 persons were afflicted, causing 46 deaths and several cases of prenatal intoxication manifesting in characteristic symptoms, e.g. motor disturbances, mainly ataxia, mental symptoms, congenital malformations (see below) and cerebral palsy as a major effect [636, 653]. In the mothers concerned no serious symptoms occurred [654, 655] which is suggested to be due to the relative ease of placental transfer of methyl mercury and its preferential concentration in the foetus [653, 656, 657]. Foetal erythrocytes contained 28% more mercury than those from the mothers [657]. The onset of tissue damage can be correlated with the concentration of mercury in the red blood cells (Fig. 12).

Postnatal intoxication involves irritation of the mucous membranes of the respiratory tract, dermatitis, and eczema upon contanct with organomercurials [625]. In systemic intoxication the latent period of weeks to months is a characteristic feature [666].

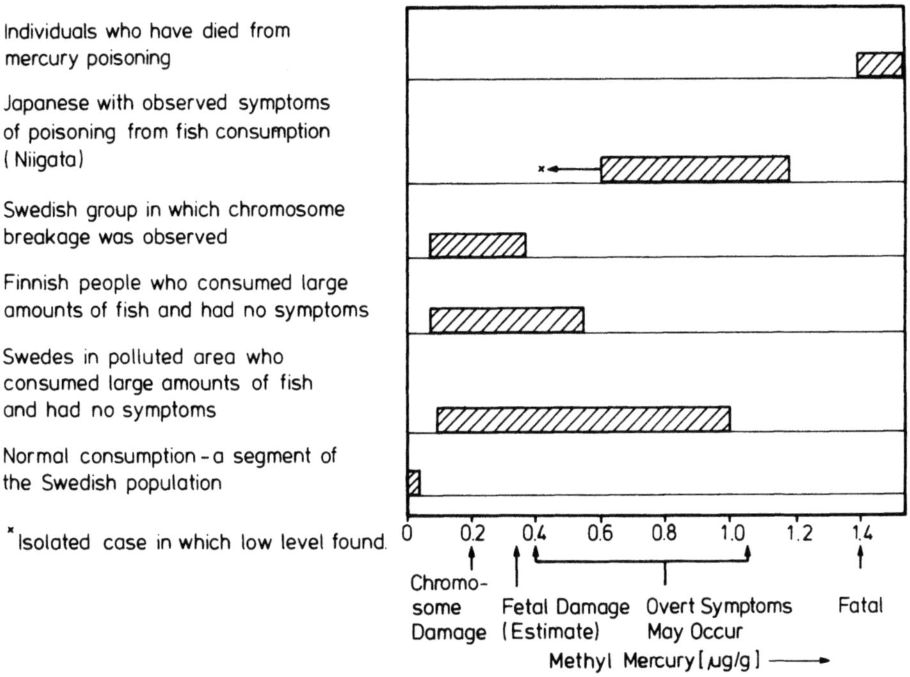

Fig. 12. Relation of methyl mercury levels in blood to physical hazards [77, 441, 658]

The classical picture contains three main symptoms:
1) Sensory disturbances in the distal parts of the extremities in the tongue and round the lips; 2) ataxia; 3) concentric constriction of the visual fields, hearing loss, symptoms from the anatomic and extrapyramidal nervous system, and mental disturbances [667].

Methyl mercury penetrates the blood brain barrier – more than mercury vapour [668] – and is distributed within the brain producing specific symptoms due to the destruction of the cells in the cerebellum and the visual and hearing centers [77]. Damaged functional nerve cells, in contrast to other types of cell, are not replaced by nerve cells produced by cell division. Their function is partly taken over by other existing nerve cells. Thus damage may be cumulative. The latent period for a manifestation of lesions can be very long [613]. The onset of tissue damage can be correlated with the concentration of mercury in the brain – assumed to be in the form of methyl mercury (Fig. 13). In the graph the practical daily intake of mercury with food [587], the acceptable daily intake (ADI) [68, 669, 670] and the practical residue limit for food set by the FAO and WHO [671] are also stated.

Maximum Allowable Concentrations (MAC-values) of Mercury and its Compounds. The MAC-value, in Germany MAK (Maximale Arbeitsplatzkonzentration), is defined as: that average concentration in the air which causes no signs or symptoms of illness or physical impairment in all but hypersensitive workers during their working day (8h/5d a week) on a continu-

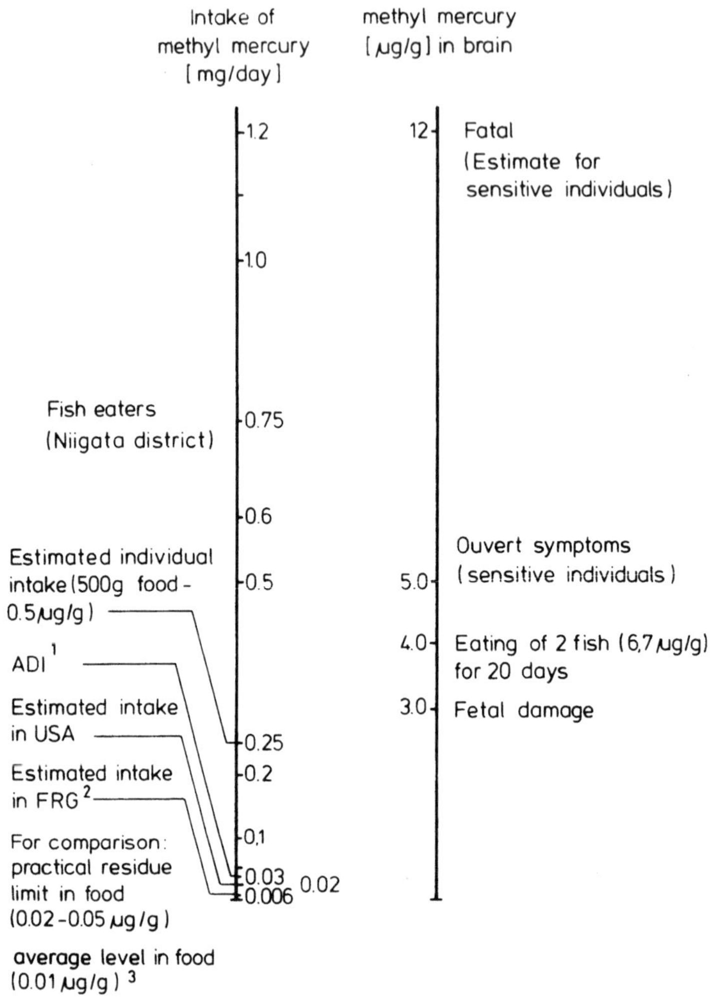

Fig. 13. Calculated relationship between methyl mercury intake and levels of methyl mercury in brain tissue [490]. 1: For a 75 kg man [669], the recommended ADI values lie between 0.03 and 0.1 mg/day [670]. 2: Evaluated total mercury from per capita consumption without consideration of beverages and fish [587]. 3: Without consideration of fish [587]. Calculations of brain tissue levels based upon: Brain distribution of 15% of total body methyl mercury at 10–15%. Continuous exposure for 1 year. With an excretion rate of 1%/day of total body mercury the indicated level will almost be reached

ing basis, as judged by the most sensitive internationally accepted test [672]. The definition is by and large, comparable with the concept of Threshold Limit Values (TLV) in the USA [673] (Table 15).

Up to now there are no MAC-values of individual mercurials. From occupational and suicidal intoxications as well as from animal experiments the quantities which exert adverse effects to human health can roughly be construed (Table 16). MAC-values should not directly be equalized with risk.

Many factors, e.g., route of entry into the body, penetration rate through the skin, and absorption rate play an important role. Sensitive persons may be affected at lower levels whereas others may tolerate much higher ones.

Table 15. Maximum allowable concentrations of mercury and its inorganic and organic compounds

Mercury/ Hg-compound	MAC (mg/m^3]		Ref.
	Western countries	USSR	
Mercury vapour	0.1		[672, 697]
Inorganic comp.	0.1	0.01	[672, 698]
Organic comp.	0.01		[672, 697]
alkyl mercury salts		0.005	[698]

Table 16. Quantities of mercury and some of its compounds at which no signs of intoxication in humans have been observed compared with some LD_{50} values

Mercury/ Mercury compound	Route of up take[a]	Quantity [g]	Ref.
Elemental Hg	or	100–1000	[693, 696]
	i.v.	<27 g	[693]
Mercury vapour	inhal.	see MAC	
Hg_2Cl_2	inhal.	<0.1-0.2	[694, 695]
	ingest.	2-3 (LD_{50})	[669]
$HgCl_2$	or	0.1-0.2	[695, 696]
		0.5(LD_{50})	
$Hg(CN)_2$	or	0.2-1 (LD_{50})	[695]
Organic Hg-comp.		n.d.	
		LD_{50} [mg/kg] body weight	
For comparison:			
CH_3HgCl (mouse)	i.p.	14	[635]
PhHg ac. (mouse)	i.p.	8	[635]
	or	26	

[a] or = orally, i.p. = intraperitoneally, i.v. = intravenously

References

1. Li Chiao P'ing: The chemical arts of old China. Amer. Chem. Soc., Easton, Pa. 1948, p. 48
2. Leicester, H.M.: The Historical Background of Chemistry, John Wiley, Inc. New York 1961, p. 58
3. Britton, R.S.: Havard J. Asiatic Studies *2*, 1 (1937)
4. Marshall, J.: Mohenjo-daro and the Indus Civilization, Vol. 2. A. Probsthain, London 1931
5. Farber, E.: The Evolution of Chemistry. Ronald Press, New York 1952, p. 21

6. Lippmann von, E.O.: Entstehung und Ausbreitung der Alchemie, Vol. I. Berlin 1919, p. 600, Verlag Chemie, Weinheim 1954
7. Gowland, W.: Archeologia *69*, 157 (1920)
8. Thompson, R.C.: Chemistry of the Ancient Assyrians. Luzac & Co., London 1925, p. 59
9. Partington, J.R.: Origins and Development of Applied Chemistry. Longmans, Green & Co., London 1935, p. 381
10. Weiskopf, A.: Z. angew. Chem. *14*, 429 (1910)
11. Lippmann von, E.O.: Entstehung und Ausbreitung d. Alchemie, Band 2. Berlin 1931, Verlag Chemie, Weinheim
12. Ruska, G.: Al'-Razi's Buch Geheimnis der Geheimnisse, Quellen und Studien, Gesch. Naturwiss. Med. Springer, Berlin 1937, p. 6
13. Calif. J. Mines Geol. *49*, 12 (1953)
14. Pauly-Wissova: Real-Encyclopädie, Vol. 9/1, Hydrargyrum. Alfred Druckenmüller, Stuttgart 1914, p. 55
15. Caley, E.R.: J. Chem. Educ. *23*, 12 (1953)
16. Aristotle: Meteorologica, Book IV, 385b, with an English translation by Lee, H.D., London-Cambridge, Mass. 1952, p. 342
17. Bennet, E.: Mining Metallurg. *29*, 6 (1948)
18. Vitruvius, M.P.: De architectura libri decem (Ed.: Rose, V.). Heinsius Kayser, Leipzig 1867
19. Stillmann, J.M.: The story of Early Chemistry. D. Appleton Comp., New York 1924, p. 30
20. Acosta de, J.: Natural and moral history of the Indies (Reprint of English translation) 1604, Hakluyt Society, London 1880, p. 211
21. Farrar, W.V., Williams, A.R.: A History of Mercury in: The Chemistry of Marcury (Ed.: McAuliffe, C.A.). McMillan Press, LTD, London 1977, p. 3
22. Goldwater, L.J.: Ann. Med. Hist. *8*, 27 (1936)
23. Needham, J.: Science and Civilization in China. Cambridge Univ. Press, Cambridge 1974, part 5, p. 2
24. Caley, E.R.: J. Chem. Educ. *3*, 1149 (1926); *4*, 979 (1927)
25. Berthelot, M.: Les origines d'Alchemie. G. Steinheil, Paris 1885, p. 207
26. Steel, R.: Isis *12*, 10 (1930)
27. Fracastoro, G.: Syphilis or the French disease (English translation). William Heineman, London 1935
28. Partington, J.R.: A History of Chemistry. McMillan Press, London 1935
29. D'Itri, P.A., D'Itri, F.M.: Mercury Contamination. A Human Tragedy. John Wiley, New York 1977
30. Anon.: Ind. Eng. Chem. *19*, 634 (1927)
31. Holmstedt, B.: The toxicology and metabolism of the different mercury compounds in: The Mercury Problem. Proc. Symp. Mercury in the Environment, Stockholm 1966, publ. as Oikos *9* (Suppl.) 25, (1967)
32. Lomonosov, M.V.: Complete Collected Works of Lomonosov. Acad. Sci., Moscow, Leningrad *3*, 377 (1952)
33. Priestly, J.: Experiments and observations of different kinds of air. J. Johnson, London *2*, 33 (1775)
34. Lavoisier, A.L.: Traité Elémentaire de chemie. Cuchet, Paris 1789, p. 33
35. Dalton, J.: A new system of chemical philosophy, Manchester. Printed by S. Russel for R. Birkerstaff, London 1827, *2*, p. 19
36. Porter, C.: Brit. Med. J. *1*, 377 (1902)
37. Howard, E.C.: Phil. Trans. Roy. Soc. *90*, 24 (1800)
38. Godenic, D., Tunell, G.: Handbook of Geochemistry (Ed. K.-H. Wedepohl). Springer-Verlag Heidelberg 1970
39. Bateman, A.M.: Economic Mineral Deposits, 2. ed. J. Wiley, New York 1955
40. Jonasson, I.R., Boyle, R.W.: Can. Inst. Mining. Met., Bull. *65*, 32 (1972)
41. Saukow, A.A.: Geochemie. Verlag Technik, Berlin 1953
42. Rankania, K., Sahama, T.G.: Mercury-Abundance and Geochemical Character, Univ. Chicago Press 1972, Chap. 39
43. D'Itri, F.M.: The Environmental Mercury Problem, Chem. Rubber Co., Cleveland, Ohio 1972

44. Lamey, C.A.: Metallic and Industrial Mineral Deposits. McGraw Hill, New York 1966, p. 302
45. White, D.E.: in: Geochemistry of Hydrothermal Ore Deposits Barnes, H.L., Ed. Holt, Rinehart, Winston, New York 1967, Chapt. 13, p. 575
46. Metallstatistik *61*, p. 54 (1974). Metallgesellschaft AG, Frankfurt/M. FRG
47. Agricola, G.: De re metallica libri XII. Basel 1621, p. 344
48. Johannsen, O., Peder Manson: Schriften über Technische Chemie und Hüttenwesen. VDI-Verlag, Berlin 1941
49. Biringuccio, V.: De la pirotechnica lib. X. Venedig 1540
50. Gould, G.J.: Mercury in: Handb. Non-ferrous metallurgy (Ed.: Lidell, D.M.) 2. Ed. Mc GrawHill, New York – London 1945, *2*, p. 473/516
51. Kaiser, W.F.: Quecksilber in: Winnacker, K., Weingartner, E., Chemische Technologie *5*, München 1953, p. 371
52. J. Chem. Educ. *33*, 607 (1956)
53. Universal Matthey Products (Deutschland GmbH) Hochreine Metalle u. Chemikalien. Köln 1978
54. Commissions Communautés Européennes, EUR 56 54 f, 1976. Le mercure: production, consommation et demande future dans le monde et dans la communautés européennes
55. Minerals Years Books, Metals, Minerals and Fuels, 1958–1973, *1*, US Bureau of Mines
56. Rauhut, A.: Quecksilber-Bilanz 1972–1976. Landesgewerbeanstalt Bayern, Nürnberg
57. Environm. Sci. Technol. *4*, 890 (1970)
58. Wallace, R.A. et al.: Mercury in the Environment: The Human Element. Oak Ridge Laboratory, USA Publication ORNL, NSF-EP-1
59. Chlorine and caustic soda; Recent Development, Noyes Development Corp., N.J., USA
60. Cook, N.E., Beitel, A.: Some aspects of other sources of mercury to the environment in: Mercury in Man's Environment. Royal Soc. Can. Sympos. Ottawa 1971, *15–16*, (2), 53
61. Osram Umweltschutz 1: Gefährdet d. Quecksilber, d. in Entladungslampen ist, unsere Umwelt? München 1972
62. Rauhut, A., Wild, L.: Metall *27*, 933 (1973)
63. Lichttechnik *29*, (9), 382 (1977)
64. Chem. Britain *7*, 310 (1971)
65. Chem. Eng. News *49*, (40), 48 (1971)
66. Miller, S.A.: Acetylene: Properties, Manufacture and Uses. Benn, London 1966
67. Kudsuna, M. (ed.): Minamata disease, Kumanoto Univ., Japan 1968
68. OECD: Mercury in the Environment, Paris 1974
69. Chem. Eng. News *48*, (14), 22 (1970)
70. Whiteley, J.: J. Oil and Colour Chemists' Assn. *43*, 842 (1960), *45*, 803 (1962)
71. Hoffmann, E., Bursztyn, B.: ibid. *46*, 460 (1963), *47*, 871 (1964)
72. O'Neill, L.A.: ibid. *46*, 425 (1963)
73. Hoffmann, E., Saraez, A.: ibid. *53*, 841 (1970)
74. Borg, W.K. et al.: Kvicksilver förgiftningar bland vilt i Sverige. Nat. Veterinary Inst., Stockholm 1965
75. Trends in Mercury Usage. Nat. Materials Advisory Boards, Washington, D.C., USA Publication NMAB-258
76. Wimer, B.A.: Lovelace Clin. Rev. *2*, 17 (1964)
77. Grant, N.: Environment *13* (4), 16 (1971)
78. Marton, J., Marton, T.: Tappi *55*, 1614 (1971)
79. Anon.: Mercury potential of the US Bureau of Mines Information Circular 8252, US Govt. Print. Off., Washington, D.C. 1965
80. Wambeke van, L.: La Géochimie du Mercure et la Pollution de l'Environnement, 1° Congr. Internat. Mercurio, Barcelona-Mayo, Vol. 2, 1974, p. 81
81. Klein, D.H.: Sources and Present Status of the Mercury Problem. Paper at the Mercury in the Western Environment Conf., Portland, Oregon 1971, Febr., p 25
82. Kuivala, A., Poijavvi: J. Erzmetall *30* (12), 555 (1977)
83. Wallace, R.A. et al.: Mercury in the Environment: The human element. Oak Ridge, Nat. Laboratory, USA Publication ORNL, NSF-EP-1
84. Bertine, U.U., Goldberg, E.D.: Science *173*, 233 (1971)

85. Man's Impact on the Global Environment. MIT Press, Cambridge, Mass. 1970
86. Breger, J.A.: in ref [84]
87. Heinrichs, H.: Naturwissenschaften *64*, 479 (1977)
88. Joensuu, O.I.: Science *172*, 1027 (1971)
89. Jonasson, I.R., Boyle, R.W.: Mercury in Man's Environment, Proc. Soc. Can. Symp. Ottawa 1971 (5), Febr. 15th
90. Schlesinger, M.D., Schults, H.: Tech. Prog. Report, US Bur. Mines, Washington D.C., Dept. of the Interior 1971, p. 43
91. Billings, C.E., Matson, W.R.: Science *176*, 1232 (1972)
92. Miners "Year Book" *1–2*, p. 377 (1968)
93. Heinrichs, H.: Z. Anal. Chem. *273*, 197 (1975)
94. Gavis, J., Ferguson, J.F.: Cycling of Mercury through the Environment, Water Res. *6*, (1972) Pergamon Press
95. Comm. Europ. Communities, Quality of Life, EUR 56 60 e, 1977
96. Comm. Europ. Communities, Europ. Colloq. EUR-5075, 1973
97. Anderson, A.A., Anderson, J.M. in: Das globale Gleichgewicht (Eds.: Meadows, D.L., Meadows, D.H.) Deutsche Verlagsanstalt, Stuttgart 1973, p. 78
98. Weiss, H.V., Koide, M., Goldberg, E.: Science *174*, 692 (1971)
99. Garrels, R.M., Mackenzie, F.T., Hunt, C.: Chemical cycles and the global environment assessing human influences. Willam Kaufmann, Inc., Los Altos, California 1975
100. Siegel, S.M., et al.: Environmental Biol. Med. *2*, 81 (1973)
101. Eshleman, A., Siegel, S.M., Siegel, B.Z.: Nature *233*, Oct. 15 (1971)
102. Lamb, H.H.: Phil. Trans., Royal Soc. London *266*, 425 (1970)
103. Aidin'yan, N.KH., Ozerova, N.A.: Sovrem. Vulkanizm *1*, 249 (1966)
104. Barnes, J.: Science *168*, 973 (1970)
105. Dickson, F.W., Tunell, G.: in: Ore Deposits of the USA 1933–1967 (Ridge J.D., ed.), Amer. Inst. Mining. Metall. Petroleum Eng. 1968, p. 1673
106. White, D.E., Hinkle, M.E., Barnes, J.: Prof. Paper US Geol. Surv., No. 713, 1970, p. 25
107. Goldberg, E.: Atmospheric Transport of Mercury From Continents to the Oceans. Ocean Sciences Board, Nat. Res. Council, Miami, Florida 1975
108. Kothny, E.L.: in: Trace Elements in the Environment, (Ed.: Kothny, E.L.) Adv. Chem. Ser. 123, Amer. Chem. Soc. 1973
109. Gmelins Handbuch d. Anorgan. Chem. Springer Heidelberg 1960, 8 ed. System Nr. 34
110. Comprehensive Inorganic Chemistry. Pergamon Press, Oxford, New York 1973
111. Roesmer, J., Kruger, P.: Radiochemistry of Mercury US Atomic Energy Commission, Washington D.C. 1961
112. Handbook of Chemistry a. Physics, Chemical Rubber Co., Cleveland, Ed. 56th (1975–76)
113. D'Ans-Lax: Taschenbuch für Chemiker u. Physiker. Springer-Verlag, Berlin, Heidelberg, New York 1967
114. Michaels, A., de Kluiver, H., Ten Soldam, C.H.: Physica *25*, 1321 (1959)
115. Warrick, P., Wewerka, E.M., Ureevoy, M.M.: J. Am. Chem. Soc. *85*, 1909 (1963)
116. Ogg. R.A., Martin, H.C., Lughton, P.A.: ibid *58*, 1922 (1936)
117. Phillips, C.G.S., Williams, R.J.P.: Inorganic Chemistry, Oxford Univ. Press 1965, Vol. 2
118. Roberts, H.L.: Adv. inorg. Chem. Radiochem. *11*, 309 (1968)
119. Hepler, L.G., Olafsson, G.: Chem. Rev., 585 (1975)
120. Chem. Soc. Spec. Public. No. 17: Stability Constants of Metal Ion Complexes, The Chem. Soc. Publ. House, London 1971, Alden & Mowbray Ltd, Oxford, Great Britain
121. Levason, W., McAuliffe, C.A.: in: The Chemistry of Mercury, (Ed. McAuliffe, C.A.) Macmillan Press, London 1977, p. 49
122. Makarova, L.G., Nesmeyanov, N.A.: Organic Compounds of Mercury. North-Holland, Amsterdam 1967
123. Reutov, O.A.: Topics Curr. Chem. *8*, 61 (1967)
124. Bloodworthy, A.J.: in: The Chemistry of Mercury (Ed.: McAuliffe, C.A.) Macmillan Press LTD, London 1977, p. 139
125. Wakefield, B.I.: Adv. Inorg. Chem. Radiochem. *11*, 341 (1968)
126. Dessy, R.E., Kitching, W.: Adv. Organometallic Chem. *4*, 267 (1966)
127. Schwarzenbach, G., Schellenberg, M.: Helv. Chim. Acta *48*, (1), 28 (1965)

128. Frankland, E.: Phil. Trans. Royal Soc. *142*, 417 (1852)
129. Buchton, G.B.: Phil. Trans. Royal Soc. *148*, 163 (1858)
130. Chem. News *12*, 213 (1865); *13*, 59 (1866); *13*, 7, 23, 35, (1866)
131. Dimroth, O.: Ber. dtsch. chem. Ges. *31*, 2154 (1898)
132. Makarova, L.G., Nesmeyanov, N.A.: in: The Organic Compounds of Mercury (Eds.: N.A. Neymeyanov and Kocheshkov, K.A.). North-Holland Press, Amsterdam 1967
133. Straub, H., Zeller, K.P., Leditschke, H.: Houben-Weyl, Meth. Organ. Chem. Georg-Thieme Verlag, Stuttgart 1977, Vol 13, 2b
134. Bonati, F., Minghetti, G.: J. Organometallic Chem. *22*, 5 (1970)
135. Allmann, R., Flatau, K., Musso, H.: Chem. Ber. *105*, 3067 (1972)
136. Budde, W., Dessy, R.E.: J. Am. Chem. Soc. *85*, 3964 (1963)
137. Kitching, W.: Organometallic Chem. Rev. *3*, 35 (1968)
138. Taylor, R.: Comprehensive Chemical Kinetics, Vol. 13, Reactions of Aromatic Compounds, (Eds.: Bamford, C.H., Tipper, C.F.H.) Elsevier, Amsterdam 1972, p. 186
139. Baker, R.A., Luh, M.D.: Mercury Analysis and Toxicity. Ind. Waste, May/June 1971, p. 21
140. Reimers, R.S., Burrows, W.D., Krenkel, P.A.: J. Water Pollut. Contr. Fed. *45*, 814 (1973)
141. Chilov, S.: Talanta *22*, 205 (1975)
142. Uthe, J.F., Armstrong, F.A.J.: in: Analyt. Asp. Mercury a. other Heavy Metals in the Environment (Eds.: Frei, R.W., Hutzinger, O.) Gordon and Breach, London 1975, p. 21
143. Lindstedt, G., Skerfving, S.: in: Mercury in the Environment (Eds.: Friberg, L., Vostal, J.) Chemical Rubber Co., Cleveland, Ohio 1972
144. Smith, R.G.: in: Environmental Mercury Contamination (Eds.: Hartung, R., Dinman, B.D.) Ann Arbor Science Publ. 1972
145. Heinonen, J., Suschny, O.: J. Radioanal. Chem. *20*, 499 (1974)
146. Fishbein, L.: Chromatography of Environmental Hazards. Elsevier, Amsterdam 1973, *II*, p. 1, 141
147. Burrows, W.D.: A review of the status of total mercury analysis. Progr. Water Technol., Suppl. Oxford 1975, p. 51
148. Valković, V.: Trace Element Analysis. Taylor and Francis, London 1975
149. Tölg, G.: Talanta *21*, 327 (1974)
150. Murphy, T.J.: in: Accuracy in Trace Analysis: Sampling, Sample Handling and Analysis, NBS Special Publ. No. 422. Washington, US Gov. Printing Off. 1976, Vol. I
151. Zief, M., Mitchell, J.W.: Contamination Control in Trace Analysis. Wiley, New York 1976, Vol. 47
152. Takeni, T.: Environmental Hg. Contamination. Ann Arbor Science 1972, p. 226
153. Kaiser, G., et al.: Z. Anal. Chem. *291*, 278 (1978)
154. Rosain, R.M., Wai, C.M.: Anal. Chim. Acta *65*, 279 (1973)
155. Sanemasa, J. et al.: Anal. Chim. Acta *87*, 479 (1976)
156. Toribara, T.Y., Shields, C.P., Koval, L.: Talanta *17*, 1025 (1970)
157. Feldman, C.: Anal. Chem. *46*, (1), 99 (1974)
158. Sanemasa, J. et al.: Anal. Chim. Acta *94*, 421 (1977)
159. Coyne, R.V., Collins, J.A.: Anal. Chem. *44*, 1093 (1972)
160. Carron, J., Agemian, H.: Anal. Chim. Acta *92*, 61 (1977)
161. Omang, S.H.: Anal. Chim. Acta *65*, 279 (1975)
162. Lo, J., Wai, C.M.: Anal. Chem. *47*, 1869 (1975)
163. La Fleur, P.D.: Anal. Chem. *45*, 1534 (1973)
164. Jensen, S. Jernölöv, A.: Nature *223*, 753 (1969)
164a. Baier, R.W., Wojnovich, L., Petrie, L.: Anal. Chem. *47* (14), 2464 (1975)
165. Matsunaga, K., Takahashi, S.: Anal. Chim. Acta *87*, 487 (1976)
166. Magos, L., Tuffery, A.A., Clarkson, T.W.: Brit. J. Industr. Med. *21*, 294 (1964)
167. Oki, S., Watanabe, W.: Japan Analyst *20*, 1204 (1971)
168. Ambe, M., Suwabe, K.: Anal. Chim. Acta *92*, 55 (1977)
169. Weiss, H.V., Shipman, W.H., Guttman, M.A.: ibid *81*, 211 (1976)
170. Christmann, D.R., Ingle, J.D. Jr.: ibid. *86*, 53 (1976)
171. Stoeppler, M., Matthes, W.: ibid. *98* (2), 389 (1978)
172. Shimamura, S., Hayashi, Y., Morita, H.: Anal. Lett. *7*, 355 (1974)
173. Bothner, M.H., Robertson, D.E.: Anal. Chem. *47* (3), 592 (1975)

174. Shimamura, S. et al.: Japan Analyst *23*, 653 (1974)
175. Warren, H.V., Delavault, R.E., Barakso, R.E.: J. Econ. Geol. *61*, 1010 (1966)
176. Pillay, K.K.S. et al.: Environmental Res. *5*, 814 (1972)
177. Pillay, K.K.S.: in: Environmental Mercury Contamination (Eds: Hartung, E., Dinman, B.D.) Ann Arbor Science, Ann Arbour. Mich. 1972, p. 137
178. Ramelov, G., Hornung, H.: Z. Anal. Chem. *293*, 80 (1978)
179. Litman, R.: Anal. Chem. *47*, 2364 (1975)
180. Lindstedt, G.: Analyst *95*, 264 (1970)
181. Mercury Analysis Working Party of the Bureau Internat. du Chlore: Anal. Chim. Acta *84*, 231 (1976)
182. Ure, A.M.: Analyst *102*, 50 (1977)
183. Knapp, G., Sadjadi, B., Spitzy H.: Z. Anal. Chem. *274*, 275 (1975)
184. Feldman, C.: Anal. Chem. *46*, 1606 (1974)
185. Sansoni, B., Kracke, W.: Z. Anal. Chem. *243*, 209 (1968)
186. Agemian, H., Chau, A.S.Y.: Analyst *101*, 91 (1976)
187. Farey, B.L., Nelson, L.A., Rolph, M.G.: Analyst *103*, 656 (1978)
188. Knapp, G.: Z. Anal. Chem. *274*, 271 (1975)
189. Agemian, H., Chau, A.S.Y.: Anal. Chem. *50* (1), 13 (1978)
190. Bethge, P.O.: Anal. Chim. Acta *10*, 317 (1954)
191. Bernas, B.: Anal. Chem. *40*, 1682 (1968)
192. Langmyhr, F.J.: Anal. Chim. Acta *45*, 358 (1970)
193. Kotz, L. et al.: Z. Anal. Chem. *260*, 207 (1977)
194. Kotz, L. et al.: Talanta *26*, 681 (1979)
195. Stoeppler, M., Backhaus, F.: Z. Anal. Chem. *291*, 116 (1978)
196. Kuehner, E.C. et al.: Anal. Chem. *44*, 2050 (1972)
197. Gladney, E.S., Owens J.W.: Anal. Chim. Acta *90*, 271 (1977)
198. Götz, D.: Dissert. Univers. Stuttgart 1977
199. Watling, R.J.: Anal. Chim. Acta *99*, 357 (1978)
200. Kaiser, G. et al.: Talanta *12*, 889 (1975)
201. Schöninger, W.: Mikrochim. Acta *1955*, 123
202. Tölg, G.: Z. Anal. Chem. *283*, 257 (1977)
203. Gleit, C.E., Holland, W.D.: Anal. Chem. *34*, 1454 (1962)
204. Kaiser, G., Tschöpel, P., Tölg, G.: Z. Anal. Chem. *253*, 177 (1971)
205. Wickbold, R.: Angew. Chem. *64*, 133 (1952)
206. Kunkel, E.: Z. Anal. Chem. *258*, 337 (1972)
207. Frimmel, F., Winkler, A.: Z. Wasser- u. Abwasserforsch. *8* (3), 67 (1975)
208. Hofer, A., Heidlinger, R.: Z. Anal. Chem. *271*, 192 (1974)
209. Nicholson, R.A.: Analyst *102*, 399 (1977)
210. Poluektov, N.S., Vikun, R.A., Zelyukovo, Y.V.: Z. Anal. Chim. *18*, 33 (1963)
211. Hatch, W.R., Ott, W.L.: Anal. Chem. *40*, 2085 (1968)
212. Toffaletti, J., Savory, J.: Anal. Chem. *47* (13), 2091 (1975)
213. Barlow, P.J. et al.: Proc. Analyt. Div. Chem. Soc. *16*, 15 (1979)
214. Yamamoto, Y., Kumamaru, T., Shiraki, A.: Z. Anal. Chem. *292*, 273 (1978)
215. Narasaki, H., Down, I.L., Ballah, R.: Analyst. *102*, 537 (1977)
216. Westöö, G.: Acta Chem. Scand. *20*, 2131 (1966)
217. Sumino, K.: Kobe, J. Med. Sci. *14*, 131 (1968)
218. Kitamura, S. et al.: Igaku Seibutsug *72*, 274 (1966)
219. Johansen, O., Steinnes, E.: Intern. J. Appl. Radiation Isotopes *20*, 751 (1969)
220. Smith, H.: Anal. Chem. *35*, 635 (1963)
221. Disam, A., Tschöpel, P., Tölg, G.: Z. Anal. Chem. *295*, 97 (1979)
222. Westöö, G.: Acta Chem. Scand, *22*, 2277 (1968)
223. Kudsk, F.N.: Scand. J. Clin. Lab. Invest. *16*, 575 (1964)
224. Kressin, J.: Talanta *19*, 197 (1972)
225. Kish, P.P. et al.: Zh. Anal. Khim *32*, 1942 (1977)
226. Patil, P.S., Shinde, V.M.: Indian J. Chem. *15A*, 1030 (1977)
227. Kuroda, R., Kiriyama, T., Ishida, K.: Anal. Chim. Acta *40*, 305 (1968)
228. Marovski, G.: Z. Anal. Chem. *253*, 267 (1971)

229. de Jong, G.J.N., Rekers, C.J.N.: J. Chromatog. *102*, 443 (1974)
230. Toribara, T.Y., Shields, C.P.: Am. Ind. Hyg. Assoc. J. *29*, 87 (1968)
231. Klimes, I., Betusova, M.: Anal. Biochem. *23*, 102 (1968)
232. Magos, L., Clarkson, T.W.: Brit. Pat. 1,130,921
233. Canic, V.D., Petrovic, S.M.: Z. Anal. Chem. *211*, 321 (1965)
234. Marchant, W.: Environment Sci. Technol. *8*, (12), 993 (1974)
235. Kurayuki, Y., Kusumoto, K.: Bunseki Kagaku *16*, 815 (1967)
236. Popper, E., Florean, E., Marcu, P.: Rev. Roum. Chim. *13*, 909 (1968)
237. Cuipe, R., Kovats, H.: Revta Chim. *27* (3), 251 (1976)
238. Zelenko, V., Kosta, L.: Talanta *20*, 115 (1973)
239. Bertoni, G. et al.: Chromatographia *11*, 55 (1978)
240. Schaller, K.H.: Z. Anal. Chem. *290*, 113 (1978)
241. Fumiaki, G., Fumie, J.: Buseki Kagaku *24*, 480 (1975)
242. Brandenberger, H., Bader, H.: Atom. Abs. Newsl. *6*, 101 (1973)
243. Kazuo, H., Akinovi, K., Takashi, T.: Japan Analyst *23*, (9), 1075 (1974)
244. Analyt. Meth. Comm., Soc. Analyt. Chem.: Analyst *90*, 515 (1965)
245. Kudsk, F.N.: Scand. J. Clin. Lab. Invest. *16*, Suppl. 77, 1 (1964)
246. Linch, A.L. Soltzer, R.F., Lefferts, D.T.: Am. Ind. Hyg. Assoc. J. *29*, 79 (1968)
247. Jansen, I.H.: Anal. Chim. Acta *92*, 71 (1977)
248. Scullman, J., Widmark, G., in: Analytical Aspects of Mercury and Other Heavy Metals in the Environment (Eds.: Frei, R.W., Hutzinger, O.) Gordon and Breach, London 1975, p. 69
249. Corte, G.L., Thomas, R.S., Lao, R.C.: Mikrochim. Acta I (5–6), 533 (1975)
250. Dogan, S., Haerdi, W.: Anal. Chim. Acta *84*, (1), 89 (1976)
251. Lech, J.F., Siemer, D., Woodriff, R. Appl. spectr. *28*, (1) 78 (1974)
252. Moffit, A.E. Jr., Kupel, R.E.: At. Abs. Newsl. *9*, 113 (1970)
253. Searingelli, F.P. et al.: Anal. Chem. *46*, 278 (1974)
254. Rawlings, G.: Dissert. Graduate Coll. Texas A & M Univers. 1974
255. Christie, A.A., Dunsdon, A.J., Marshall, B.S.: Analyt. *92*, 185 (1967)
256. Eberling, C. et al.: Air Pollution Measurement Techn., Spec. Environmental Rep. No. 10, World Meteorol. Organ. Geneva 1977
257. Andersen, D.H. et al.: Anal. Chem. *43*, 1511 (1971)
258. Schullmann, J., Widmark, G.: Int. J. Environm. Anal. Chem. *2* (1), 29 (1972)
259. Dürr, W., Hausmann, K.: Z. Anal. Chem. *283*, 337 (1977)
260. Hemeon, W.L.C., Haines, G.F. jr.: Am. Int. Hyg. Assoc. J. *22*, 75 (1961)
261. Peirson, D.H. et al.: Nature *241*, 252 (1973)
262. Henriques, A., Isberg, J., Kjellgren D.: Chemica Scripta *4*, (3), 139 (1973)
263. Braman, R.S., Johnson, D.L.: Environm. Sci. Technol. *8*, 996 (1974)
264. Matheson, D.H.: Inland Water Directorate. Canada Centre for Inland Waters, Burlington, Ont. 1977
265. Hartley, A.M., Devine, D.J.: Illinois Inst. Environmental Quality, Document 76/04, 1976
266. Stock A.: Naturwissenschaften *19*, 499 (1931)
267. Stock, A., Cucuel, F.: Naturwissenschaften *22*, 390 (1934)
268. Iwantscheff, G.: Dithizon und seine Anwend. in d. Mikro- u. Spurenanalyse, 2. Aufl., Verlag Chemie/Weinheim, 1972
269. Sawaya, T., Ishii, H., Odashima, T.: Bunseki Kagaku *22*, 318 (1973)
270. Willis, J.B.: Anal. Chem. *34*, 614 (1962)
271. Kirkbright, G.F., West, T.S., Wilson, P.J.: Analyst *98*, 49 (1973)
272. Hadeishi, T., MacLaughlin, N.: Science *174*, 404 (1971)
273. Manning, D.C.: Atomic Absorption Newsl. *9*, 97 (1970)
274. Helsby, C.A.: Mikrochim. Acta 1976 I, 307
275. Mittelbauer, H.M.: Atom. Absorp. Newsl. *9*, 134 (1970)
276. Kokot, M.L.: Erzmetall *28* (5), 251 (1975)
277. Nakamura, Y., Moriki, H.: Eisei Kagaku *20*, (5), 300 (1974)
278. Alberts, J.J. et al.: Anal. Chem. *46* (3), 434 (1974)
279. Fitzgerald, W.F., Lyons, W.G., Hunt, C.D.: Anal. Chem. *46*, 1882 (1974)
280. Simpson, W.R., Nickless, G.: Analyst *102*, 86 (1977)
281. Dittrich, K., Mueller, H.: Chem. Anal. (Warsaw) *23*, 81 (1978)

282. Soo-Loong Tong: Anal. Chem. *50*, 412 (1978)
283. Braun, R., Husbands, A.P.: Spectrovision *26*, 2 (1971)
284. Harsanyi, E.G. et al.: Hung. Sci. Instrum. 36, 13 (1976)
285. Nelson, K.H., Brown, W.D., Staruch, S.I.: in: Analyt. Aspects of Mercury and Other Heavy Metals in the Environment (Eds.: Frei, R.W., Hutzinger, O.) Gordon and Breach, London, New York, Paris 1975, p. 77
286. Agemian, H., Chau, A.S.Y.: Anal. Chim. Acta *75*, 297 (1975)
287. Abbas, A.E., Miller, Carter, M.J.: Anal. Chem. *48* (1), 110 (1976)
288. Skare, I.: Analyst *97*, 148 (1977)
289. Bailey, B.W., Lo, F.C.: Anal. Chem. *43*, 1525 (1971)
290. Tsujii, K., Kuga, K.: Anal. Chim. Acta *101* (1), 199 (1978)
291. Velghe, N.: Atomic Absorpt. Newsl. *17* (2), 37 (1978)
292. Agemian, H., V. Cheam: Anal. Chim. Acta *101*, (1), 193 (1978)
293. Kirkwood, D.: Lab. Practice, March 1967, p. 233
294. Velghe, N.: Atom. Absorpt. Newsl. *16* (1), 28 (1977)
295. Ramelow, G.I., Balkas, T.I.: Atom. Absorpt. Newsl. *15*, 55 (1976)
296. Marei, S.A. et al.: J. Radioanal. Chem. *43*, 101 (1978)
297. Jirka, A.M., Carter, M.J.: Anal. Chem. *50*, 91 (1978)
298. Lidums, V.: Chemica Scripta *2*, 159 (1972)
299. Chapmann, J.F., Dale, L.S.: Anal. Chim. Acta *101*, 203 (1978)
300. Matthes, W., Flucht, R., Stoeppler, M.: Z. Anal. Chem. *291*, 20 (1978)
301. Aldous, K.M. et al.: Anal. Chim. Acta *54*, 233 (1971)
302. Lichte, F.E., Skogerboe, R.K.: Anal. Chem. *44*, 1321 (1972)
303. Watling, R.J.: Anal. Chim. Acta *75* (2), 281 (1975)
304. Gilbert, T.R.: Anal. Lett. *10* (7+8), 599 (1977)
305. Kozechowski, J.: Anal. Chim. Acta *99* (2), 293 (1978)
306. Atsuya, Ikuo et al.: Anal. Biochem. *77* (1), 208 (1977)
307. Kirkbright, G.F., Ward, A.F., West, T.S.: Anal. Chim. Acta *64*, 353 (1973)
308. April, R.W., Hume, D.N.: Science *170*, 849 (1970)
309. Knight, D.M., Pyzyno, M.K.: Appl. Spectrosc. *28*, 372 (1974)
310. Browner, R.F., Dagnall, R.M., West, T.S.: Talanta *16*, 75 (1969)
311. Muscat, V.J., Vickers, T.J., Andren, A.: Anal. Chem. *44*, 218 (1972)
312. Muscat, V.J., Vickers, T.J.: Anal. Chim. Acta *96*, 771 (1971)
313. Walsh, P., Hamrick, P., Underwood N.: Rev. Scient. Instr. *44* (8), 1019 (1973)
314. Mancini, C.: Isotopenpraxis *11*, 417 (1975)
315. D'Silva, A.P., Fassel, V.A.: Anal. Chem. *44* (11), 2115 (1972)
316. Clechet, P. et al.: Analusis *5*, 366 (1977)
317. Kalbus, G.E., Wesley, R.D., Kalbus, L.H.: Analyst *96*, 488 (1971)
318. Overman, R.F.: Anal. Chem. *43*, 616 (1971)
319. Rohm, T.J., Purdy, W.C.: Anal. Chim. Acta *72*, 177 (1974)
320. Kabanova, O.L., Zalogina, E.N.: Zh. Analit. Khim. *26* (4), 714 (1971)
321. Okuno, I., Wilson, R.A., White, R.E.: J. Assoc. Off. Ana. Chemists *55*, 96 (1972)
322. Weiner, I.M., Müller, O.H.: J. Pharmacol. Exptl. Therap. *113*, 24 (1955)
323. Kambara Tomihito, Hara Minora: Bull. Chem. Soc. Jap. *44* (4), 1040 (1971)
324. Sontag, G., Kerschbaumer, G., Kainz, G.: Mikrochim. Acta II, 411 (1976)
325. Usatenko, Yu. I., Sitalo, L.I.: Zh. Analit. Khim. *29* (1), 35 (1974)
326. Linden, van der, W.E.: Z. Anal. Chem. *264*, 353 (1973)
327. Stock, J.T.: in: Amperometric Titrations. Interscience, New York 1965
328. Southworth, B.C., Hodecker, J.H., Fleischer, H.D.: Anal. Chem. *30* (6), 1152, (1958)
329. Ehrlich-Rogozynski, S., Sperling, R.: Anal. Chem. *42* (9), 1089 (1970)
330. Fukai, R., Huynh-Ngoc, L.: Anal. Chim. Acta *83*, 375 (1976)
331. Kendall, D.R.: Anal. Lett. *5* (12) 867 (1972)
332. Allen, R.A., Johnson, D.C.: Talanta *20*, 799 (1973)
333. Taddia, M.: Microchem. J. *23* (1), 64 (1978)
334. Roizenblat, E.M., Verentina, G.N.: Zh. Analit. Khim. *29* (12), 2376 (1974)
335. Miwa, T., Toyota, T., Mizuike, A.: Bunseki Kagaku *25*, 402 (1976)
336. Andrews, R.W., Larochelle, J.H., Johnson, D.C.: Anal. Chem. *48* (1), 212 (1976)

336a. Kritsotakis, K., Laskowski, N. Tobschall, H.J.: Intern. J. Environ. Anal. Chem. *6*, 203 (1979)
336b. Sipos, L. et al.: Fresenius Z. Anal. Chem. *298*, 1 (1979)
337. Rannev, G.G., Valkova, V.S., Murtazaev, A.M.: Zavodsk. Lab. *36*, 1446 (1970)
338. Filby, R.H. et al.: Mikrochim. Acta 1130 (1970)
339. Kosta, L., Byrne, A.R.: Talanta *16*, 1297 (1969)
340. Kim, J.T., Sanoko, S.R.: J. Radioanal. Chem. *16*, 257 (1973)
341. Erdmann, G., Aboulwater, O.: Z. Anal. Chem. *270* (1), 1 (1974)
342. Krämer, N.J., Neidhart, B.: J. Radioanal. Chem. *37*, 835 (1977)
343. Zmijewska, W.: J. Radioanal. Chem. *35*, 389 (1977)
344. Wysocka, B.: Dissnes Pharm. Warsaw *17*, 99 (1965)
345. Wysocka, B., Bronisz, H.: Acta Pol. Pharm. *22*, 491 (1965)
346. Luckow, V., Rüssel, H.A.: J. Chromatogr. *150*, 187 (1978)
347. Dressman, R.C.: Paper 19th Ann. Detroit Anachem. Conf. Detroit, Mich., 1971, Oct. 31st
348. Gage, J.C.: Analyst *86*, 457 (1961)
349. Hartung, R.: in: Environmental Mercury Pollution (Eds.: Hartung, R. Dinman, B.D.). Ann Arbor Science, Mich. 1972, 157
350. Krivan, V.: Z. Anal. Chem. *253*, 192 (1971)
351. Rohm, T.J., Purdy, W.C.: Anal. Chim. Acta *72*, 177 (1974)
352. Mealor, D., Townshend, A.: Talanta *15*, 747 (1968)
353. Weisz, H., Ludwig, H.: Anal. Chim. Acta *62*, 125 (1972)
354. Hutzinger, O., Jamieson, W.D., Safe, S.: in: Analyt. Asp. Mercury a. Other Heavy Metals in the Environment (Eds.: Frei, R.W., Hutzinger, O.) Gordon and Breach, London, New York, Paris 1975, p. 55
355. Takeshita, R. et al.: J. Chromatog. *51*, 283 (1970)
356. Gage, J.C., Warren, J.M.: Am. J. Occupational Hyg. *13*, 115 (1970)
357. Clarkson, T.W., Greenwood, M.R.: Anal. Biochem. *37*, 236 (1970)
358. Hey, H.: Z. Anal. Chem. *256*, 361 (1971)
359. Longbottom, J.E.: Anal. Chem. *44*, 1111 (1972)
360. Dressman, R.C.: J. Chromat. Sci. *10*, 167 (1972)
361. Dagnall, R.M., Pratt, S.I., West, T.S.: Talanta *17*, 1009 (1976)
362. Bache, C.A., Lisk, D.I.: Anal. Chem. *43*, 950 (1971)
363. Talmi, Y.: Anal. Chim. Acta *74*, 107 (1975)
364. Jones, P., Nickless, G.: J. Chromatog. *89* (2), 201 (1974)
365. Zarnegar, P., Mushak, P.: Anal. Chim. Acta *69* (2), 389 (1974)
366. Cappon, Ch.J., Smith, I.C.: Anal. Chem. *49*, 365 (1977)
367. Frimmel, F., Winkler, H.A.: Wasser 1975, *45*, Verlag Chemie, Weinheim
368. Ohkoshi, S., Takahashi, T., Sato, T.: Japan Analyst *22*, 593 (1973)
368a. Kritsotakis, K., Rubischung, P., Tobschall, H.J.: Fresenius Z. Anal. Chem. *296*, 358 (1979)
369. Mac Crehan, W.A., Durst, R.A., Bellama, J.M.: Anal. Letter, *10* (14), 1175 (1977)
370. Jernelöv, A., Landner, L., Larsson, T.: J. Water Pollut. Control Fed. *47*, 810 (1975)
371. Lockeretz, W.: Water, Air, Soil Poll. *3*, 179 (1974)
372. Gardner, D.: Marine Poll. Bull. *6*, 43 (1975)
373. Lamb, H.H.: Phil. Trans. Royal Soc. London *266*, 425 (1970)
374. Goldschmidt, V.M.: Geochemistry (Ed.: A. Muir) Oxford, Clarendon Press 1954
375. Mc Carthy, J.M. jr., et al.: US Geol. Surv. Prof. Paper 713 (1970)
376. Willistone, S.H.: J. Geophys. Res. *73*, 7051 (1968)
377. Rissanen, K., Erkawa, J. Miettinen, J.K.: Food and Agriculture Organization, FAO FIR MP/70/E-61, Oct. 15th, 1970, Rome
378. Wilkins, P.E., Bressan, D.J.: Nuclear Techniques in Environmental Pollution, Proc. Ser., Vienna, Int. Atom. Energ. Author., 1971, p. 297
379. Mac Intyre, F.: Scient. Amer. *230*, 62 (1974)
380. Boutron, C., Echevin, M., Lowius, C.: Geochem. Cosmochim. Acta *36*, 1029 (1972)
381. Erikson, E.: Tellus *12*, 63 (1960)
382. Jenne, E.A.: in: Mercury in the Environment, Geol. Survey, Prof. Paper 713, US Government Printing Office, Washington 1970
383. Hinkel, M.E., Learned, R.E.: US Geol. Survey, Prof. Paper, 650 D, D. 251, US Government Printing Office, Washington 1969

384. Goldberg, E.D.: in: McGraw Hill Yearbook of Sci. Technol. 1975, p. 65
385. Hickel, W.J., Pecora, W.T.: US Geol. Survey, Prof. Paper 713, US Government Printing Office Washington 1970
386. Goeij de, J.J.M.: Chem. Eng. News, June 7th, p. 6 (1971)
387. Wroblewski, S.C. Splitter, T.M., Harrison, P.R.: J. Air. Poll. Contr. Assoc. *24*, 778 (1974)
388. Johnson, D.L., Braman, R.S.: Environ. Sci. Technol. *8*, 1003 (1974)
389. Harrison, P.R.: J. Air Poll. Contr. Assoc. *21* (9), 563 (1971)
390. Natusch, D.F.S., Wallace, J.R., Evans, C.A.: Science *183*, 202 (1974)
391. Murgatroyd, R.J.: Phil. Trans. Royal Soc. London, *A 265*, 273 (1979)
392. Tonani, F.: in: Trace Elements in the Environment (Ed.: Kothny, E.L.) Adv. Chem. Ser. 123, Amer. Chem. Soc. Washington 1973
393. Willistone, S.H.: ibid
394. Brosset, C., Svedung, I.: Swedish Water and Air Pollution Res. Lab. Gothenb. 1977, B 378
395. Gibbs, G.P.: Science *180*, 71 (1973)
396. Hem, J.D.: US Geol. Survey, Prof. Paper, 713, p. 19 (1970)
397. Alberts, J.J., Schindler, J.E., Miller, R.W.: Science *184*, 895 (1974)
398. Krauskopf, K.B.: Geochim. Cosmochim. Acta *9*, 1 (1956)
399. Ginzburg, I.I.: Principles of geological prospecting, techniques of prospecting for non-ferrous ores and rare metals Pergamon Press, Oxford 1960, p. 311
400. Koksoy, M., Bradshaw, P.M.D.: Colorado School Mines Quart. *64*, (1), 333 (1969)
401. James, C.H.: London Imp. College, Sci. Technology, Techn. Communication *41*, 42 (1962)
402. Sergeev, E.A.: Internat. Geol. Rev. *3*, 93 (1961)
403. Dall'Aglio, M.: in: Origin and distribution of the elements (Ed.: Ahrens, L.H.) Pergamon, New York 1968, p. 1065
404. Sayre, W.W.: IAEA, Vienna, Techn. Rep. 1973, Ser. No. 145
405. IAEA, Vienna, 1973, Techn. Rep. No 45, Tracer Techniques in Sediment Transport
406. Ruivo, M.: Marine Pollution and Sea Life, London, Fishing News Ltd. 1972
407. Weidemann, H., Sendner, H.: in: Marine Pollution and Sea Life, (Ed.: Ruivo, M.) London Fishing News Ltd. 1972, p. 115
408. Talbot, J.W.: Ref. [406], p. 122
409. Klein, D.H., Goldberg, E.D.: Environment Sci. Technol. *4*, 765 (1970)
410. Jernelöv, H.: Nord. Hyg. Tidskr. *50*, 174 (1969)
411. Wood, J.M., Rosen, C.G., Kennedy, S.F.: Nature *220*, 173 (1968)
412. Jernelöv, A.: Limnol. Oceanography *15*, 958 (1970)
413. Wood, J.M.: Science *183*, 1049 (1974)
414. Kaiser, G., Tölg, G., Schlichting, E.: Natürl. u. anthropogene Quecksilberanreicherg. in: Filderböden, Sendereihe Umwelttagung Univ. Hohenheim 1978, Heft 22
415. Frissel, M.J., Poelstra, P., Klugt, van der, N.: Geologie en Mijnbouw *53*, (4) 1974
416. Lagerwerff, J.V.: in: Micronutrients in Agriculture (Ed.: Morvedt, J.J.) Soil Sci. Soc. Amer., Madison, Wisc. 1972
417. Erikson, E.: Mercury in Nature: The Mercury Problem. Oikos, Suppl. *9*, 13 (1967)
418. Jonasson, I.R.: Geol. Surv. Canada Pap. 1970, 70
419. Anderson, A.: Grundförbättring *95*, 3 (1967)
420. Warren, H.V., Delavault, R.E.: Oikos *20*, 537 (1969)
421. Toribara, T.Y., Shields, C.P.: J. Am. Ind. Hyg. Ass. *29*, 87 (1968)
422. Norseth, T., Clarkson, T.W.: Arch. Environ Health *21*, 717 (1970)
423. Strohal, P., Huljev, D.: Nucl. Tech. Environ. Poll. Proc. Symp. IAEA, Vienna 1970
424. Bisogni, J.J., Lawrence, A.W.: Water Resour. Mar. Sci. Cent., Cornell Univers. Ithaca, N.Y. 1973, Rep. 63
425. Shcherbakov, V.P., Dvornikov, A.G., Zakrenichnaya, G.L.: Dopov. Akad. Nauk Ukr. RAR, Ser. B *32*, 126 (1970)
426. Trost, P.B., Bisque, R.E.: in: Environ. Mercury Contam (Eds.: Hartung, R., Dinman, B.D.) Ann Arbor Sci., Ann Arbor 1972, p. 178
427. Commiss. Europ. Communities Eur. 5683 e, Ispra, Italy 1977
428. Tonomura, K., Maeda, K., Futai, F.: J. Ferment. Technol. *46*, 685 (1968)
429. Tonomura, K., Kanzaki, F.: Biochem. Biophys. Acta *184*, 1969
430. Rogers, R.D.: J. Environ. Quality *5* (4), 454 (1976)

431. Rogers, R.D.: ibid *6* (4), 463 (1977)
432. Jernelöv, A.: Chemical Fall-out, Current Res. Persistent Pesticides (Eds.: Miller, M.W., Berg, G.G.) Charles C. Thomas, Springfield, Ill. 1969, p. 68
433. Greeson, P.E.: Mercury in the Environment. Geolog. Survey Prof. Paper 713, Washington 1970
434. Fuyita, M., Hashizume, K.: Chemosphere *1*, 203 (1972)
435. Ishikura, S. Shibuya, C.: Eisei, Kagaku *14*, 228 (1968)
436. Windom, H. Taylor, F., Stickney, R.: J. Conc. int. Explor. *Mer 35* (1), 18 (1973)
437. Borg, K.: Nord. Hyg. Tidskr. *50*, 9 (1969)
438. Wanntorp, H. et al.: Nord. Vet. Med. *19*, 474 (1967)
439. Borg, K. et al.: Environ. Pollut. *1*, 91 (1970)
440. Keith, J.A., Gruchy, I.M.: Proc. Roy. Soc. Can. Symp. *91*, Feb. 15–16 (1971)
441. Berglund, F. et al.: Nord. Hyg. T. Suppl. 4, 1971, Suppl. 3, 1970
442. Dolar, S.G., Keeney, D.R., Chesters, G.: Environ. Lett. *1*, 191 (1971)
443. Stahl, Q.R.: Prelim. Air Poll., Survey of Mercury and its Compds., Department of Health Education and Welfare, Public Health Service, National Air Pollution Control Administration Raleigh, N.C. US, 1969
444. Booer, J.R.: Ann. Appl. Biol. *31*, 340 (1944)
445. Pickard J.H., Grainger, J.: Ann. Rep. Agric. Hort. Res. Stat., Long Ashton, Bristol 1962, p. 65
446. Smart, N.A.: J. Sci. Food, Aric. *15*, 102 (1964)
447. Pickard, J.A., Martin, J.T.: Ann. Rep. Agric. Hort. Res. Sts., Long Ashton Bristol 1961, p. 106
448. Blanton, C.J. et al.: Paper 9th Ann. Conf. Trace Substances in Environ. Health, Columbia, Miss. 1975, June 10–12
449. Miller, E.J.: Plant Pathol. *5*, 119 (1959)
450. Ross, R.G., Stewert, D.K.R.: Can. J. Plant Sci. *40*, 117 (1960)
451. Saxena, J., Howard, P.M.: Adv. Appl. Microbiol. *21* 185 (1977)
452. Jensen, S., Jernelöv, A.: Intern. Atom. Energ. Author., Wien, Techn. Rep. 1972, Series No. 137, Chaps. 4, 43
453. Kothny, E.L.: R. Ass. Geol. Argentina *25*, 63 (1970)
454. Fagerström, T., Jernelöv, A.: Water Res. *5*, 121 (1971)
455. Furukawa, K.T., Suzuki, T., Tonomura, K.: Agric. Biol. Chem. *33*, 128 (1969)
456. Brunker, R.L., Bott, T.L.: Appl. Microbiol. *27*, 870 (1974)
457. Wood, J.M.: in: Biochem. and Biophys. Persp. in Marine Biology, (Eds.: Malins, D.C. and Rogers, J.R.) Academic Press, 1975–6, III, 407
458. Vonk, I., Sjipersteinsteijn, A.: Antonie van Leeuwenhoek *39*, 505 (1973)
459. Jernelöv, A.: Environ. Mercury Contam. (Eds.: Hartung, R., Dinman, B.D.) Ann Arbor Science, Ann Arbor 1972, p. 167
460. Werner, J.: Sved. Kem. Anal. Lung, 1967, cited in: Jernelöv, A. in: Chemical Fallout. Thomas Springfield, Illinois 1969, p. 68
461. Kiemeneij, A.M., Klosterboer, J.G.: Anal. Chem. *48* (3), 575 (1976)
462. Imura, M. et al.: Science *172*, 1248 (1971)
463. Bertilsson, L., Neujahr, N.Y.: Biochem. *10*, 2805 (1971)
464. De Simone, R.E.: Chem. Soc. London, J. Chem. Commun. *13*, 780 (1972)
465. Wood, J.M. Scott, K.F., Rosen, C.G.: Nature *220*, 173 (1968)
466. Jensen, S., Jernelöv, A.: Nordforsk. Biocidinformation *12*, 3 (1968)
467. Beckert, W.F. et al.: Nature *249*, 674 (1974)
468. Rogers, R.D.: Proc. Internat. Conf. Electric Power Res. Institute, Environmental Protection Agency, Institution for Environmental Studies, Nat. Inst. of Occupational Safety and Health, WHO in 4 parts (Heavy Metals Environment) 1975, Paper C-218, Toronto, Canada
469. Landner, L.: Nature *230*, 452 (1971)
470. Spangler, W.J. et al.: Science *180*, 192 (1973)
471. Hamdy, M.K., Noyes, O.R.: Appl. Microbiol. *30*, 424 (1975)
472. Gomer, R., Noyes, W.A.: J. Am. Chem. Soc. *71*, 3390 (1949)
473. Slemr, F.: private Communication, MPI für Chemie, Abt. Chem. d. Atmosph. u. Physikal. Chemie d. Isotope, Univ. Mainz
474. Ganther, H., Sunde, M.: J. Food Sci. *39*, 1 (1974)
475. Schrauzer, G.: Progr. Chem. Organ. Natural Prod. *31*, 563 (1974)

476. Segall, H., Wood, J.: Nature 248, 456 (1974)
477. Clarkson, T.W.: J. Occup. Med. *10*, 351 (1968)
478. Nordberg, G.F., Skerfving, S.: in: Mercury in the Environment. Chemical Rubber Co., Cleveland Ohio 1972, p. 29
479. Cole, H.N., Schreiber, N., Sollmann, T.: Arch. Dermatol. *21*, 372 (1930)
480. Lang, E.R. et al.: J. Pharmacol. Exp. Ther. *89*, 52 (1947)
481. Berlin, M.H., Nordberg, G.F., Serenius, F.: Arch. Environm. Health *18*, 42 (1969)
482. Fraser, A.M., Melville, K.I., Stehle, R.L.: J. Ind. Mealth *16*, 77 (1934)
483. Gage, J.C.: Brit. J. Ind. Med. *18*, 287 (1961)
484. Rothstein, A., Hayes, A.D.: J. Pharmacol. Exp. Ther. *130*, 166 (1960)
485. Magos, L.: Brit. J. Ind. Med. *25*, 315 (1968)
486. Tennant, R., Johnston, H.J., Wells, J.B.: Conn. Med. *25*, 106 (1961)
487. Bornmann, G. et al.: Arch. Toxicol. *26*, 203 (1970)
488. Viola, P.L., Cassano, G.B.: Med. Lavaro *59*, 437 (1968)
489. Morrow, P.E., Gibb, F.A., Johnson, L.: Health Phys. *10*, 543 (1964)
490. Prickett, C.S., Lang, E.P., Kunze, F.M.: Proc. Soc. Exp. Biol. Med. *73*, 585 (1950)
491. Clarkson, T.W.: Food Cosmet. Toxicol. *9*, 229 (1971)
492. Sollmann, T., Schreiber, N.E.: Arch. Intern. Med. *57*, 46 (1936)
493. Turk, J.L., Bakev, H.: Brit. J. Dermatol. *80*, 623 (1968)
494. Frithz, A.: Acta Derm. Venerol. *50*, 345 (1970)
495. Wahlberg, J.E.: Arch. Environ. Health *11*, 201 (1965)
496. Friberg, L., Skog, E., Wahlberg, J.E.: Acta Derm. Venerol. *41*, 40 (1961)
497. Swensson, A.: Acta Med. Scand. *143*, 365 (1952)
498. Hagen, U.: Arch. Exp. Pathol. Pharmacol. *224*, 193 (1955)
499. Östlund, K.: Acta Pharmacol. Toxicol. *27*, Suppl. 1 (1969)
500. Miettinen, J.K. et al.: Ann. Clin. Res. *3*, 116 (1971)
501. Ulfvarson, U.: Int. Arch. Gewerbepath. *19*, 412 (1962)
502. Itsuno, Y.: Minamata Disease (Ed.: Kutsuna, M.), Study Group of Minamata Disease, Kumamoto University, Kumamoto-shi 1968, p. 338
503. Wahlberg, J.E.: Acta Dermatovener. *45*, 397 (1965)
504. Berlin, M., Ullberg, S.: Arch. Environ. Health *6*, 610 (1963)
505. Trenholm, H.L. et al.: Toxicol. Appl. Pharmacol. *18*, 97 (1971)
506. Snyder, R.D.: New Eng. J. Med. *284*, 1014 (1971)
507. Bakulina, A.V.: Sovet Med. *31*, 60 (1968)
508. Lang, E.P., Kunze, F.M.: J. Pharmacol. Exp. Ther. *95*, 460 (1961)
509. Abrams, J.D., Majzoub, U.: Brit. J. Opthal. 54, 59 (1970)
510. Ellis, R.W., Fang, S.C.: Toxicol. Appl. Pharmacol. *11*, 104 (1967)
511. Berlin, M., Ullberg, S.: Arch. Environ. Health, *6*, 589 (1963)
512. Suzuki, T. et al.: Ind. Health *5*, 149 (1967)
513. Clarkson, T.W., Gatsy, J., Dalton, C.: Div. Radiation Chem. Toxicol., Univ. Rochester, Atom Energ. Project, Rochester, N.Y. 1961, UR-582
514. Magos, L.: Environ. Res. *1*, 323 (1967)
515. Berlin, M., Fazackerly, J., Nordberg, G.: Arch. Environ. Health, *18*, 719 (1969)
516. Nielson Kudsk, F.: Acta Pharmacol. *27*, 149 (1969)
517. Platonov, N.: Health Rev. *20*, 9 (1968)
518. Irukayama, K. et al.: Japan J. Hyg. *20*, 11 (1965)
519. Gage, J.C.: Brit. J. Ind. Med. *21*, 197 (1964)
520. Donelly, W.J.C.: M.S. Thesis, Mich. State Univers. 1965
521. Östlund, K.: Nord. Hyg. Tiskr. *50*, 82 (1969)
522. Takeda, Y. et al.: Toxicol. Appl. Pharmacol. *13*, 156 (1968)
523. Takeda, Y., Ukita, T.: Toxicol. Apll. Pharmacol. *17*, 181 (1970)
524. Takahashi, T. et al.: Jap. J. Hyg. Chem. *17*, 93 (1971)
525. Miller, V.L., Klavano, P.A., Conka, E.: Toxicol. Appl. Pharmacol. *2*, 344 (1960)
526. Daniel, J.W., Gage, J.C.: Biochem. J. *122*, 24 (1971)
527. Gage, J.C.: in: Mercury, Mercurial, and Mercaptans (Eds.: Miller, M.W., Clarkson, T.W.) 4th Rochester Conf. on Environ. Toxicity, 1971, June 17–19, Springfield, Charles C. Thomas Ill.

528. Daniel, J.W., Gage, J.C., Lefevre, P.A.: Biochem. J. *121*, 411 (1971)
529. Anghileri, L.J.: Nucl. Med. *4*, 193 (1964)
530. Norseth, T.: Thesis, Dpt. Radiation Biol. Biophys. Univ. Rochester, Rochester, N.Y. 1969
531. Åberg, B. et al.: Arch. Environm. Health *19*, 478 (1969)
532. Seifert, P., Neudert, H.: Zbl. Arbeitsmedizin *4*, 129 (1954)
533. Jernelöv, A., Lann, H.: Environm. Science and Technol. *7*, 712 (1973)
534. Suzuki, T., Farakawa, K., Tokomura, U.: Rep. Fermentation Res., Inst. Chiba, Japan 1969 (1–9), p. 36
535. Spangler, W.J. et al.: Science *180*, 192 (1973)
536. Kimura, Y., Miller, V.L.: Agr. Food. Chem. *12*, 253 (1964)
537. Yamada, M., Dazai, M., Tonomura, K.: Hakko Kogaku Zasshi *47*, 155 (1969)
538. Tonomura, K. et al.: Rep. Ferm. Res. Inst. *32*, 25 (1967)
539. Tonomura, K. et al.: J. Ferment. Technol. *46*, 506 (1968)
540. Furukawa, K., Suzuki, T., Tonomura, K.: Agr. Biol. Chem. *33*, 128 (1969)
541. Tonumura, K.: Biochem. Biophys. Acta *182*, 227 (1969)
542. Tonomura, K., Kanzaki, F.: ibid. *184*, 227 (1969)
543. Tonomura, K. et al.: Nature *217*, 644 (1968)
544. Harres, J.O., Eisenstark, A., Dragsdorf, R.D.: J. Bacteriol. *68*, 745 (1954)
545. Sode, R.L.: Water Qual. Off. Dpt. Interior, Washington D.C. 1970
546. Ball, R.C., Bahr, T.G., Kevern, N.R.: Water Qual. Office, Dpt. Interior, Washington D.C., (Mich. State Univ.) 1971
547. Jernelöv, A.: Internat. Conf. Environ. Mercury Contamination. Ann Arbor, Mich., Sept. 1970
548. Jernelöv, A.: Environmental Mercury Contamination. Ann Arbor Sci. Publ. Mich., Ann Arbor, Mich. 1972, p. 174
549. Olsson, M.: Nord. Hyg. Tidskr. *50*, 179 (1969)
550. Lann, H.: Rep. Swed. Environmental Protect. Board, No. 060370, 1970
551. Wallace, R.A. et al.: Mercury in the Environment – The Human Element, ORNL-NSF–EP1, Oak Ridge, Nat. Laboratory, Oak Ridge, Tenn., Jan. 1971
552. McCarthy, J.H. et al.: Mercury in the Environment. Geolog. Survey Prof. Paper, 713, 1970, p. 37
553. Jepsen, A.F. in: Trace Elements in the Environment, see [108]
554. Brune, D.: Anal. Chim. Acta *44*, 15 (1969)
555. Cambray, R.S., Jefferies, D.F., Topping, G.: Environ. Med. Sci. Divis. AERE, Harwell, Oxfordshire 1975
556. Commission Europ. Communities: V/F/1966/74e Vol. 2 and 3, Health Protection Directorate, Luxembourg 1974
557. Fitzgerald, W.F., Lyons, W.B.: Nature *242*, 452 (1973)
558. Hosohara, K.: Nippon Kagaku, Zasshi *82*, 1107 (1961)
559. Weiss, H.V., Yamamoto, S.: Environ. Sci. Technol. *6*, 644 (1972)
560. Rat d. Europ. Gemeinschaften, Vorschläge, Richtlinie über Qualitätsanforderungen an Oberflächenwasser für die Trinkwassergewinnung in den Mitgliedstaaten, Nov. 1974
561. Winkler, H.A.: Chem.-Ing.-Techn. *47*, 659 (1975)
562. DVGW – Regelwerk (Deutscher Verein von Gas- u. Wasserfachmännern) Frankfurt, Arbeitsblatt W 151, May 1975
563. Young, D.R. et al.: Nature *244*, 273 (1973)
564. Förstner, U., Müller, G.: Schwermetalle in Flüssen und Seen. Springer-Verlag, Berlin, Heidelberg, New York 1974
565. Aston, S.R. et al.: Nature *241*, 16 (1973)
566. Thomas, R.L.: Canad. J. Earth, Sci. *9*, 636 (1972)
567. Derryberry, O.M.: Environmental Mercury Contamination, Ann Arbor Sci. Publ., Mich. 1972
568. Rucker, R.R., Amend, D.F.: Prog. Fish, Culturist *31*, 197 (1969)
569. Hamilton, A.: Royal Soc. Canada, Int. Symp. Mercury in Man's Environment, Ottawa, Canada 1971, Feb. 15–16, p. 87
570. Johnels, A.G.: Oikos *18*, 323 (1967)
571. Johnels, A.G. et al.: Fauna Flora, Uppsala *63*, 169 (1968)

572. Hannerz, L.: Rep. Inst. Freshw. Res. Stockholm Sweden *48*, 120 (1968)
573. D'Itri, F.: Proc. 162nd Amer. Chem. Soc. Meet. Washington, D.C. 1971
574. Bäckström, J.: Acta Pharmacol. Toxicol. *27* (3) (1969)
575. Keckes, S., Miettinen, J.K.: Proc. FAO Conf., Rome 1970, Dec. 9–18
576. Miettinen, V. et al.: FAO Techn. Conf. Marine Pollution and its Effects on Living Resources, Rome 1970, Dec. 9–18
577. Wobeser, G. Nielson, N.O., Dunlop, R.H.: J. Fisheries Res. Board, Canad. *27*, 830 (1970)
578. Ackefors, H., Löfroth, G., Rosen, C.–G.: Oceanogr. Mar. Biol., Ann. Rev. *8*, 203 (1970)
579. Vostal, J. in: Mercury in the Environment (Eds.: Friberg, L., Vostal, J.) CRC Press, Cleveland 1972
580. Shacklette, H.F., Boerngen, J.G., Turner, R.L.: Geolog. Survey Circular 644, US Geolog. Survey, Washington, D.C. 1971
581. Jernelöv, A., Wallin, T.: Atmospheric Environment *7*, (1973)
582. Anderson, W.L., Smith, K.E.: Environ. Sci. Technol. *11*, 75 (1977)
583. Frissel, M.J., Poelstra, P., van der Klugt, N.: Geol. en mijnbouw *53* (4) (1974)
584. Heinrich, H., Mayer, R.: J. Environ. Qual. *6*, (4), 402 (1977)
585. Kloke, A., Schenke, H.D.: in: Problems of the Contamination of Man and His Environment by Mercury and Cadmium CEC, 5075, Luxemburg 1973
586. Pickard, J.A., Martin, J.T.: Ann. Rep. Agric. Hort. Res. Stat. Long Ashton Bristol 1959, p. 93
587. Schelenz, R., Diel, J.F.: Z. Lebensmittel Unters.-Forsch. *151*, 369 (1973)
588. Loon, J.C.: Environmental Lett. *3* (6), 211 (1974)
589. Shacklette, H.T.: Prof. Pap. US Geol. Surv., No. 713, 1970, p. 35
590. Löfroth, G., Duffy, M.E.: Environment *11*, 10 (1969)
591. James, P.E., Lagerwerf, J.V., Dudley, R.F.: Proc. Intern. Symp. Identificat. and Measurement of Environmental Pollutants, Ottawa 1971
592. Westöö, G.: Nord. Hyg. Tisdkr. *50*, 67 (1969)
593. Lagervall, M., Westöö, G.: Var Föda *21*, 9 (1969)
594. Rauter, W.: Z. Lebensmittel Unters.-Forsch. *162*, 1 (1967)
595. Wheeler, B.E.J., Pickard, J.A., Martin, J.T.: Ann. Appl. Biol. *51*, 403 (1963)
596. Stewart, D.K.R., Ross, R.G.: Canad. J. Plant Sci. *42*, 370 (1962)
597. Ross, R.G. and Stewart, D.K.R.: Canad. J. Plant Sci. *44*, 123 (1964)
598. Stone, H.M., Jacks, H.: N.Z.J. Sci. Technol. *1*, 373 (1958)
599. Seeger, R.: Z. Lebensmittel Unters.-Forsch. *161*, 115 (1976)
600. Westöö, G.: Acta Chem. Scand. *21*, 1790 (1967)
601. Tejning, S.: Oikos *18*, 334 (1967)
602. Smart, N.A.: Residue Rev. *23*, 1 (1968)
603. Ellis, R.W., Fang, S.C.: Toxicol. Appl. Pharmacol. *11*, 104 (1967)
604. Cember, H.: Am. Ind. Hyg. Ass. J. *30*, 367 (1969)
605. Sumari, P. et al.: Nord. Hyg. Tidskr. *50*, (2), 97 (1969)
606. Tejning, S.: Rep. No. 67 02 06, Dept. of Occup. Med., Univ. Hospital Lund, Sweden 1967, p. 3
607. Tejning, S.: Rep. No. 67 08 31, Dept. of Occup. Med., Univ. Hospital Lund, Sweden 1967, p. 30
608. Skerfving, S.: Toxicity of methyl mercury with special reference to exposure via fish, Thesis, Karolinska Institute, Stockholm 1972
609. Reiniger, P. et al.: Nucl. Techn. in Environ. Poll., IAEA, Vienna 1971, p. 407
610. Takizawa, Y. et al.: Acta Med. Biologica *19*, 193 (1972)
611. Tomizawa, C.: Shokuhin. Eisergaku Zasshi, *7*, 26 (1966)
612. Iyengar, G.V., Kollmer, W.E., Bowen, H.J.M. in: Elemental Composition of Human Tissues and Body Fluids. Verlag Chemie, Weinheim 1978
613. Skerfving, S. in: Mercury in the Environment. CRC, Cleveland, Ohio 1972, p. 109
614. Saha, J.G., Mc Kinlay, K.S.: in: Anal. Aspects of Mercury and other Heavy Metals in the Environment (Eds. Frei, R.W. and Hutzinger, O.) Gordon and Breach, London, New York, Paris 1975, p. 1
615. Borg, K. et al.: J. appl. Ecol., Suppl. *3*, 171 (1966)
616. Wood, J.M.: Environm. *14* (1), 33 (1972)
617. Hopmann, A.: Bull. Hyg. (London) *3*, 585 (1928)

618. Miettinen, J.K. et al.: Proc. 9th Japanese Conf. Radioisotopes B/II–17 474, Jap. Ind. Forum, Inc. Tokyo 1969
619. Miettinen, J.K., Heyraud, M., Kekes, S.: Proc. FAO Conf. (Food and Agriculture Organization) FAO-FIR/MP:70/ E–90, Rome 1970, Nov. 5th
620. Matthes, F.T. et al.: Pediatrics *22*, 675 (1958)
621. Miller, V.L. et al.: Toxicol. Appl. Pharmacol. *3*, 459 (1961)
622. Takahata, N. et al.: Folia Psychiatr. Neurol Jap. *24*, 59 (1970)
623. Beidas, A.S., Higgans, D.J.: J. Sci. Food Agric. *8*, 597 (1957)
624. Joselow, M.M., Goldwater, L.J., Weinberg, S.B.: Arch. Environ. Health *15*, 64 (1967)
625. Dunlap, L.: Chem. Engng. News *49*, (27), 22 (1971)
626. Birke, G.: Arch. Environmental Health *25*, 77 (1972)
627. Slemr, F., Seiler, W., Schuster, G.: Ber. Bunsenges. Phys. Chem. *82*, 1142 (1978)
628. Kussmaul, A.: Untersuch. konstitutionellen Mechanismus u. sein Verhältnis z. konstitutionellen Syphilis, Stahel, Würzburg 1861
629. Frachtenberg, I.M.: The chronic action of mercury on the organism, current aspects of the problem of mercurialisms and its prophylaxes, Zdorovija, Kiev 1969
630. Skerfving, S., Vostal J. in: Mercury in the Environment (Eds.: Friberg, L., Vostal) CRC Press, Cleveland, Ohio 1972, p. 93
631. Hazards Group of Mercury Study Group: Environmental Res. *4*, 1 (1871)
632. Takeuchi, T. in: Environmental Mercury Contamination (Eds.: Hartung, R., Dinman, B.D.) Ann Arbor Science, Publ. 1972, p. 247
633. Bidstrup, P.L.: Elsevier Monogr. Toxic Agents (Ed. Browning, E.) Elsevier New York 1964, p. 114
634. Friberg, L., Nordberg, F. in: Mercury in the Environment (Eds.: Friberg, L., Vostal, J.) CRC, Press, Cleveland, Ohio 1972, p. 113
635. Skerfving, S.: ibid, p. 141
636. Kurland, L.T., Faro, S.N., Siedler, H.: World Neurology *1*, 370 (1960)
637. Ramel, C. in: Mercury in the Environment (Eds.: Friberg, L., Vostal, J.) CRC Press, Cleveland, Ohio 1972, p. 169
638. Hughes, W.L. jr.: Ann. N.Y. Acad. Sci. *65*, 454 (1951)
639. Takeda, Y., Ukita, T.: Toxicol. Appl. Pharmacol. *17*, 181 (1970)
640. Webb, J.L.: Enzyme and Metabolic Inhibitors, Vol. 2, Chapt. 7, Academic Press, New York 1966
641. Plunkett, E.R.: Handb. Ind. Toxicology, Chemical Publishing Co., New York 1966
642. Salvaterra, P., Lown, B., Massaro, E.: Acta Pharmac. Tox. *32*, 111 (1975)
643. Mc Crehan, W.A., Durst, A.R., Bellama, J.M.: Anal. Lett. *10* (14), 1175 (1977)
644. Mackay, D., Wolhoff, A.: Curr. Res. Environ. Sci. Techn. *7*, No. 7 (1973)
645. Hartung, R. in: Environmental Mercury Contamination (Eds.: Hartung, R. and Dinman, B.D.) Ann Arbor Sci. 1972, p. 172
646. Landner, L.: Restoration of Mercury Contaminated Lakes and Rivers Swedish Water and Air Pollution Research Laboratory Rep. B 76, 1970
647. Friberg, L., Hammarström, S., Nyström, A.: Arch. Ind. Hyg. Occup. Med. *8*, 149 (1953)
648. Atkinson, W.S.: Amer. J. Ophtal. *26*, 685 (1943)
649. Fanconi, G., Botszstejn, A., Schenker, P.: Helv. Pediat. Act *2*, (4), 3 (1947)
650. Sollman, T.: A Manual of Pharmacology, 8th ed. Saunders, Philadelphia 1957, p. 1310
651. Mc Alpine, D., Araki, S.: Lancet *2*, 629 (1958)
652. Takeuchi, T. et al.: A pathological Study of Minamata Disease in Japan, Acta Neuropath., (London) *2*, 40 (1962)
653. Harada, Y. in: Minamata Disease (Ed.: Kutsuna, M.) Study Group of Minamata Disease, Kumamoto, Univers., Kumamoto, Japan, 1968, p. 93
654. Irukayama, K.: Adv. Pollution Res. *19*, 153 (1969)
655. Clegg, D.J.: Proc. Royal Soc. Canada, Symp. Man's Environment, Ottawa, Canada 1971, Feb. 15th
656. Snyder, R.D.: New England, J. Med. *284*, 1014 (1971)
657. Tejning, S.: Rep. 68 02 20, Dept. Occupat. Med., Univ. Hospt., Lund, Sweden, p. 5
658. Löfroth, G.: Swedish Nat. Sci. Res. Council, Stockholm Sweden, 2nd Ed., Sept. 1970
659. Fiskesjo, G.: Hereditas *62*, 314 (1969)

660. Ramel, C.: Nord. Hyg. Tisdkr. *50*, 135 (1969)
661. Ramel, C., Magnuson, J.: Hereditas *61*, 231 (1969)
662. Skerfving, S., Nanson, K., Lindstein, J.: Arch. Environ. Health *21*, 133 (1970)
663. Ramel, C. in: The Mercury Problem Oikos (Suppl.) *9*, 35 (1967)
664. Löfroth, G.: Ecological Res. Comm. Bull. 4, Swedish Nat. Sci. Res. Council 1969
665. Cohen, M.M.: Md. Med. J. *7*, 236 (1958)
666. Höök, O., Lundgreen, K.-D., Swensson, A.: Acta Med. Scand. *150*, 131 (1954)
667. Tsubaki, T. in: Mercury in Man's Environment. Proc. Royal Soc. Can. Symp. 1971, *131*, Feb., p. 15
668. Takahata, N. et al.: Folia, Psych. Neuro. Japan *24*, 59 (1970)
669. Morrison, A.B.: Canadium Approach to Acceptable Daily Intakes of Mercury in Foods, Spec. Symp. on Mercury in the Man's Environment, Ottawa, Canada, Feb. 1971
670. Berglund, F., Berlin, M.: in: Chemical Fallout (Eds.: Miller, M.W., Berg, G.G.) Charles C. Thomas Publ., Springfield, Ill. 1969
671. FAO Working Party on: Pesticide Residues and WHO Expert Comm. on Pesticides Residues, WHO Techn. Rep. Ser. No. 370, FAO Agricult. Studies, No. 73, FAO, Rome 1967
672. Arch. Environm. Health *19*, 891 (1969)
673. Threshold Limit Values of Air-Borne Contaminants for 1968: Recommended a. Intended Values, Adapted at the 30th Ann. Meet. Amer. Conf. Gov. Ind. Hyg., St. Louis, 1968, May 13, (Eds.: Americ. Conf. of Gov. Ind. Hyg.) Cincinnati
674. Sergeyev, Ye.: Proc. First All-Union Conf. Geochem. Meth. of Prospecting for Ore Deposits Moscow 1957
675. Fujimura, Y.: Japan J. Hyg. *18*, 10 (1964)
676. Neville, G.A.: Canad. Chem. Educ. *3* (1), 4 (1967)
677. Ganther, H., Sunde, M.: J. Food Sci. *39*, 1 (1974)
678. Schelenz, R., Diehl, J.-F.: Z. Lebensmittel-Untersuch. Forsch. *154*, 160 (1974)
679. Krausche K. in: [43], p. 51
680. Konrad, J.G. in: Environmental Mercury Contamination (Eds.: Hartung, R. and Dinman, B.D.) Ann Arbor Sci. Publ., Inc., Ann Arbor, 1972, p. 52
681. Vernet, J.P., Thomas, R.L.: Eclogae, Geol. Helv. *65* (2), 307 (1972)
682. Takeuchi, T. [42]
683. Interlaboratory Comparative Analysis of the DFG: Trace Metals in Rhine Sediment, Bundesanstalt für Gewässerkunde, Koblenz, FRG, 1976
684. Wilander, L.: Grundforbattring *21* (4), 151 (1968)
685. Brune, D., Landström, D.: Radiochem. Acta *5*, 228 (1966)
686. Heide F., Lerz, H., Böhm, G.: Naturwissenschaften *44*, 441 (1957)
687. Aidin'yan, N. Kh., Ozerova, N.A., Gipp, S.K.: Akad. Nauk. SSR, Trudy Inst. Geol. Rudn. Mestorozhd. Petrogr., Mineral. Geokhim. *99*, 5 (1963)
688. Chester, R. et al.: Mar. Poll. Bull. *4*, 28 (1973)
689. Leatherland, T.M. et al.: Nature *232*, 112 (1971)
690. Carr, R.A., Hoover, J.B., Wilkniss, P.E.: Deep-Sea Res. *19*, 747 (1972)
691. Turkeim, H.J.: Lancet *257*, 1150 (1949)
692. Kutsuna, M.: Minamata Rep. (Study Group of Minamata Disease) Kumamoto Univ., Japan 1968
693. Braun, W., Dönhardt, A.: Vergiftungsregister, Georg Thieme Verlag, Stuttgart 1970
694. Jordi, A.: Z. Unfallmed. Berufskrankh. *36*, 136 (1943); Schweiz med. Wschr. *1947*, 621
695. Moeschlin, S.: Klinik u. Therapie d. Vergiftungen, 5th Ed., Georg Thieme, Stuttgart 1972
696. Bader, E.W.: Gewerbekrankheiten, 4 Ed., Urban & Schwarzenberg, München, Berlin 1954
697. Deutsche Forschungsgemeinschaft, Maximale Arbeitsplatzkonzentrationen 1979, Senatskommission z. Prüf. gesundheitsschädl. Arbeitsstoffe, Mitt. XIV, Harald Boldt, Boppard, FRG
698. Korbakova, A.I.: Maximum Allowable Conc. Mercury in the USSR Working Paper, Stockholm Symp., Nov. 1968 (paper can be obtained from the Institute of Hyg. Karolinska, S-10401 Stockholm 60, Sweden)
699. Parizek, J. et al. in: New Trace Elements in Nutrition (Eds.: Mertz, W., Cornatsev, W.E.) Macel Dekker Inc., N.Y. 1971
700. Kosta, L., Byrne, A.R., Zelenko, V.: Nature *254*, 238 (1975)

Cadmium

U. Förstner
Institut für Sedimentforschung der Universität Heidelberg
D-6900 Heidelberg, Federal Republic of Germany

Introduction

Cadmium is regarded as one of the most toxic metals, although there is no rigid order of toxicity of trace metals in the environment [1, 2]. The acute toxicity of cadmium upon inhalation or ingestion was recognized long ago and its chronic toxic effect on workers exposed to dust and vapor has been known approximately since 1948 [3]. Pollution by cadmium in aquatic systems appears to be less widespread than that by mercury, but has nonetheless had hazardous effects on humans. During 1947 an unusual and painful disease was recorded as of a "rheumatic nature" in the case of 44 patients from villages on the banks of the Jintsu River, Toyama Prefecture, Japan. It became later known as the "itai-itai" disease (meaning "ouch-ouch") in accordance with the patients' shrieks resulting from painful skeletal deformities; it is estimated that approximately 100 deaths occurred due to the disease until the end of 1965 [4]. However, the cause of itai-itai disease was completely unknown until 1961 when sufficient evidence led to the postulation that cadmium plays a role in its development [5]. Based on the findings of further government-supported studies, the Japanese Ministry of Health and Welfare declared in 1968: "The itai-itai disease is caused by chronic cadmium poisoning, on condition of the existence of such inducing factors as pregnancy, lactation, imbalance in internal secretion, aging, deficiency of calcium, etc." [6].

After the first alarming suspicion and initial diagnosis of the itai-itai disease, numerous detailed investigations have been carried out in many countries. Heavy cadmium pollution in aquatic systems (without indications of acute toxic effects on humans, however) has been recorded in the Hudson River Estuary, New York (nickel-cadmium battery factory), the Hitachi area near Tokyo/Japan (braun tube factory), from Palestine Lake, Indiana/USA (plating industry), Sörfjord/Norway and Derwent Estuary, Tasmania (smelting emissions), from the Neckar River/FRG (pigment factory), and the Vesdre River/Belgium. The accumulation of cadmium in agricultural soils and its increased uptake by plants is presently of world-wide concern.

Although analytically valid long-term human balance studies are not available to enable precise estimates of the rate of cadmium absorption and excretion, it has been shown that cadmium originating in different pathways can constitute a health hazard to parts of the populations. Good examples of such pathways are food (e.g., crops grown on contaminated soil), air pollution and cigarette smoking. Summaries and discussions of the environmental impact and of dose-response relationships of cadmium have been given by Friberg et al. [6], by the Panel of Hazardous Trace Substances of the US National Institute of Environmental Health [7], by the Task Group on Environmental Health Effects of Cadmium, World Health Organization [8], by the Subcommittee on the Toxicology of Metals of the International Association of Occupational Health [9] and by the Working Group of Experts for Cadmium, prepared for the Commission of the European Communities [10].

Mining and smelting companies in various parts of the world established the Cadmium Association in the UK and the Cadmium Council in the US in 1976 to collect and disseminate information on cadmium to consumers, governments, and other concerned groups. These two organizations publish *Cadmium Abstracts*, a quarterly summary of current world literature on the properties and uses of cadmium, its alloys, and compounds. The First International Cadmium Conference, organized jointly by the two bodies and the International Lead and Zinc Research Organization, was held in San Francisco in January 1977 [11], and the second conference in Cannes/France in February 1979. A handbook on the biogeochemistry of cadmium is presently edited by J. O. Nriagu [12]. This work includes data on cadmium in natural waters, in the atmosphere, in soils, in biota, and in sediments, on occupational exposure to cadmium, metabolism and toxicity in organisms, human health (including immunological, cellular, teratogenic, mutagenic, and pathological aspects).

Production, Consumption and Use

Production

Cadmium[1] is basically recovered as a by-product from the smelting and refining of zinc concentrates at a rate of approximately 7–8 lbs/t primary zinc. No ores are mined and processed exclusively to obtain cadmium.

Data for 1977 show that cadmium production in western industrial countries was approximately 14,000 t, 7% higher than in the year before. This rate is also 17% higher than that in 1975, the year when a distinct slump occurred in demand for cadmium in most of the western countries (Fig. 1). The majority of cadmium producing countries, particularly Japan and Western Europe,

1 According to Mining Ann. Rev. *1978*, p. 101–102

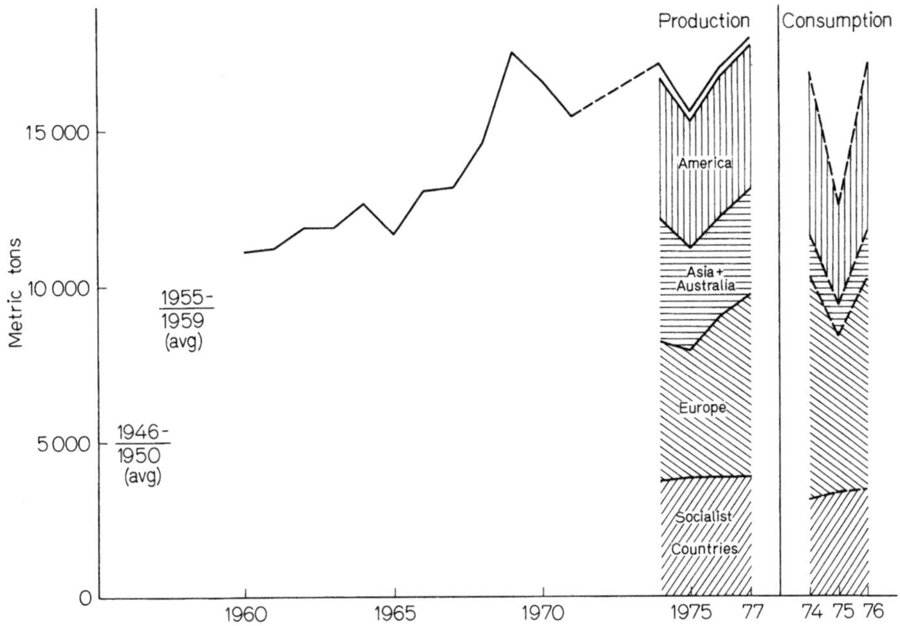

Fig. 1. Production and consumption of cadmium (metric tons). (Data from [7] and [13])

recorded significant increases in production during 1977; this, of course, reflects expansion of zinc output (Table 1). Against a background of a 2% increase in zinc production in 1977, the 7% increase in cadmium production raised the cadmium/zinc recovery ratio. This apparently means that stocks of semi-refined material were economically converted to saleable metal boosting somewhat smelter revenues [13].

An important source of cadmium is the processing of secondary raw sources (recycling materials). These are partially added to the cadmium production process with primary raw materials. Secondary raw materials are cadmium-containing end products which have become unuseable (such as nickel-cadmium batteries), cadmium-containing by-products that are subsequently reprocessed using special procedures (dust particles, cadmium-containing muds, etc.), and all types of materials whose reprocessing has become economically feasible or necessary by legislative enactment. In the Federal Republic of Germany in 1975 as much as 20% of the total processed cadmium came from recycled material.

Consumption

Consumption of cadmium in Western industrial countries rose at an average compound rate of about 3.5% between 1964 and 1974. In 1975, however, a slump in consumption of 34% occurred, but consumption rapidly picked up in 1976 and almost regained the previous levels of around 14,000 t. In 1977 cadmium demand in some countries was down again with the consumption in

Table 1. Major producers and consumers of cadmium (t) [13]

	Metal production				*Metal consumption*		
	1974	1975	1976	1977[a]	1974	1975	1976
Europe							
Belgium	1,043	950	1.176	1,445	1,682	1,036	1,426
Bulgaria[b]	200	220	220	220	-	-	-
East Germany[b]	18	18	18	18	450	450	460
Finland	156	217	426	519	-	-	-
France	644	602	586	782	1,460	825	958
West Germany	1,338	923	1,454	1,535	2,016	1,139	2,100
Italy	505	421	439	435	430	360	360
Netherlands	75	272	398	300	-	-	-
Poland[b]	305	370	400	400	230	240	200
Spain	178	205	246	300	-	-	-
Sweden	-	-	-	-	170	264	403
United Kingdom	273	275	190	285	1,441	1,032	1,414
USSR[b]	2,900	2,950	2,900	2,900	2,000	2,200	2,300
Africa							
Zaire	272	264	266	160	-	-	-
America							
Brazil	-	-	-	-	198	200	200
Canada	1,152	1,140	1,368	1,416	49	38	50
Mexico	348	577	680	894	-	-	-
Peru	182	144	174	-	-	-	-
USA	2,970	2,055	2,150	2,142	4,584	2,893	5,081
Asia							
India	59	53	34	36	160	120	96
Japan	3,084	2,741	2,566	2,794	968	444	1,181
Australia	759	543	648	608	165	200	174

Sources: World Bureau of Metal Statistics, US Bureau of Mines. [a] Preliminary figures; [b] Estimate

the US 9% less to 4,626 t, in the UK to 1,242 t (a loss of 12%), and in Japan to 771 (a loss of 34%). In West Germany consumption rose only slightly, whereas in France an increase in demand was reported at 25% to 1,200 t.

In constant values, the price of cadmium has declined over the past decade – the US producer price was 3 dollars/lb in 1964 when zinc was 10–12 cents/lb. Now it is less than 2 dollars/lb, while zinc is around 30 cents. In the past years, price fluctuations have had the greatest influence on cadmium consumption, but environmental considerations have now assumed greater importance [13].[2]

Use

Electroplating is the largest single use of cadmium amounting to about 34% of all cadmium consumption. Presently it requires about half the cadmium used in the US and about 40% of that in the UK, but it is of less importance in other European countries. became insignificant when legislation strongly reduced

2 Beginning in July 1980, Swedish government will ban the use of cadmium in electroplating, as a stabilizer and as a coloring agent; it is also working to reduce the level of cadmium impurities in phosphorous fertilizer (ES&T, Dec. 1979, p. 1447)

the limit for cadmium in plating effluents. Legislation has forced a number of European and US electroplaters to find alternatives to cadmium. However, strong traditional demands from the aerospace and telecommunication industries, requiring the superior resistance of cadmium to alkaline and humid atmospheres is expected to continue, although due to recent developments in zinc plating processes, this metal has replaced cadmium in some instances [13]. Small amounts of cadmium are used for the production of fungicides [16], control rods for nuclear reactors, fluorescent lamps, phosphors for television picture tubes, luminescent dials, compounds used in photography, lithographs, etc. [10]. 23% of cadmium consumed is in the form of *pigments*. In the plastic and ceramic industries cadmium sulfide and cadmium selenide are used; these have bright clear colors in the yellow-orange-red range and are light-fast up to 600 °C. With the increase in plastics production is is reasonable to expect a similar increase in the demand for cadmium-based pigments.

Nickel-Cadmium batteries, originally introduced in the 1920's, account for about 15% of total consumption. Growth in use of some 6% per year is expected because of wider application of rechargeable batteries in hand-held electronic devices and in portable domestic electrical appliances. The most rapidly expanding application for this type of battery is in calculators [15, 10]. Use of cadmium as a *stabilizer* in PVC presently accounts for some 15% of consumption and 8% of cadmium consumption goes for the production of solders, brazing materials, and other alloys. The low melting point cadmium alloys are used in fire prevention devices, and Cu-1% Cd is widely used for overhead conductors for trains, trams and trolleybuses, and occasionally for overhead telephone wires [13]. Current consumption trends have been determined by recent detailed studies in the FRG [14]. Investigations of sources provide results which are quite variable (Table 2). Since 1973, there has been a steady decrease in the production of stabilizers, batteries, galvanized products and cadmium-containing glass. The production of pigments increased slightly in 1974, but fell off and even decreased in 1975. A similar development was determined also for alloys and rectifiers, whereby the 1975 values for alloys remained 39% above that for 1973. The use of cadmium in galvanizing processes decreased markedly in 1975 [11].

Table 2. Consumption of Cd in the Fed. Republic Germany according to useage [14]

	1973 t Cd	%	1974 t Cd	%	1975 t Cd	%
Pigments	719	100	771	107	424	59
Stabilizers	349	100	248	71	207	59
Batteries	335	100	150	45	150	45
Galvanizing	386	100	354	92	196	51
Glass products	30	100	22	73	13	43
Alloys	54	100	87	161	75	130
Rectifiers	15	100	24	160	18	107
By-products	10	100	28	280	22	220
Cadmium salts	72	100	25	35	13	18
Impurities in Zn	59	100	30	51	21	36

General Chemistry, Mineralogy, Geochemistry, Aquatic Chemistry

Chemistry

Cadmium has the atomic number 48, an atomic weight of 112.40 and consists of eight stable isotopes of the following relative abundances [17]:
^{106}Cd = 1.22%, ^{108}Cd = 0.88%, ^{110}Cd = 12.39%, ^{111}Cd = 12.75%,
^{112}Cd = 24.07%, ^{113}Cd = 12.26%, ^{114}Cd = 28.86%, ^{116}Cd = 7.58%.

Its specific gravity is 8.65g/cm³, melting point 320.9 °C and boiling point 767 °C. Cadmium, a white metal with a bluish tinge, was discovered in 1817 by Strohmeyer in Germany. It is soft, easily worked and is of considerable ductivity [10]. Like zinc and mercury, cadmium is a transition metal in Group IIb of the periodic table of elements. Cadmium and zinc, however, differ from mercury in that the latter has 14 additional electrons in the fourth orbital, which probably accounts for the high stability of compounds of mercury-carbon bonds, whereas the similar alkyl-cadmium compounds are extremely unstable and react rapidly with water and moist air under normal environmental conditions [7]. As a result they are not expected to be of importance as environmental pollutants [10].

Cadmium and zinc show only valence +2 in their compounds. These metals are also generally similar in reactivity, zinc being the more reactive, and cadmium showing a slightly greater tendency to form covalent bonds, especially with sulfur. The ionic radius of Cd^{2+} has been found to be 1.03 Å (Goldschmidt, 1926) and 0.97 Å (Pauling, 1927); the observed coordination number of Cd in its compounds are usually four and six, in a few cases also five, seven, eight, nine, and twelve. Cd-compounds are often isotypic with the corresponding compounds of Zn^{2+}, Mg^{2+}, Fe^{2+}, Co^{2+}, Ni^{2+} and in some cases of Ca^{2+} [18].

In air, cadmium vapor oxidizes quickly to cadmium oxide, and the metal dissolves in weak dilute acids, a property which has been responsible for acute oral intoxication in man. The sulfide (CdS), the carbonate ($CdCO_3$), the oxide (CdO), and the hydroxide ($Cd(OH)_2$) are insoluble in water (negative logarithms of solubility products – pH = 7 at 25 °C – are 27.8 for CdS, 11.3 for $CdCO_3$, and 14.4 for $Cd(OH)_2$ [19]. Cadmium sulphide is decomposed by acids with the liberation of hydrogen sulfide gas. The fluoride, chloride, bromide, iodide, nitrate, and sulfate of cadmium are relatively soluble compounds. Cadmium forms also a wide variety of soluble complexes, notably with cyanides and amines [10].

Mineralogy

The common cadmium minerals are greenockite, hexagonal CdS; hawleyite = cubic CdS; otavite = $CdCO_3$; monteponite = CdO, and cadmoselite, hexagonal CdSe. Significant amounts of naturally occurring cadmium are found only in association with zinc ores, in which the amount of cadmium varies considerably. Usually 0.1–0.5% (maximum 5%) is present in zinc blende (sphalerite) and calamine (zinc spar, smithonite). Greenockite occurs usually as an earthy coating on zinc minerals, especially sphalerite; crystals found in amygdaloidal cavities in basic igneous rocks are rare [20].

Geochemistry

Cadmium is a strongly chalcophilic element, i.e., it is concentrated in sulfide deposits together with zinc and mercury, and to a much lesser extent with lead and copper [7]. The abundance of cadmium in the earth's crust is generally estimated to be 0.11 ppm [21]. Its concentration is low in all igneous rocks and shows no clear relation to any major element, not even to zinc; the ratio Zn/Cd varies widely in all types of igneous rocks, with recorded extremes of 27–7,000 [7].

An accumulation of cadmium takes place in the sedimentary environment, in addition to the amount contributed by rock weathering. If the shale + clay concentration (Table 3) is taken to be representative of the average abundance of cadmium in sedimentary rocks, a 2.4-fold increase is measured as compared to magmatic rocks of the upper continental crust. This accumulation has been explained as due to a degassing of the earth [21]. A further slight increase of cadmium concentration is observed in pelagic clay, which is partly explained as the high adsorptive capacity of sedimentary iron and manganese compounds [22]; Mn-nodules from the North and South Pacific were found to contain 8.40 and 5.06 ppm Cd, respectively [23]; in Fe/Mn concretions from Quinte Bay, Ontario, Cd concentrations of 0.8–6.4 – average 3.0 ppm Cd – were measured [24]. Compared to magmatic rocks and shales, concentrations

Table 3. Cd in magmatic rocks, sediments, phosphorites, coal and oil [33]

Example	Cd (ppm)	Ref.
Granite rocks	avg. 0.075–0.100	[21]
Basaltic rocks	0.130	[21]
Ultramafic rocks	0.026	[21]
Shales	0.300	[21]
Pelagic clays	0.405	[21]
Sandstones	0.020	[26]
Limestones	0.035	[25]
Limestones	0.048	[26]
Limestones	0.090	[21]
Diatomaceous ooze	0.39 (n = 5)	[23]
Globigerina ooze	0.42 (n = 3)	[23]
Radiolarian ooze	0.45 (n = 4)	[23]
Red clay	0.56	[23]
Green mud	0.27	[23]
Calcaerous ooze	0.57	[23]
Organic mud	0.39	[23]
Phosphorites	~ 15	[27]
Coal	0.2–30	[30]
	avg. ~ 1	[32]
Oil	0.01–16	[7]

of cadmium in sandstones and limestones are diluted (Table 3). Due to the crystal chemical properties, the cadmium concentrations in limestones are expected to be controlled by their clay fraction [21]; this explains the relatively large differences in the average values for cadmium in limestones shown in Table 3 [25, 26]. Generally, cadmium seems to be enriched in some organic-rich sedimentary rocks such as dark shales, while depleted in others such as red shales relative to igneous and metamorphic rocks and the crust. Such enrichment occurs primarily through the adsorption and/or complexation of cadmium onto organic matter followed by the accumulation of organic debris in the depositional environment. Since such an environment will also be reducing, formation of cadmium sulfides during or after deposition can be expected [22].

Recent data on unpolluted environments are rather controversial. Different sediment types in pelagic deposits do not seem to affect cadmium concentrations (Table 3). Significant enrichment has only been observed from marine phosphorites [27]. For lake sediments higher concentrations of cadmium are reported from the organic fractions [28, 29]. It is assumed that gelatinous colloidal substances, which are formed from dissolved organic acids, spores, pollen and decayed leaves take up the metal ion from water. Cd concentrations in fossil organic substances are highly variable. Cadmium content of coal ranges between 0.04 and 30 ppm, with an average concentration of approximately 1 ppm [30–32]. Even larger variations are found for oil, but no average value can be given accurately.

Aquatic Chemistry

Reliable data on the distribution of cadmium in natural waters have only recently become available. Typical depth profiles from mid-Pacific sampling stations [34] show that cadmium together with phosphate and silicate is depleted in the surface water (approx. 0.01 µg/kg seawater) relative to the deeper ocean water (~0.07 µg Cd/kg). Such a distribution indicates uptake by organisms at the surface and regeneration from sinking biologic debris deeper in the water column. The particularly high covariance of Cd with phosphate suggests that cadmium occurs in a shallow cycle like the labile nutrients, rather than deeper in the ocean in silicates [34]. Natural Cd concentrations in freshwater are about the same as in deeper seawater: 0.07 µg/l were determined from Amazon River water [34], and 0.1 µg/l in water samples from the lower Mississippi River [35].

Differentiation of chemical species of metals has been performed by analytical procedures (see Section on Analytical Methods) and by computation of equilibrium models. Calculations of equilibrium solubilities with $Cd(OH)_2$ or $CdCO_3$ indicate minimum solubility at pH 9.0–10.0 [36]. The pH/E_h systems for $Cd+S+CO_2+H_2O$ is given in Fig. 2 after Hem [36]; the activity of dissolved Cd is $10^{-7.05}M$, which is equivalent to 10 µg/l, and those of dissolved carbon dioxide and sulfur species are 10^{-3} M. In the system as defined, cadmium solubility is below the actual standard limit (5 µg/l or 10 µg/l; last section) only at high pH (between pH 8.9 and 10.7) or in reduced

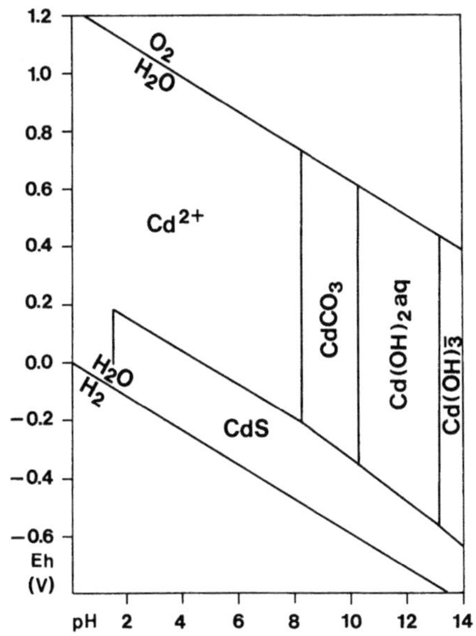

Fig. 2. Fields of stability of solids and predominant dissolved cadmium species in the system $Cd+CO_2+S+H_2O$ at 25 °C and 1 atm pressure in relation to Eh and pH. Dissolved cadmium activity $10^{-7.05}$ moles/l; dissolved carbon dioxide and sulfur species 10^{-3} moles/l (Hem [36])

systems in which oxygen is severely depleted [36]. Most natural waters are unsaturated with respect to hydroxide or carbonate; about 20% of the waters have carbonate contents in reasonable agreement with values calculated assuming $CdCO_3$ as the equilibrium solid phase [7]. Computer models have provided the following data for inorganic speciation of Cd in aquatic systems (listed according to increasing pH):
Cd^{2+}, $CdCO_{3(s)}$, $Cd(OH)_{2(s)}$, accounting for more than 90%,
$CdSO_4$ and $CdCl^+$, accounting for a few percent [37].
Dissolved Cd species in aerated seawater ($CdCO_3$ as solid species) are calculated as follows [38]:
$CdCl^+ = 56\%$, $CdCl_2^0 = 15\%$, $CdCl_4^{2-} = 10\%$, $CdCl_6^{4-} = 9\%$
other calculations: $CdCl_2^0 = 38\%$, $CdCl^+ = 29\%$, $CdCl_3^- = 28\%$ [39]; $CdCl_2^0 = 50\%$, $CdCl^+ = 40\%$, $CdCl_3^- = 6\%$ [40].

In reducing marine environments $Cd(HS)^0$ is the predominant dissolved species [37]. In oxidizing fresh waters sulfato- (45%), carbonato- (42%), and chlorospecies (13%) represent the dissolved fraction of Cd, whereas in the respective reducing environment free (aquo- = 88%) and chloro-species (11%) are calculated as the dominant dissolved species [37]. However, organic ligands can play a major role in natural waters by complexing trace metal ions and keeping them in solution [41]. In addition, organic ligands are likely to mediate large interactions among metal ions; for example, the presence of NTA couples the free concentrations of copper and cadmium [37].

Laboratory experiments on natural samples suggest that "labile forms" (free cationic species and complexes with low stability constants, both organic and inorganic) of Cd predominate in natural and polluted freshwater systems [42, 43]. Because of the tendency of these species to be sorbed on particulates, there is the need to determine the cadmium content of the suspended matter and sediments for assessing the sources, distribution, and fate of Cd contamination in aquatic systems [6, 33].

Analytical Methods

Determination of Cd in air, water, food, and organisms has been performed by different methods. Some major methods are colorimetry (dithizone method), emission spectroscopy, atomic absorption spectrophotometry (flame and flameless), electrochemical methods (anodic stripping voltametry, polarography), X-ray fluorescence methods, neutron activation analysis, isotope dilution, spark source mass spectrometry, and fluorimetry. Some of these methods have been summarized by Fleischer et al. [7], and are partially reproduced here in Table 4 (for references, see original paper by Fleischer). Currently,

Table 4. Cadmium analyses [7]

Material	Method
Granite and diabase rocks	Atomic absorption, mass spectrometry, neutron activation, optical spectrography, polarography, spectrophotometry
Water	Atomic absorption, spectrophotometry, polarography, stripping voltammetry, x-ray fluorescence
Air	Low-temperature ashing of glass fiber filters, atomic absorption
	Neutron activation
	Anodic stripping voltammetry
Food	Dry ashing, atomic absorption
Shellfish	Homogenization, wet digestion, atomic absorption
Blood	Wet ashing, dithizone extraction, atomic absorption
	Dry ashing, dithizone extraction, optical spectrography
Blood, tissues, and hair	Wet digestion, anodic stripping voltammetry
Renal tissue	Drying, neutron activation
	Dry ashing, atomic absorption
Urine	Dithizone, atomic absorption

atomic absorption spectrometry is probably the most widely used method for Cd analyses [44]. Anodic stripping voltammetry is extremely sensitive and is especially useful for determinations in natural waters and its use may yield information on the nature of binding of cations in water ([7]; see section on Aquatic Chemistry). A neutron activation method has been developed for Cd determination *in vivo* for organs, mainly the liver [10, 45]. *Pretreatment* of the samples is often required either for concentration (to improve sensitivity or to eliminate interferences, such as for NaCl matrices) or for extraction of distinct solid or aqueous species. For such differentiation in natural waters analytical schemes have recently been proposed [42, 43], which, however, need further development and standardization (e.g., Chelex-100 resin for the differentiation of labile and more stable species of transition metals).

Digestion of biological samples and sediments is often performed with conc. nitric acid [46] or $HCl\text{-}HNO_3 = 1{:}1$ [47]. The *aqua regia* digestion ($HCl{:}HNO_3 = 3{:}1$) is commonly used in the extraction of metals from polluted sediments. The fact that more volatile elements such as cadmium are not boiled as is the case for hydrofluoric acid in combination with nitric, perchloric or sulfuric acids is a major advantage in their analysis. A compilation of laboratory methods which have been proposed to measure Cd in soil, sewage sludge, and sediments, with special emphasis on methods for the assessment of cadmium available to plants, has been given by Symeonides and McRay [48] and is partially reproduced in Table 5. Other extraction processes for determining chemical associations of cadmium in particulate phases will be discussed in the section on chemical reactions.

Table 5. Suggested methods for assessing Cd in soil [48, 49]

Extraction solution	Strength	Ref.
Acetic acid	2.5% (0.41 N)	[50]
Hydrochloric acid	1 N	[51]
	0.1 N	[52]
Nitric acid	1 N	[53]
	2 N	[54]
Ammonium acetate	1 N, pH 7	[51, 53]
Ammonium acetate-acetic acid buffer	1 N, pH 4.8	[53]
Ammonium nitrate	1 N, unbuffered	[48]
Ammonium chloride	1 N	[54]
Calcium chloride	0.05 M, pH 6	[49]
	0.5 M, pH 6.5	[55]

Sources, Pathways, and Reservoirs in the Environment

With the attempt to determine the material flow of cadmium in the environment it becomes clear that complete knowledge of the processes involved is lacking. In the following section we refer to the detailed survey compiled in respect to the fluxes in US and adjacent marine waters by A.F. Sarofim in the framework of the Subpanel on Cadmium, US National Institute of Environmental Health [7]. Estimates by Davis et al. [56] on atmospheric emissions and data from Chizikhov [57] on Cd fluxes in smelters and refineries are the basic information source for this study. The most important data is compiled in Table 6.

Table 6. Estimated rates of emission of cadmium during production and disposal of cadmium products for 1968 in the USA [7]

				Losses during use and disposal			
	Mining and ore concentration, t/yr	Primary cadmium production, t/yr	Electroplating, pigment and plastic formulation, t/yr	Coal and oil combustion t/yr	Cadmium-plated metals, t/yr	Pigment, plastics, and miscellaneous, t/yr	Alloys and batteries, t/yr
Air contamination		955		120	500	90	40
Water contamination	3,000	240	300				
Soil contamination (140: Phosphate fertilizers)							
Accumulation in service					1,420	2,080	380
Land disposal (dumps, land fills, slag pits, mine tailings)	300	310		360	500	490	220

Sources

Coal and Oil Combustion, Cement Production, Incineration. If it is assumed that for the 455 million tons of coal consumed in 1968, the average particulate collection efficiency was 80% and the average cadmium concentration was 1.0 ppm (Table 3), an estimated 100 t of cadmium was emitted for that year [56]. Cement manufacture will provide approximately 3 t/yr, assuming that 80% of the emissions were captured in particulate collection devices, and approx. 100 t of cadmium were released during incineration of waste products in the same year [56]. With the closing of open-burning dumps and stricter air pollution regulations on incineration, this source of emission should be significantly reduced [7].

Industrial Emissions. Of the major industries employing cadmium, *electroplating* shops, *pigment* plants and producers of *alloys* and *batteries* can be expected to be major sources of cadmium pollution. The contribution of Cd to New York City wastewater treatment plants (approximately 25 t year) is shared by the following activities [60]:
electroplaters 33%,
other industrial sources 6%,
stormwater runoff 12%,
residential waste 49%.

From non-metallurgical industries the following metal concentrations in wastewaters (μg/l) were recorded in the same study [60]:
laundry 134, for dressing and dyeing 115, ice cream production 31, textile dyeing 30, miscellaneous chemicals 27, car washing 18, fish processing 14, meat processing 11 (mass fluxes are not noted).

Sewers. Cd input to sewers stems both from industrial processing and domestic sources (see above). Major domestic sources are atmospheric fallout on residential areas – which can produce particularly adverse effects by shock load during stormwater events [61] – and corrosion of zinc-containing roof fittings and household pipes [62]. On the average, approximately 50% of the cadmium discharge in the waste stream is retained within the sewer [63].

Phosphate Fertilizers. Significant amounts of Cd will reach the agricultural soils during application of phosphate fertilizers, which, on the average, contain approx. 5 ppm Cd [64]. For the entire US an input of 140 t yr has been estimated ([7] Table 6). The cadmium contents in superphosphate fertilizer is normally in the range of 2–50 ppm [10], but concentrations as high as 50–170 ppm have been determined [64]. At moderate Cd levels (5–10 ppm) in phosphate fertilizers no significant correlation between the rates of application and cadmium concentrations in the plow layer soil was found after 20 years [65].

Pathways

Major transport of Cd from the various sources to the reservoirs (see appropriate sections) and eventually to organisms can be via atmosphere, water, suspended sediment, land application (of sewage material and phosphate fertilizer) and waste dumping.

Table 7. Selected examples of Cd concentration in the atmosphere (see [10])

Northern Norway	0.10 ng/m^3
Swiss Alps	0.28 ng/m^3
Erlangen (FRG)	1.50 ng/m^3
Munich (FRG)	6.90 ng/m^3
Tokyo (Japan)	10-53 ng/m^3
El Paso (USA)	120 ng/m^3

Atmosphere (Table 7). Natural background values lie below 0.1 ng Cd/m^3 air. In urban areas an average of approximately 2 ng Cd/m^3 has been determined [7]. Under unfavorable conditions these values may increase by an order of magnitude. Even higher concentrations, up to 500 ng Cd/m^3 and more have been recorded in the vicinity of Zn smelters [10].

Water. A compilation of water data was made by Fleischer [7]; more recent examples from freshwater bodies are given in Table 8. The highest concentration of cadmium in surface waters is usually found in areas of high

Table 8. Selected references for Cd concentrations in river waters

Amazon River	0.07 µg/l	[34]
Mississippi River	0.10 µg/l	[35]
Missouri River	0.80 µg/l	[69]
Lower Rhine River (FRG)	5.5 µg/l	[70]
Meuse River (Belgium)	10.4 µg/l	[71]
Neckar River (FRG)	max. 220 µg/l	[72]
Coeur d'Alene R. (Idaho)	max. 450 µg/l	[73]

population density [66]. In a study sponsored by CIPS in Belgium, the following distribution of cadmium concentration in 480 surface water samples has been determined [10]: <1 µg/l–35.4‰, 1–5 µg/l–22.5‰, 6–10 µg/l–10.8‰, 11–20 µg/l–13.8‰, 21–30 µg/l–7.7‰, 31–40 µg/l–6.2‰, 41–50 µg/l–3.6‰. With an upper limit of 5 µg Cd/l for drinking water and water for irrigation purposes approximately 40% of these cannot be used. This allowable limit is also valid for surface waters reclaimed by conventional methods, such as bank filtration and artificial groundwater recharge.

The average concentration of cadmium in water from public water supplies in 7 large cities of the European Communities was 1.1 µg/l with a range of 0.2–4.0 µg/l [67]. These values, however, apply mainly to the water quality at the pumping station and the actual concentration of cadmium at the tap can be expected to be higher [10]. The galvanized pipes which are sometimes used in plumbing are a potential source of cadmium in drinking water [68]. If the water is soft and somewhat acid the cadmium can conceivably remain in solution.

Suspended Sediment. While the concentrations of Cd in water seem to be largely unaffected by the water discharge (despite flushing effects at the onset of stormwater runoff), this parameter is of great influence for the transport behavior of Cd (and other trace metals) in suspended particles. Figure 3 gives an example from the middle section of the Rhine River near Koblenz; values for the spring/summer period are marked by open circles, those for the autumn/winter period by dots. For both categories, a clear depencency on the river flow can be seen. However, even the moderate Cd-concentrations in the winter period exceed the maximum values for the spring/summer period, despite the much higher water discharge. Such a development is due to the suspended sediments rich in cadmium being held back in periods of low flow, e.g., in Germany mainly in the summer period in the lock-regulated Neckar and Main Rivers; in autumn/winter the tendency is reversed when the material is carried by high water flow into the Rhine in increased quantities.

Data on Cd concentrations on particulates from natural and polluted aquatic systems are listed below ("sediments").

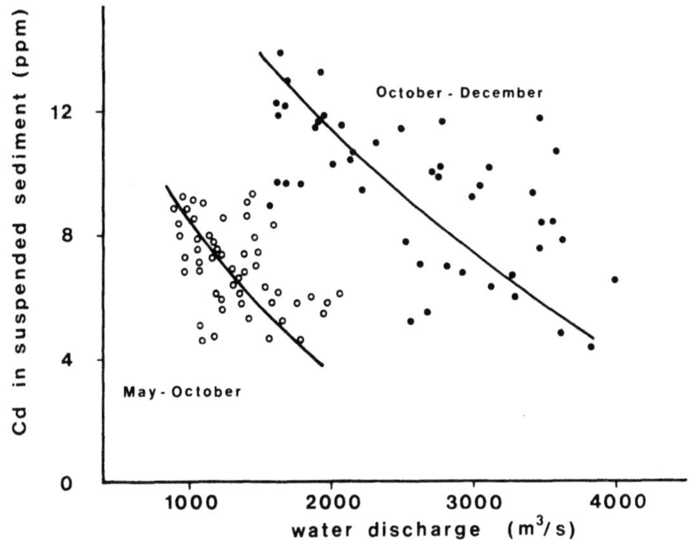

Fig. 3. Water discharge *vs.* cadmium concentration in suspended sediments of the Rhine River near Koblenz (Schleichert [74])

Sewage Sludge. Despite large local variations, which are mostly due to industrial input, the average (median) values of Cd in sewage sludges from different countries, such as Sweden [75, 76], USA [77, 78], and the FRG [79, 80] have a narrow range of 7–15 ppm. It has been calculated that annual application of a few tons of sewage sludge containing 20 ppm (and more) to unpolluted agricultural soils will raise the concentration of the ploughed layer of the soil to levels between 1.2 and 6 ppm Cd [81, 10].

Waste Dumping. Disposal of metal-bearing waste material may affect aquatic biota and groundwater quality. Investigations on soils beneath sewage sludge, effluent disposal and stormwater retention ponds showed movement of zinc, cadmium, copper, and chromium in various manners [82, 83]. It was shown that the distribution of metals with depth was closely related to changes in chemical oxygen demands, suggesting that the metals moved as soluble metal-organic complexes (see Section on Chemical Reactions). Waste waters infiltrated into the substratum under a zinc processing plant at Nievenheim in the lower reaches of the Rhine River led to an increase of Cd concentration of up to 600 μg/l [84]. Similar effects were reported from sanitary landfill leachates [85, 86].

Waste material is discharged to the sea via sewer outfalls and barges. From an annual input of approx. 50 t of cadmium into New York Bight, 82% is delivered by barges, 5% from municipal wastewater effluents, 2% stems from atmospheric sources, and approx. 10% is introduced by surface runoff [87]. Similarly, approximately 50 t of cadmium are annually discharged into the Southern California coastal zone from the five largest municipal effluents; about two-thirds of that reach the sea via Hyperion's Joint Water Pollution Control Plant [86]. Maximum factors of enrichment of dissolved trace metals in the main outfall plume, as compared with the surface water composition, are 13 for lead and zinc, 8 for copper, and 4 for cadmium [89]. To show what quantities of cadmium and other heavy metals from anthropogenic sources enter this relatively confined area, it is pointed out that the Mississippi River carries in its suspended substances approx. 50 t of cadmium – mainly from geochemical sources – into the Gulf of Mexico each year [90].

Reservoirs

The various reservoirs for cadmium contamination can be characterized by the *mean residence times* and the *concentration*. τ_{Cd} has been evaluated for natural environments as follows [91]:
air, 20–30 days;
river water, a few days;
lake water (example: Lake Washington), 1–2 yr [92];
humans, 20–30 yr [6];
soil (example: upper Thames valley) 280 yr [91];
ocean water, 250,000 yr;
pelagic sediment, $2–5 \times 10^8$ yr [93].

Concentrations of cadmium have been measured:
air, 0.1–500 ng/m^3;
water, 0.01–42,000 μg/l;
aquatic, organisms 0.001–1120 mg/kg;
soil, 0.01–500 ppm;
sediments, 0.01–50,000 ppm.

It has been proved effective in making conclusions as to the origin of cadmium enrichment in a certain reservoir to refer to the Cd/Zn ratio. This ratio in contaminated samples is frequently below that in geological reserves, due to the selective vaporization of cadmium during incineration processes. This is particularly valid for smelter emissions. Values for smelter recovery efficiencies of 75% for cadmium and 89–97.5% for zinc have been quoted [94], suggesting that, for an ore with a Zn/Cd ratio of 200, the ratio in the unrecovered portions (losses in the atmosphere and the slag) will range from 20 to 88 [7]. As the level of cadmium increases in a distinct compartment, generally the Zn/Cd ratio decreases. Among the major reservoirs for cadmium contamination which may be available to organisms, soils, stagnant waters, and sediment are considered now in more detail.

Soil. A selection of representative data on the cadmium content of soils is given by Fleischer [7]. Recent analysis of contaminated soil profiles indicate that the normal, average content of cadmium in soils is about 0.4 ppm. There is a clear increase near highways [95]. Typical regional differences are shown from investigations performed by Klein [96] in the Grand Rapids, Michigan area: average Cd concentrations in *residential areas* was 0.41 ppm, in an *agricultural area* 0.57 ppm, in a *industrial area* 0.66 ppm and near an *airport* 0.77 ppm Cd. In these experiments each sample was collected from the top 5 cm of soil.

Table 9. Cadmium content and Zn/Cd ratios in uncultivated soil surrounding East Helena Stack. (From Miesch and Huffmann [97], modified by [7])

	1.8 km from stack		*3.6 km from stack*		*7.2 km from stack*	
Depth of soil, cm	Cd ppm	Zn/Cd	Cd ppm	Zn/Cd	Cd ppm	Zn/Cd
0–2.5	68	16	17	14	4	12
5–10	30	33	7	25	2	15
15–25	3	70	2	42	1	33

The largest contamination originates from smelters and metallurgical plants, and it has been shown that a major fraction of the emissions from smelters accumulates in the surrounding soil [97]. Data from an extensive study of the pollution in areas near lead and zinc smelters in East Helena, Montana are summarized in Table 9. A significant decrease of Cd concentrations in all three profiles is recorded. Simultaneously the Zn/Cd ratios increase, but are lower than those in geological reserves. Low Zn/Cd ratios were also observed in settled dust and suspended particulate, supporting the postulation that the relatively low Zn/Cd ratio in the soil reflects the low ratio at the source and not a selective leaching of zinc from the soil [7].

The effects of phosphatic fertilizers are obviously less important than generally thought. Kloke [16] has estimated that the increase in cadmium concentrations of soil resulting from the application of phosphatic fertilizers (50 kg P_2O_5/ha) would be at the most 0.016 ppm per year ([10]; see also the chapter on availability). On the other hand, the application of sewage material and polluted stream sediments can significantly effect the Cd-content of soils.

Water. Obviously, the physico-chemical conditions in normal surface waters prevent large-scale dispersion of dissolved cadmium, even in cases where strong contamination is observed from sediment studies (see below). Only in acidic waters, e.g., from mine tailing effluents, has significant enrichment been found at greater distances from the source. The Cd concentrations in most lake waters studied are relatively less enriched, even for examples with significant overall pollution effects. Average concentration of Cd in Lake Ontario has been determined as 0.09 μg (range: 0.03–0.15 μg/l) by Chau and colleagues [98]; for Lake Michigan 0.30 μg/l Cd has been recorded [99]. It seems that during periods of higher biologic activities the Cd-concentrations may fall below the general geochemical background values.

At the same time, the input from polluted rivers can significantly affect the Cd contents of coastal waters. Investigations performed by Abdullah and co-workers [100] on the distribution of transition metals in Welsh rivers clearly reflect the influence of mineralization zones. The rivers and lakes in regions where no mineral deposits are known show cadmium levels ranging between 0.1 and 0.6 μg/l, whereas the annual average cadmium levels in rivers of the mineralized regions are found to range between 1.2 and 4.7 μg/l, with the highest recorded concentration being 20 μg/l. These waters characteristi-

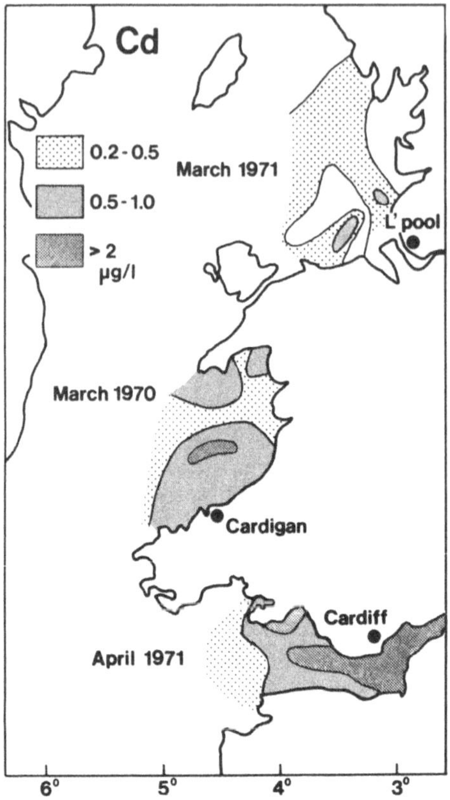

Fig. 4. Cd concentrations in surface waters of the Severn Estuary, Cardigan Bay, Liverpool Bay, and the adjacent Irish Sea [101]

cally influence the composition of the adjacent Irish Sea (Fig. 4). The highest concentrations of cadmium in the Bristol Channel are probably derived from industrial effluents entering the area from the Avon and Severn Estuary. In Cardigan Bay, however, which is relatively free from industrial effluent, little

Table 10. Mean values of Cd in coastal waters of Great Britain [102]

English Channel	0.06 μ/l	[103]
Atlantic Ocean (Iceland-Faroes Ridge)	0.07 μ/l	[104]
Irish Sea (offshore)	0.11 μ/l	[103]
Liverpool Bay	0.27 μ/l	[101]
North Sea (nearshore)	0.5 μ/l	[105]
Firth of Clyde	0.5 μ/l	[106]
Conway Bay (nearshore)	0.76 μ/l	[107]
Cardigan Bay	1.11 μ/l	[101]
Bristol Channel	1.94 μ/l	[108]

domestic waste is produced due to a low population density; runoff from the mineralized zones and sites of former mining activity is the main source of cadmium and of other trace metals [101]. A compilation of data from coastal waters off Great Britain (Table 10) shows particularly strong enrichment of dissolved cadmium in the Firth of Clyde, Conway Bay, Liverpool Bay, Cardigan Bay and Bristol Channel, which amounts to more than a 10-fold increase compared to the normal values from the open ocean or even from the Irish Sea [102–108].

Sediment. Analyses of sediment are useful when selecting critical sites for routine water sampling. Their profiles (cores) often uniquely preserve the historical sequence of pollution intensities, and lateral distributions (quality profiles) are used to determine and evaluate local sources of pollution [2]. A general description of sediment properties is given in Volume 1 of the Handbook; a chapter on "Cadmium in Polluted Sediments" has been prepared for

Table 11. Cd contamination in natural and polluted sediments

Investigation site	Cd value (ppm)	Sources	Ref.
Hudson River, Foundry Cove (N.Y.)	max. 50,000	Cd-Ni battery	[111]
Palestine Lake (Indiana)	3.2–2,640	Electroplating	[112]
Derwent Estuary (Tasmania)	0.8–862	Zinc smelter	[113]
Los Angeles River (Calif.)	max. 860		[114]
Sörfjord (Norway)	16–850	Pb-Zn smelter	[115]
Vesdre River (Belgium)	max. 430		[71]
Hitachi area, NE Tokyo (Japan)	max. 368	Braun tube factory	[116]
River Tawe (Wales)	max. 355	Metal processing	[117]
Neckar River (F.R.Germany)	max. 340	Pigment industry	[118]
Takahara River (Kamioka Mine, Japan)	4.1–238	Mine effluents	[119]
Meuse River (Belgium)	max. 230		[71]
Tennessee River (Tennessee)	max. 227	(Suspended solids)	[120]
South Esk River (Tasmania)	max. 153	Mine effluents	[121]
Main River (F.R.Germany)	max. 151	(Suspended solids)	[122]
Milwaukee River (Wisconsin)	max. 149		[123]
New Bedford Harbor (Massachusetts)	max. 130	Metal processing	[124]
Corpus Christi Harbor (Texas)	2–130		[125]
River near Himeji City (Osaka, Japan)	max. 129		[126]
Gadura River (Israel)	max. 123	Battery plant	[127]
Stola River (Poland)	max. 116	Zn smelter	[128]
River Conway (Wales)	3–95	Mine effluents	[129]
Ginsheimer Altrhein (F.R.Germany)	2–95		[130]
Coeur d'Alène River (Idaho)	max. 80	Mine effluents	[131]
Upper Rhône River (Switzerland)	0.1–73		[132]
Voglajna River (Yugoslavia)	max. 66		[133]
Santa Monica Canyon (Calif.)	max. 65		[134]
Unpolluted lakes in South America, Asia, Africa, and Australia (n = 72)	avg. 0.35 (0.04–0.84)		[109]
Very slightly influenced sea sediments (Japan)	avg. 0.45		[110]
Mississippi River (clayey sediment; n = 4)	avg. 0.49		[90]

J. O. Nriagu's "Cadmium in the Environment" [12, 33]. Table 11 lists examples of less polluted sediments (background data) from lacustrine, marine, and fluviatile environments and examples of sediment investigations on 25 of the most heavily polluted aquatic systems. Various sources – effluents from mine tailings, processing plants, smelters, the pigment industry, battery plants, the electroplating industry, and from a Braun tube factory – are responsible for very strong accumulations of cadmium in aquatic sediments.

Core sediment studies from moderately polluted lacustrine and coastal marine environments indicate that next to mercury, concentrations of cadmium generally prove to be the most important metal enrichment of all heavy metals under investigation [33]. The order of enrichment in anthropogenically influenced sediments

$$Cd > Pb > Zn > Cu$$

found in these studies corresponds to the accumulation of metals in fossil fuel residues [135], and the above sequence is also reflected in the metal enrichment of air-borne particulates. The pollution by cadmium in many aquatic systems is apparently still increasing. By analyzing sediments collected since 1922, Salomons and DeGroot [136] were able to trace the development of metal pollution in the Rhine River in the Netherlands over a period of more than 50 years (Fig. 5): the samples taken from the flood plain at the beginning of this century showed already anthropogenic influences of lead, cadmium, and mercury, as can be shown by comparison with sediment data from polders reclaimed in the 15th and 18th centuries. Between 1920 and 1958 all trace element concentrations studied have increased in the sediment from the Rhine River. Whereas the concentrations of mercury and lead decreased between

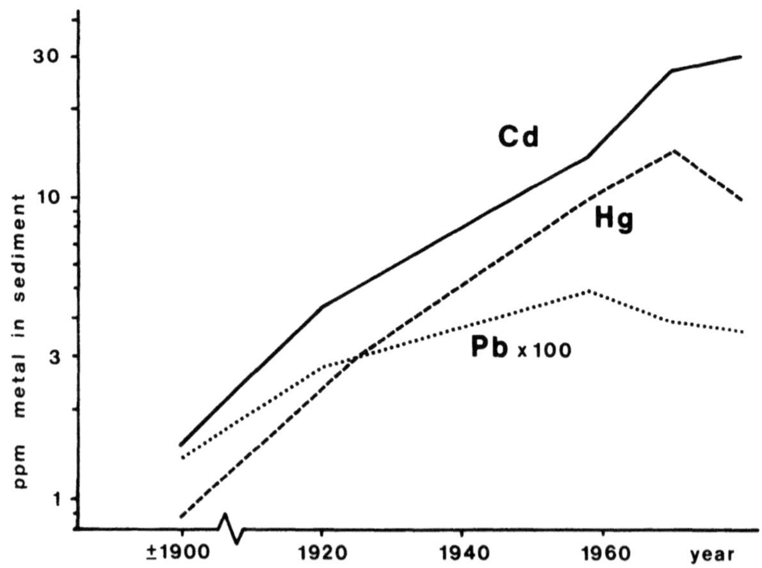

Fig. 5. History of trace metals in sediments from the Dutch Rhine River. (After Salomons and De Groot [136])

1958 and 1975, cadmium continued to increase up to a level 100 times greater than that in sediments of Rhine polders reclaimed in 1977.

Biota. In order to determine the major compartments of an enrichment of cadmium, biological systems cannot be ignored. Because of their generally greater biomass, plants are of special interest; their capacity for storing cadmium is normally an order of magnitude greater than the corresponding

Table 12. Estimates of cadmium concentrations in some plants and plant parts [7]

Plant or plant part	Reported or estimated cadmium concentrations or ranges in concentrations; ppm in dry material	
	Environments presumably having normal cadmium levels	Environments having greater than normal cadmium levels
Marine algae	0.1 -1	n.d.
Mosses (bryophytes)	0.7 -1.2	8-340
Lichens (fruticose type)	0.1 -1.4	1
Grasses	0.03-0.3	0.6-40
Alfalfa	0.02-0.2	0.2-2.4[a]
Grains		
Corn (Zea mays)	0.1	2
Rice (polished)	n.d.	0.5
Barley, wheat, and oats	0.1 -0.5	0.1-1.5[b]
Vegetables		
Asparagus	n.d.	4
Beet root	0.05	0.24
Cabbage leaves	0.05	6-12
Carrots	<0.35	8
Chinese cabbage	n.d.	41
Eggplant fruit	n.d.	8
Kale	1	n.d.
Leafy vegetables used as pot herbs or salads	0.3 -0.5	3-50
Leeks	n.d.	14
Lettuce	0.3 -0.5	4-16
Potatoes	0.05-0.3	6-20
Spinach	0.6 -1.2	n.d.
Turnip, roots	n.d.	5
leaves	n.d.	15
Tomatoes	n.d.	2
Trees, deciduous		
leaves	0.1 -2.4	4-17
stems (branches)	0.1 -1.3	0.03-1.5
Epiphytes (Spanish moss)	0.1	1
Floating aquatic plants (duckweed)	n.d.	17
Marine flowering plant *(Zostera marina)*	0.23	n.d.

n.d. = no data; [a] = Original data given in wet weight; converted to concentration in dry material ba assuming 25% water in original sample reservoir capacity of animal organisms (see section on cycling of cadmium in natural systems). The latter will be discussed in more detail in connection with aspects of metabolism and food chain behavior. A detailed review of the cadmium concentrations in plants is given by Shacklette [137] and Shacklette and Nisbet in Fleischer et al. [7]; Table 12 is excerpted from the latter study. Important observations are summarized as follows:

(i) Where environmental cadmium levels are low, cadmium concentrations in plant tissue frequently vary more with species than with soil type.
(ii) Where cadmium content of soils are higher than background amounts, the cadmium content of plant tissue tends to increase with increased concentrations of soil cadmium.
(iii) The cadmium contents of many species of plants reflect above-normal amount of cadmium that are introduced into the environment both from natural sources [138] and from cadmium pollution of soils [139], water [4] and air [140]. Cadmium from these sources may be absorbed by the plant through roots or leaves or both, and thus be incorporated into the tissues.

Airborne particulate matter containing cadmium may be deposited on the surface of leaves ([7]; see section on biological uptake and accumulation).

Cycling of Cadmium in Natural Systems

Examples for the transfer pathway of cadmium in the environment are reproduced from results of investigations undertaken by Windom et al. [141] in a coastal marine ecosystem. The annual rate of input of cadmium by nine major river systems emptying into the Georgia Embayment was determined based on analyses of river water samples collected bimonthly and by integrating the metal concentrations with flow rates [142]. Once these metals are

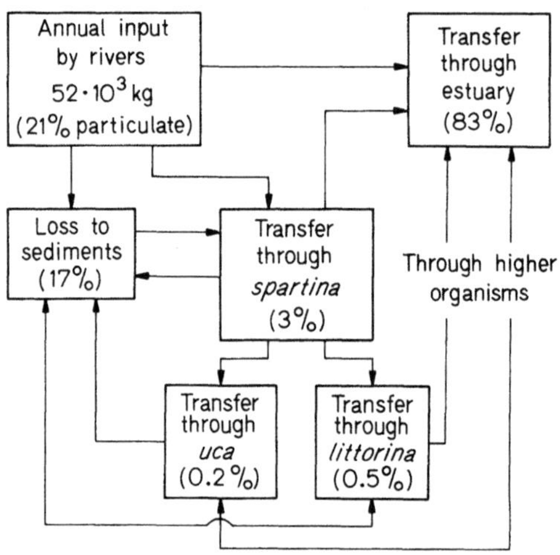

Fig. 6. Transfer pathways of cadmium through the estuarine zone of Georgia Embayment [141]. Estimated percentages of the total input passing through biological compartments are given in parentheses

delivered to the estuarine zone they follow various pathways within the environment entering major biologic and nonbiologic reservoirs (Fig. 6). Once the *sedimentation rates* in the salt marshes are known, the rate of loss of the metals to the sediment can be determined. After the *production rates of major biological components* of the estuarine zone and their metal concentration is found it is possible to determine to what extent the biota transfers metals through this interface. For example, the annual production rate of *Spartina alterniflora*, the major primary producer in the estuarine zone, is approximately 700 g/m^2 dry weight; approx. 3% of the total cadmium annually delivered by rivers is transferred through marsh grass. A similar approach can be taken using the major primary consumers (*Littorina* and *Uca*), since data exist on their annual production [141]; these transfer routes are less important than for the primary producer (0.5% and 0.2%, respectively). The data show that the loss of cadmium to sediments is roughly equivalent to the total amount of Cd supported by rivers in particulate form, whereas the soluble fraction is ultimately transferred through the estuarine zone either in solution or in biological compartments. It has been demonstrated by Windom et al. [141] that the importance of the biological cycles of cadmium in the estuarine system is much less than for mercury; transfer of Hg through *Spartina* was found to be 17% of the total mercury input into the system.

Chemical Reactions: Sorption and Release of Cd on Particulates

The availability of trace metals for metabolic processes is closely related to the chemical species involved, both in solution and in particulate matter. The type of chemical association between metals and particulates has therefore become of interest in connection with problems arising from the disposal, i.e. land application, of sewage sludge and contaminated dredged sediments. In this respect cadmium presently seems to pose the greatest problems of all metals [33].

Leaching Methods

In pedology, for the determination of plant-available concentrations of cadmium and other heavy metals (e.g. Cu, Ni, and Zn) numerous leaching tests have been introduced. Table 5 lists several reagents that are important for the assessment of plant-available cadmium in soils. Recently, attempts are made to standardize extraction procedures with respect to the use of sludge/soil mixtures [143]. Chumbley [144] proposed that 0.5 N HOAc be used to assess zinc and nickel availabilities, and 0.05 *M* EDTA (ethylene-diamine-tetraacetic acid) to assess copper availability for regulating sludge application to soils. 0.05 *M* EDTA has been used for cadmium extraction, for example, by Webber [145], Davies [146, 147] and Symeonides and McRae [48]. Similarly, 0.005 *M* DTPA (diethylene-triamine-penta-acetic acid) has been used to determine the solubilities in soils of both nutrient metals [148, 149] and non-nutrient metals, such as Cd [143, 150, 151]; the proposed DTPA-solution by Lindsay and

Norvell [148] contains 0.005 *M* DTPA, 0.01 *M* triethanol amine, and 0.01 *M* $CaCl_2$ adjusted to pH 7.3. The 0.01 *M* $CaCl_2$ also is used separately as neutral salt solution and seems to provide an index of metal solubility in the soil solution, which, at least for zinc, is related to plant uptake [152]. Increased levels of $CaCl_2$-extractable cadmium from the sludge/soil mixtures with heavily contaminated sludges indicate a potential movement of this metal into groundwater [143]. In a comparative study on the relationship between Cd in soils and radish plants it was shown [48] that the most sensitive of several possible indices to Cd uptake by plants is the amount extracted by a 1-h shaking with 1 N ammonium nitrate solution at a soil/solution ratio of 1:10 (wt/vol).

Generally, however, these correlations do not give information from which chemical compound the metals are preferably adsorbed into the plants. Plant-available metal concentrations in particulates may be investigated by determining the chemical association of cadmium by successive chemical leaching processes. A number of these methods are summarized in Table 13.

Both extraction procedures were done, for comparison, on a heavily Cd-contaminated sediment sample (14.8 ppm Cd) from the Neckar River

Table 13. Extraction of Cd-forms in particulate matter (examples)

Chemical fraction	Leaching method	Ref.
Soluble fraction	H_2O; elutriate test (disposal site water)	[153]
Soluble organics/ soluble free cations	Supernatant solution is passed through cation exchange resin	[154]
Exchangeable cations + easily extractable phases	0.2 *M* $BaCl_2$ triethanolamine pH 8.1	[155, 156]
	1 *M* sodium acetate, pH of suspension 1 N ammonium acetate, adjusted to sediment pH	[157, 158]
Reducible phases	0.3 N HCl	[159]
Carbonate fraction	CO_2-treatment	[156]
	Acidic cation exchange resin	[160]
Easily reducible fraction (carbonate, Mn-oxide, amorphous Fe-oxide)	0.1 *M* hydroxylamine hydrochloride + 0.01 M HNO_3	[161]
	0.15 *M* oxalic acid + 0.25 *M* ammonium oxalate	[162, 154]
Moderately reducible phases (hydrous Fe-oxide)	Sodium dithionite/citrate	[163]
	1 *M* hydroxylamine hydrochloride + 25% acetic acid ("acid-reducible agent")	[164]
Oxidizable phases	30% H_2O_2 (95 °C), extraction with 1 N ammonium acetate	[158]
Humic acids	0.5 N NaOH, 0.1 N NaOH	[165, 166]
Lipids, asphalt	Benzene/dichloromethane/methanol	[167]
Organic residues	Sodium hypochlorite, citrate-dithionite extraction	[168]

[169]. The determination of the compound phases by *successive* chemical leaching according to the methods compiled in Table 13 had the following results:

Soluble fraction	disposal site water	0.2%
Cation exchange	0.2 N $BaCl_2$-triethanolamine	28%
Carbonate fraction	acidic cation exchangers	48%
Easily reducible fraction	0.1 *M* $NH_2OH \cdot HCl$ + 0.01 *M* HNO_3	12%
Moderately reducible fraction	0.3 N HCl	4%
Residual fraction	$HF/HClO_4$	8%

The pedological experiments (simultaneous) showed the following Cd amounts being released:

1 *M* ammonium nitrate unbuffered pH 4	2.5%
1 *M* ammonium acetate acetic acid pH 7	17%
Mixed reagent 0.005 *M* DTPA, 0.1 *M* TEA, 0.01 *M* $CaCl_2$	38%

The exact amounts of metal actually being extracted and the preferred chemical phases from which the plant-available Cd originates have not yet been determined. From the available data, however, it can be clearly seen that there is a general decrease of the residual bonding forms of cadmium (and of other heavy metals), i.e., the predominantly inertly fixed cadmium content as the anthropogenic metal enrichment increases [170, 171]. There is a characteristic affinity of cadmium for organic substances and sulfides in polluted sediment at lower carbonate concentrations. At higher carbonate contents the association with carbonate minerals – either as discrete carbonate or as coprecipitates with calcite seems to provide the major process of immobilization of elevated Cd concentrations in the effluents [171, 172].

Remobilization Processes

Trace metals temporarily immobilized in the bottom sediments and suspended matter of aquatic systems may be *released as a result of physicochemical changes* such as:

(i) increased salinity,
(ii) lowering of pH
(iii) increased input of organic chelators,
(iv) microbial activity, and
(v) change in the redox conditions.

Increased salinity in a water body leads to competition between dissolved cations and adsorbed trace metal ions and can result in partial replacement of the latter. Such effects can be expected particularly in the estuarine environment [173, 174]. Experiments performed by Van der Weijden [175] with artificial seawater indicate desorption of trace metals from particulate matter, presumably by inorganic complex formation, which was highest for cadmium. From experiments on desorption of metals from sludge material diluted in seawater Rohatgi and Chen [176] found that 93% of the original Cd-content of sewage particles were released during 4 weeks treatment.

A lowering of pH leads to the dissolution of carbonate and hydroxide minerals and – as a result of hydrogen ion competition – to an increasing desorption of metal cations. Long-term changes of the pH conditions have been observed from waters poor in bicarbonate ions, which are effected by atmospheric SO_2 emissions. Significant increases of cadmium were reported from water of the Sudbury mining area [177]. Cd-enrichment in acidic mine effluents by factors of 1,000 and more, in respect to normal surface waters, has been observed [178, 119, 121]. On the other hand, Cd-availability is greatly reduced in soils rich in carbonate [179, 199, 300].

Significant impacts on remobilization from polluted sediments may result from the growing use of synthetic *complexing agents* (e.g. NTA, nitrilotriacetic acid) in detergents replacing polyphosphates. Experiments performed by Banat et al. [180] with polluted river sediments indicate a high percentage of cadmium mobilization. The results of a test made by Chau and Shiomi [181] in various NTA metal complexes in natural waters of Lake Ontario show a very delayed degradation of Cd chelates. Thus, a potential danger seems to arise for drinking water obtained from bank filtration or artificial recharge processes [182].

Oxygen deficiency in sediments leads to an initial dissolution of manganese oxides followed by that of hydrous iron oxides. Since these metals are readily soluble in their divalent states, any coprecipitates with metallic coatings become partially remobilized. However, the presence of sulfide under anoxic conditions will precipitate toxic metals. These are released by conversion of sulfide to sulfate under oxidizing conditions [38]. Isotope studies performed by Gambrell et al. [154] with Mississippi River sediment material indicated that exchangeable ^{109}Cd levels are strongly pH-redox-potential dependent. A

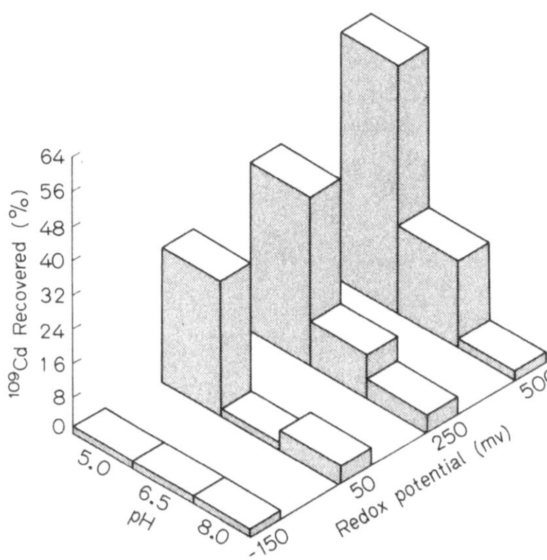

Fig. 7. Effects of pH and redox potential on exchangeable ^{109}Cd in Mississippi River sediment suspensions [154]

much greater proportion of the incubated cadmium isotope was recovered in the readily bioavailable forms than for any other potentially toxic heavy metal studied. Figure 7 indicates the influence of pH and redox potential on exchangeable ^{109}Cd in the Mississippi River sediment suspensions. It is suggested that considerable cadmium release to relatively mobile forms may occur as cadmium-contaminated sediment is transported from a near-neutral pH, reducing environment to a moderately acid, oxidizing environment. Under these conditions, cadmium levels of subsurface drainage water from upland disposal of dredged materials may be increased, and cadmium availability to plants growing on the material enhanced [154].

The burial of Cd-rich sediments under succeeding layers, whereby the sediments become anaerobic, may be very effective mechanisms to fix Cd onto solid phases [183, 154]. With log $K_{sp} = 27.8$ [19], CdS is one of the least soluble metal sulfides (following HgS and CuS). For example in San Francisco Bay dredged sediments [184], about 92% of the total cadmium was found in the organic and sulfide phases. It has therefore been proposed by Jackson [185] that sewage together with heavy metal effluents should be introduced into settling ponds to achieve an effective method for preventing heavy metal pollution of natural waters; such mechanisms involve the stimulation of algal blooms with attendant H_2S production. Dredging activities and other physical perturbations of the surface sediment layers in contaminated deposits may have adverse effects on both water quality and aquatic biota due to the release of cadmium [2, 186]. Holmes et al. [125] reported that the cadmium introduced into Corpus Christi Bay Harbor from industrial effluents in the summer when the harbor water was stagnant reach the surface sediment with the sulfide ions and precipitate as CdS. In the winter months, however, the increased flow of oxygen-rich water into the bay results in the desorption of some of the precipitated metal [154]. Older data suggesting insignificant effects of dredging and other activities on the release of heavy metals [187–191) should be reexamined with respect to the behavior of cadmium [33].

Biological Uptake and Accumulation of Cadmium in Organisms

Cadmium is not an essential trace element for organisms, but rather a typical contaminant [68, 192]. Its concentration and the typical log-normal distribution in organisms [193] is influenced by the concentration in the environment.

Uptake in Plants

Table 12 reproduces cadmium concentrations in some plants and plant parts. In Cd-contaminated environments, very high enrichment rates are found in mosses, cabbage, carrots, radishes, lettuce, potatoes, and turnip roots. Obviously cadmium is readily taken up by roots and distributed throughout the plant [194]. The amounts of Cd in plants grown in contaminated soils lie in the same order of magnitude as the cadmium concentrations in the substrates.

The amount of uptake is influenced by soil factors such as cation exchange capacity [195, 196], pH [195], phosphorous levels [195], fertilizers [197], other heavy metals [195], soil temperature [195], and organic matter [195].

The accumulation of Cd varies with the plant tissue and the species of plant investigated [194]. Corn roots contained 2–4 times as much Cd as did shoots when grown in solution culture for 12 days [198]. Uptake and translocation into shoots from a loamy sand soil was 48 mg/kg, while uptake from a silty clay loam soil having a high cation exchange capacity was 8.4 mg/kg [194]. Foliar and root uptake of Cd are equally effective [195]; in both methods of application the quantity of Cd distribution had the sequence: stems > leaves > pots > beans [194]. Uptake into crops is significantly higher from acid than from calcareous soils [199].

Uptake, Absorption, Storage, and Excretion in Animals

The removal of Cd from the water by organisms occurs by external adsorption as well as internal uptake through organs such as the gills. Heavy metals appear to be accumulated by ion-exchange processes involving organic molecules such as proteins, e.g. in phytoplankton, seaweed, etc. [200]. These processes are responsible for the typical distribution of Cd in vertical ocean water profiles [34, 201]. Agents such as moulted exoskeletons and faeces of zooplanktonic animals affect the vertical distribution of metals in the sea mainly in coastal areas where nutrients for high biological productivity are available from upwelling of the ocean water or from runoff from the land [202, 200].

The following are factors for the *uptake of metals from solution* (Prosi, in [2]): temperature and oxygen content [203], water hardness [204], pH values [204], salinity, and the concentration of organic compounds [205]. With regard to the latter factor, it has become evident that the environmental impact of a particular metal species may be actually more important than the total metal concentration. Organic ligands, such as fulvic acids, NTA and EDTA, can inhibit the uptake of metals and raise the toxic threshold [206, 207]. Free ion activity, i.e. Cd $(H_2O)_x^{2+}$ is considered as an approximate measure for toxic effects of metals [208], especially in respect to phytoplankton [209]. Experiments of Ramamoorthy and Kushner [210] indicate that the metal affinity toward the different microbial growth media largely follows the availability of free cations, i.e.

$$Cd^{2+} \gg Cu^{2+} \gg Pb^{2+} > Hg^{2+}$$

(the reverse of the Irving-Williams series of stability constants of metals to organic ligands). In addition to these factors, age of the organism plays a role in the metal concentration, as well as a number of species-specific effects, that are, however, little known as yet.

Particularly in large animals, the *adsorption of heavy metals from food* may be very important. In oysters, for example, metals such as Zn are obtained from ingested particles rather than from solution; differences in the availability of metals in foodstuffs depend on factors such as the facility with which the material is digested, the chemical form of the metal, and the relative binding capacities of the animal and the products of digestion in its gut [200].

Excretion of abnormal concentrations of heavy metals can take place in a number of ways [200]: through the gills such as in the crab and in rainbow trout; in a particulate form from the mantle edge via the byssus gland, such as in mussels; into the gut such as in the cyprid larva of barnacles; and removal in the faeces, such as in most of the higher organisms. Liver and kidney usually are the major *storage* organs. In the bivalve mollusc *Pecten maximus* high concentrations of Fe, Cd, and Cu are found in the liver, whereas Zn, Mn, and Pb are stored in the kidney. Storage proteins such as metallothionein for Cd, Zn, and Cu have been found in terrestrial mammals as well as in aquatic animals [200].

The behavior of Cd compared to that of Zn is interesting especially in respect to the fact that zinc is more readily removed from sea water, probably because it is better regulated – as an essential element – by organisms than Cd [200]. Bryan and Hummerstone [211] have shown that in the polychaete *Nereis diversicolor*, Cd is adsorbed from solution more slowly than Zn, but that with increasing levels of both metals the rate of adsorption of Cd increases more rapidly than that of Zn. Investigations of Peden et al. [212] on the carnivorous gastropod *Nucella* suggest that once having been adsorbed, Cd is less readily excreted than Zn, so that ultimately a higher concentration factor is present for Cd [200].

Table 14. Geometric mean concentrations of cadmium in different groups of organisms [200]

	mg Cd/kg dry weight
Seaweed (all types)	0.5 mg/kg
Phytoplankton	2 mg/kg
Filter-feeding groups	
Zooplankton (copepods)	4 mg/kg
Tunicates (mainly ascidians)	1 mg/kg
Bivalve molluscs[a]	2 mg/kg
Oysters	10 mg/kg
50:50 carnivorous and herbivorous or particulate feeders	
Gastropod molluscs	6 mg/kg
Echinoderms	2 mg/kg
Basically carnivorous groups	
Decapod crustaceans	1 mg/kg
Coelenterates	1 mg/kg
Cephalopod molluscs	5 mg/kg
Fish	0.2 mg/kg

[a] Excludes *Pectinidae*

Data on cadmium concentrations in animals are summarized, among other authors, by Prosi in [2], in the contribution of Shacklette and Nisbet in the "Subpanel Report on Cadmium" [7] and by Bryan [200]. The geometric mean concentrations of cadmium in different groups of marine organisms from the latter work are given in Table 14.

Food Chain Effects

In field investigations dealing with heavy metal enrichment in organisms, it is imperative to group the organisms according to their habitat and ecologic behavior, i.e., feeding habits (phytophageous, carnivorous, omnivorous, filter feeding, sediment feeding, detritus feeding, etc.), life cycle, life history, sessility and wandering [213]. In addition, the physiological response of various organisms towards metal pollution may be different with respect to organ distribution of the metal, synergistic or antagonistic effects of other metals on metal uptake, heavy metal resistance, etc. [214]. When all these factors are considered in respect to heavy metal amplification in the food chain, it becomes clear that, in many cases, elevated heavy metal concentrations in higher trophic levels do occur but not necessarily in the classical sense of *food chain enrichment* [213].

In an urban-influenced river section, Prosi [215] determined a significant increase of Cd in the food web of benthic invertebrates compared to fish (Fig. 8): It was generally found that according to feeding habits, sediment-depen-

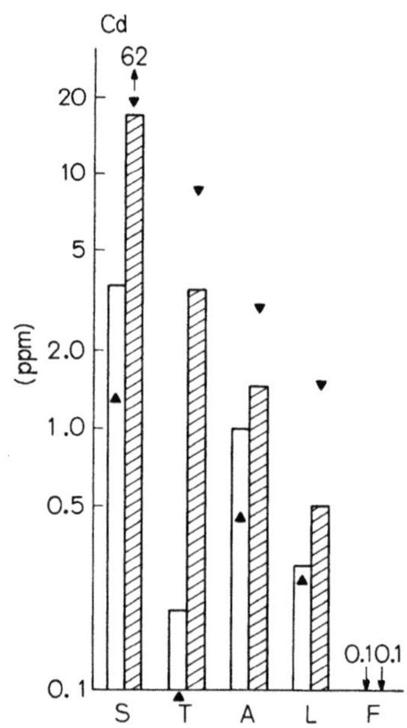

Fig. 8. Cd distribution in two sections of the Elsenz River (light column: rural; shaded: urban-industrial influenced) at different trophic levels. *S* = sediment < 2 μm, *T* = tubificid worms, *A* = isopods (*Asellus aquaticus*), *L* = leeches, *F* = fish (roaches, sticklebacks). Mean concentrations (dry weight); arrows indicate minimum and maximum values [215]

dent organisms (Tubificidae) has greater metal concentrations than other biota. Metal contents of the benthic food web, sludgeworms *(Tubifex tubifex* and *Limnodrilus hoffmeisteri)*, isopods *(Asellus aquaticus)*, and leeches *(Herpobdella octoculata)* constantly decrease, so that the lowest concentrations appear in the fish.

Investigations performed by Butterworth et al. [216] in the Severn Estuary demonstrate the effects of pollution on the concentrations of Cd in aquatic organisms. Coastal waters bordering the southern shore of the Bristol Channel contain abnormal amounts of cadmium, zinc and lead, which are probably introduced from the Bristol area via the River Avon. In the water samples the effects of the pollution have been traced as far away as Harland Quay, some 150 km to the west from Avonmouth into the Bristol Channel. Table 15

Table 15. Cadmium concentrations in water, seaweeds and shore animals of four collecting stations on the southern side of Severn Estuary and Bristol Channel [216]

Collecting point	Distance from Avonmouth	Seawater µg Cd/l	*Fucus* mg Cd/kg	*Patella*	*Thais*
Portishead	4 km	5.8	220	550	-
Brean	25 km	2.0	50	200	425
Minehead	60 km	1.0	20	50	270
Lynmouth	80 km	0.5	30	50	65

indicates that the contamination in the water by cadmium is obviously transmitted to the living material inhabiting the shore – at relatively low levels in seaweed *Fucus* (the producer), at higher levels in limpets *Patella* (a primary consumer), and greatest concentrations in the dog whelk, *Thais* (secondary consumer). There are significant differences of the Cd contents of different tissues. Mullin and Riley [23] found that in molluscs, levels of Cd were of the order of 1.5 mg/kg in muscle, and up to 550 mg/kg in digestive glands and renal organs. Brooks and Rumsby [217] found that in oysters cadmium was strongly concentrated in the gills, visceral mass, and the heart. The same authors found 2000 mg Cd/kg dry weight in the liver of the scallop *Pecten novae-zelandiae*. Analyses made by Bryan [200] from *Pecten maximus* revealed 32 mg Cd/kg dry weight for the whole animal, 321 mg Cd/kg for the liver, 79 mg Cd/kg for the kidney and 2.2 mg Cd/kg for the muscle and for other tissue. Schroeder and Balassa [197] found that in lobster, levels of cadmium were 14 times higher in the digestive gland than in muscle; analyses from Topping [218] on the lobster *Homarus gammarus* reveal 0.3 mg Cd/kg for the abdominal muscle, 17 mg Cd/kg for the gills and 12 mg Cd/kg dry weight for the liver. Fish tissues from teleost *Scombresox saurus* [219] contain 0.05 mg Cd/kg dry weight in the muscle and 0.62 mg Cd/kg in the liver. Jaakkola et al. [220]

analysed pike from polluted and other areas in Finland; whereas the Cd content in muscle was similar for both areas (0.026 mg/kg dry weight in polluted areas, 0.041 in other areas), there is a significantly higher concentration of Cd in the kidney of fish from the polluted area (1.52 mg/kg) compared to those from other, less polluted areas (0.95 mg/kg).

Indicator Organisms

Due to their wide distribution in the marine environment *Mytilus sp.* (especially *Mytilus edulis*) and oyster species *(Ostrea edulis* and *Crassostrea sp.)* have proved to be especially useful indicator organisms[3]. In his "mussel watch" Goldberg [222] has even suggested that as a long-term indicator, bivalves can

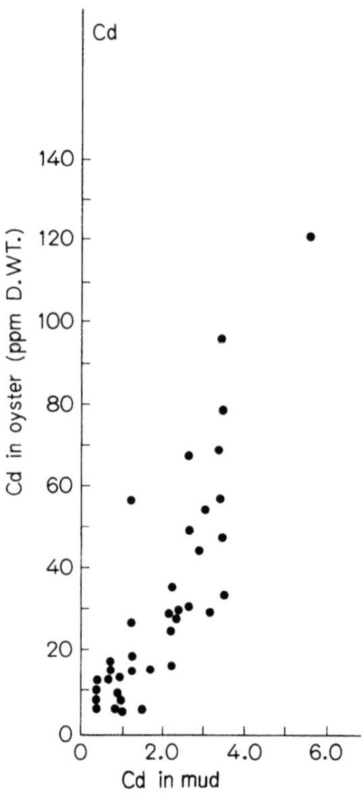

Fig. 9. Cadmium concentrations in dried mud and in oysters *(Crassostrea gigas)* in Tamar Estuary, Tasmania. Values in ppm [223]

3 According to Bryan [200], poor regulators, i.e., organisms having very little ability to regulate the total concentration in their body and which tolerate metals in the tissues or their storage in an inactivated form are suitable for use as biological indicators. Phillips [221] has proposed that the best-studied indicator types to date are the bivalve molluscs and the macroalgae. Among the former group, "*Mytilus edulis* may be the appropriate candidate because of its extensively-studied physiology, its world – wide distribution in temperate waters and the amount of accumulated knowledge concerning its uptake of metals and its metal content in various waters"

make certain water and sediment sampling procedures unnecessary. There is a distinct straightline relationship for cadmium in molluscs and sediment, as shown for the example from the Tamar Estuary in Tasmania (Fig. 9 [223]). Values of Cd concentrations in mussels and oysters from both less and more strongly contaminated examples are listed in Table 16. The higher Cd concentrations reached in some mussels and oysters are suspected of being dangerous for humans upon consumption. According to Ratkowsky et al. [234] cases of nausea and vomiting in consumers who had eaten oysters from the Derwent Estuary in Tasmania was probably caused by the contamination of these bivalves. The admissible daily intake of cadmium of 100 μg/70 kg of body weight (see below) is reached with approximately 50 g of oyster (wet weight) from a moderately polluted area, and only 10 g of oyster from some parts of Derwent Estuary and several other areas listed in Table 16.

Table 16. Cadmium concentrations in mussels and oysters (mg/kg dry weight)

Bivalve mussels		
Mediterranean Sea	1.9	[224]
NW coast France/Italy	(0.4–5.9)	
SW Spain/Portugal	1.7–3.6	[225]
Mediterranean Sea		
Trondheimsfjorden	2	[226]
Norway	(1–5)	
Irish Sea	5.1	[227]
Tasman Bay, New Zealand	10	[217]
Bristol Channel	18	[228]
	(4–60)	
Derwent Estuary,	18.6	[113]
Tasmania	(4.3–38)	
Port Phillip Bay, Australia	24.6 ± 21.9	[229]
Oysters		
San Antonio Bay, Texas	3.2	[230]
Knysna Estuary, R.S.A.	3.7	[231]
SW England Estuaries	2.2–26.7	[232]
Severn Estuary, U.K.	17–40	[233]
Tamar Estuary, Tasmania	33.2	[223]
Tasman Bay, New Zealand	35	[217]
	(10–43)	
Port Phillip Bay, Australia	91.6 ± 73.1	[229]

Human Intake, Absorption, and Excretion of Cadmium

Food Concentrations

Relative to the data on the concentration of cadmium in plants and animals, a short summary is given here on the contents of cadmium in several foodstuffs relevant for human nutrition. Characteristic data are excerpted from the

CEC-Study on Cadmium, which is one of the most recent and up-to-date compilations[4] in that respect (Table 17) [10].

Table 17. Cd contents in major foodstuffs (examples) [10]

Cereals and vegetables	mg/kg dry weight	Country	Ref.
Wheat flour	0.029–0.108	Sweden	[6]
Wheat flour	0.05 –0.10	Canada	[235]
Potatoes	0.02 –0.05	New Zealand	[236]
Potatoes	0.039	F.R.Germany	[237]
Carrots	0.016–0.088	F.R.Germany	[238]
Tomatoes	0.015	F.R.Germany	[237]
Cabbage	0.022–0.094	F.R.Germany	[238]
Radishes	0.011–0.027	F.R.Germany	[238]
Rhubarb	0.010–0.057	F.R.Germany	[238]
Lettuce	0.031–0.198	F.R.Germany	[238]
Spinach	0.055–0.063	F.R.Germany	[238]
Onion	0.018–0.040	New Zealand	[236]
Fruit			
Apples	0.005–0.027	F.R.Germany	[238]
Prunes	0.014–0.067	F.R.Germany	[238]
Dairy products			
Milk	0.010–0.076	F.R.Germany	[238]
Butter	0.02	New Zealand	[236]
Eggs, whole	0.04	New Zealand	[236]
Meats			
Beef	0.02 –0.10	New Zealand	[236]
Pork	0.03 –0.10	New Zealand	[236]
Chicken	0.03 –0.08	New Zealand	[236]
Kidney (beef)	0.17 –0.27	New Zealand	[236]
Kidney (pork)	0.07 –0.18	New Zealand	[236]
Kidney (beef)	0.27	F.R.Germany	[239]
Kidney (beef)	4.10	F.R.Germany (Stolberg)	[239]
Kidney (elk)	8.0	Finland (Poorvool)	[220]
Liver (elk)	1.5	Finland (Poorvool)	[220]
Liver (horse)	7.5	UK (industrial)	[240]
Seafood			
Muscle of various fish	0.08 –1.67	UK	[212]
Oyster	0.1 –7.8	USA (eastern)	[241]
Oyster	0.2 –2.1	USA (western)	[241]
Oyster (canned)	3.31	New Zealand	[230]
Crab	5.0 –33.1	UK	[212]
Crab	22	Europe	[71]
Molluscs	2 –50	Europe	[71]
Various seafish	0.1 –0.6	Europe	[71]
Freshwater fish	0.2 –1.2	Europe	[71]

4 A review of the effects of cadmium in mammalian systems has just been published: J.H. Mennear (ed.) Cadmium Toxicity, Marcel Dekker, Inc. New York, 224 p. (1979)

Most foodstuffs from less contaminated areas contain less than 0.1 mg Cd/kg, whereas liver, kidney and shellfish can show much higher concentrations. Investigations on cattle from southern Germany [242] show average values of <0.005 mg Cd/kg for meat parts, whereas the contents in the liver increase to 0.08 mg Cd/kg (0.005–0.3 mg/kg) and in the kidneys to 0.9 mg Cd/kg (0.04–1.41 mg/kg). Some vegetables and cereals concentrate cadmium when cultivated in polluted soil [10]. It is suggested that upon conditions of general air contamination or through the accumulative effect of fertilizers there should be a significant tendency for the Cd concentrations in foodstuffs to increase [243].

Intake from Food, Water, and Air

Representative studies on the dietary intake of cadmium and other noxious substances were first carried out in the United States and later in many other countries (basically involving the analyses of samples representative of food at the point of ingestion [244]). Such studies usually reflect the composition of the diet of the average person. Table 18 shows the data of the cadmium intake

Table 18. Cadmium in United States and Canada market basket survey [245, 246]

	USA (1968/69)		Canada (1969)	
	Range mg Cd/kg	μg daily intake	Range mg Cd/kg	μg daily intake
1. Milk and dairy products	0.01–0.09	5	<0.02–0.06	15
2. Meat, fish and poultry	0.01–0.06	4	0.05–0.08	19
3. Grain and cereal	0.02–0.08	14	<0.02–0.14	13
4. Potatoes	0.02–0.13	7	<0.03–0.22	19
5. Leafy vegetables	0.01–0.23	5	<0.02–0.05	2
6. Legumes	0.01–0.03	1	<0.02–0.06	1
7. Root vegetables	0.01–0.08	1	0.03–0.09	3
8. Garden fruits	0.01–0.38	5	<0.02–0.06	3
9. Fruits	0.01–0.38	2	<0.01–0.02	1
10. Oils and fats	0.01–0.13	2	0.03–0.07	1
11. Sugar and adjuncts	0.01–0.07	1	<0.02–0.03	3
12. Beverages	0.01–0.04	4	<0.01–0.04	2
		50 μg		80 μg

in the US Market Basket Study for June 1968 to April 1969 [245] and for Canada in the Hull-Ottawa area during 1969 [246]. Similar data were calculated for West Germany (48 μg/person/day [237]), for Romania (38–64 μg/person/day [248]), for Czechoslovakia (59 μg/day [248]), for Japan (59 μg/day [249] – for Japan there are other figures ranging from 25–120 μg/person/day [250, 251]) – for New Zealand (21 μg/person/day [252]), for France

(20–30 μg/day [253]), for the UK (15–30 μg/day – the average individual consuming 1.5 kg of food per day [244]), and for Sweden (10–17 μg/person/day [254, 255]). In areas where the soil has been found highly contaminated by cadmium, oral intake (mainly through contaminated rice) has been calculated to be as high as 600 μg/day [10]. If these data are excluded, the total daily intake in non-polluted areas ranges between 6 and 94 μg/person/day with a median of approx. 43 μg/day from food and 3 μg/day from water [10]. From air intake the following figures have been calculated [10], differentiating smokers from non-smokers (assuming a daily inhalation of 20 m^3 at 25% deposition equal to 3 μg/day from 40 cigarettes):

	Non-smokers	Smokers
Rural areas	0.0005 - 0.215	3.0005 - 3.215 μg/day
Urban areas	0.01 - 3.5	3.01 - 6.5 μg/day
Industrial areas	0.05 -25	3.05 -28 μg/day

It can be seen that most of the cadmium intake originates from foodstuffs. Cadmium intake from water on the average is equivalent to that inhaled during smoking. In industrial areas approx. 30% of the total intake of cadmium stems from air pollution.

Absorption

Inhalation. The effect of cadmium absorption by the lungs depends on the amount retained (Q deposited – Q rapidly eliminated via the upper respiratory tract) and probably also on the chemical form of the retained particles [10]. From experiments with dogs it has been demonstrated that cadmium oxide dusts and cadmium chlorides were more readily absorbed than cadmium sulphide [256, 257]. Male non-smokers (up to 60 yr old) were found to have an average 6.6 mg Cd in their kidney, liver, and lungs, whereas smokers (one pack of cigarettes per day for 40 yr) showed 14 mg in these organs [258]. Assuming that 64% of the cadmium deposited can be absorbed, this means that in the general environment 13–19% of the inhaled cadmium is absorbed; with regard to residents in urban areas the amount of Cd absorbed has been calculated as 0.006–2.24 μg/day for non-smokers and 1.9–4.2 μg/day for smokers [10].

Absorption by the Gastrointestinal Tract. Experiments on human volunteers (19 to 50 yr old) given labelled Cd orally, indicate that the absorption rate ranges between 4.7 and 7% [259]. Age may be important in the rate of gastrointestinal absorption of cadmium in humans [10], since the uptake rate is higher for adolescents than for adults. Assuming 6% absorption, the total amount of Cd absorbed approx. 2.6 μg/day from food and 0.2 μg/day from water [10].

Body Distribution

Normal concentration of Cd in blood averages below 1 μg/100 ml, with a large variation between 0.06–15.9 μg/100 ml [10]. Smokers have higher blood cadmium concentrations than non-smokers [260]. In workers no correlation was found between cadmium in blood and exposure time [261] and all present results would suggest that cadmium in blood is probably not a reflection of the *body burden*, but is rather influenced mainly by current exposure [10]. Cadmium accumulates with age (at least until age 50) and about 50% of the accumulated cadmium is found in the kidney and liver. In these tissues this toxic metal is mainly bound to metallothionein, a protein of low molecular weight (10,000–12,000), very rich in cysteine residues, and deficient in aromatic amino acids [10, 262]. It is suggested that cadmium acts as a highly specific inducer of metallothionein [263] and that Cd toxicity occurs when available metallothionein is insufficient to bind all the cadmium [6, 107]. When the cadmium-metallothionein complex is synthesized within the cell it may protect temporarily against cadmium toxicity [10]. However, the protective role of metallothionein against acute toxicity of cadmium has been questioned by recent work of Goyer et al. [264].

The Cd contents in the kidney, liver, and lungs of 172 US adults were for non-smokers 4.16 mg, 2.28 mg, and 0.36 mg, respectively, and for smokers 10.28 mg, 3.06 mg, and 0.81 mg [265]. The average body burden for adult non-smokers in the USA has been estimated at 19.2 mg, for smokers 32.4 mg [266]. Friberg et al. [6] have estimated the total body burden of a 50-year-old adult to be approx. 15–20 mg in the UK and Sweden and 80 mg in non-polluted areas of Japan. The total body burden of cadmium in workers could exceed 1200 mg [6]. It is possible, however, to find lower cadmium concentrations in kidney of exposed workers and Itai Itai patients than in normal individuals because when renal damage is present, urinary cadmium excretion is increased and therefore renal levels may decrease [10, 267, 268]. Decreased cadmium contents in the kidney cortex with age has been explained, among other hypotheses [10], with the fact that 70% of the total world production of cadmium has occurred within the last 20 yr [6].

Excretion

In normal adults the amount of Cd excreted daily via the urine is probably below 2 μg/day (range 0.2–3.1 μg/l); in workers exposed to cadmium urinary Cd excretion can reach several hundred μg/day (compilation by Friberg et al. [6]). It is still not known conclusively whether Cd concentration in urine reflects body burden or current exposure (see discussion in [10]): "at low exposure levels the amount of cadmium absorbed may be insufficient to saturate all the body binding sites (e.g. induced metallothionein) and urinary excretion does not increase proportionally to the exposure levels; in high exposure conditions (e.g. workers, adult itai-itai patients) and in the absence of renal lesion the urinary concentration would be more a reflection of exposure levels – all the binding sites are now saturated" [10, 261].

Biological Half-Time in Humans

Information on the various aspects of biologic half-time of cadmium in organisms is summarized by Friberg et al. [6]. With regard to the determination of these parameters for humans, calculations have been performed by Sudo and Nomiyama [269] from urine samples of former cadmium workers with proteinuria. A value of about 200 days was determined. From whole blood data among workers exposed longer to cadmium a half-time value was roughly calculated to be about 6 months [6]. Tsuchiya [270] estimated Cd half-life in workers to be about 1 yr. Summarizing a large number of whole body burden data involving measurements of uptake and excretion, the biological half-time is suggested to correspond to values of 13–47 yr [6]. Theoretical models of cadmium metabolism imply that the half-time in whole body are 9–18 yr [271]. With inclusion of liver accumulation into these models, the biological half-time for the human kidney is estimated to be 17.6 yr on an average and for the liver 6.2 yr [272].

Toxicological Aspects of Cadmium Pollution

Summaries on the toxicology of cadmium have been given, among others, by Flick et al. [273], Fulkerson and Goehler [65], Friberg et al. [6], Nordberg [9], in the CEC Study [10], and during the First International Cadmium Conference in San Francisco ([11; 17 articles on pp. 167–255]). Review articles in the German language were compiled by Rosmanith [192] and Ohnesorge [274], in French by Godfrain et al. [275]. The human-toxicological aspects of cadmium will be shortly summarized here in accordance with the description by the Commission of the European Communities' study "Criteria (Dose/Effect Relationships) for Cadmium" [10].

Toxic Effects on Aquatic Organisms

As for most other heavy metals, the toxicity of cadmium towards organisms is generally associated with the inhibition of enzyme systems. Cd interferes severely with metalloproteins, metalloenzymes, metallothioneins and phospholipids [276].

There is a wide range of Cd concentrations in the water phase which can be tolerated or which may show lethal effects for individual groups of organisms. Examples of the lethal toxicity data for marine organisms were given in Table 19 (from Bryan [200]). It should be noted that the toxicity may change according to influences of the metal's form in water, by the presence of other metals or poisons, by physiological factors, such as temperature, pH, dissolved oxygen, light and salinity, by the condition of the organism, and by special behavioral responses [200].

Data on sublethal effects of cadmium are still rare with the exception of investigations on morphological changes [277]. Further parameters to be considered include inhibitory effects on growth, settlement, reproduction, and

Table 19. Lethal toxicity of Cd on marine organisms [200]. References see original paper of Bryan [200]

		LC_{50} (ppm)			Minimum observed LC_{50}		Conditions		
Group	Species	24 h	43 h	96 h	LC_{50}	Time (h)	salinity	°C	Form
Fish	*Fundulus heteroclitus*	140	-	49	23	264	S‰ = 20	20	Chloride
	Agonus cataphractus	-	33–100	33	-	-	-	15	Chloride
Molluscs	*Mytilus edulis*	>200	165	25	-	-	S‰ = 20	20	Chloride
Bivalves	*Mya arenaria*	>200	50	2·2	-	-	S‰ = 20	20	Chloride
	Cardium edule	-	10–33	3·3	-	-	-	15	Chloride
Crustaceans	*Crangon crangon*	-	3·3–10	1·0	-	-	-	15	Chloride
Shrimps	*Crangon septemspinosa*	2·4	0·5	0·32	-	-	S‰ = 20	20	Chloride
Crabs	*Pagurus longicarpus*	>200	3·7	0·32	-	-	S‰ = 20	20	Chloride
	Carcinus maenas	100	16·6	4·1	-	-	S‰ = 20	20	Chloride
	Uca pugilator	-	-	47	9·5	240	S‰ = 20	20	Chloride
Echinoderm	*Asterias forbesi*	12	1	0·82	-	-	S‰ = 20	20	Chloride
Annelids	*Nereis virens*	25	25	11	-	-	S‰ = 20	20	Chloride
	Nereis diversicolor	-	-	-	10	816	50% S.W.	13	Sulphate
	Ophryotrocha labronica	-	-	8	1	410	-	20	Sulphate

metabolism and on behavioral processes such as feeding, learning, and swimming activities [200].

Toxic Effects on Humans [10]

In humans the two main target organs of *acute and short-term* cadmium exposure are the gastrointestinal tract after ingestion and the lungs after inhalation. *Long-term exposure* mainly affects these two organs; the critical organ is the kidney:

Lung. Repeated or prolonged inhalation (no-effect levels are probably close to 2 μg/m³ –24 h/day exposure for 70 yr) by exposed workers may produce an obstructive pulmonary syndrome and emphysema, probably from direct action of cadmium on the lungs.

Kidney. Kidney lesions (mainly on the proximal tubules) have been observed in workers exposed to airborne cadmium (usually preceding lung damage) and in persons that have ingested contaminated food. Epidemiological surveys in Japan suggest that a continuous oral daily intake of 200 μg cadmium could cause an increased prevalence of kidney damage in persons over 50 yr of age. It is estimated that such an intake corresponds to an average urinary excretion of 6 μg Cd/l (see discussion in [10]). In humans renal damage is likely to occur when the cadmium concentration in the renal cortex exceeds 200 mg/kg wet weight. Evidence of renal damage is shown by mild proteinuria, in which mainly low molecular weight globulins and some albumin, glycosuria, amino-aciduria are lost; reduced ability to concentrate urine or to excret acid also occur. A high phosphate clearance with hypercalcuria and stone formation has been likewise found [278].

Cardiovascular System. Controversial results have been reported regarding the hypertensive action of cadmium. A significant increase in blood pressure could be measured in rats fed with a diet containing 0.56–0.63 mg Cd/kg [279]. It has been suggested that the cadmium to zinc ratio in the kidney plays a more important role in the development of hypertension than the concentration of cadmium [68].

Bones. Osteomalacia and osteoporosis with a tendency to fracture and bond deformation accompanied by lumbar pains, leg myalgia and pains on bone pressure as well as gait have been described in itai-itai patients, principally in women after menopause who have born several children [280]. Whether cadmium exerts a direct toxic action on bone tissue (which can even precede kidney damage) or is due to disturbed calcium and phosphorous metabolism secondary to the kidney lesion is a point of discussion (see [11], pp. 211–250).

Hematopoietic System. Slight hypochromic anaemia has been observed in most itai-itai patients as well as among workers exposed to cadmium.

Carcinogenic, Mutagenic, and Teratogenic Effects. Insufficient epidemiological data are available on the potential effects of these influences on humans [10]. In animals, cadmium and its compounds have been shown to induce sarcoma at injection sites [281–285], whereas on oral administration cadmium does not seem to act as carcinogen [279, 286]. Likewise, no data relevant to industrial societies (e.g. from Japan) is available to suggest that non-occupational exposure to cadmium constitutes a carcinogenic hazard. Chromosomic anomalies of peripheral leucocytes of workers exposed to lead and cadmium suggest a synergistic effect between both metals [287].

Regulations

Guidelines and standards for environmental levels of cadmium summarized here may be subject to changes.

Food. Approximately 200 mg Cd/kg wet weight has been proposed as a tentative critical concentration in the human kidney cortex. If the total intake of cadmium does not exceed 1 µg/kg body weight per day, it is likely that the levels of cadmium in the renal cortex won't exceed 50 mg/kg, assuming an absorption rate of 5% and a daily excretion of only 0.005% of the body load (reflecting the long half-life of cadmium in the body). The Joint FAO/WHO Expert Committee of Food Additives [288] has therefore proposed a provisional *tolerable weekly intake of 400–500* µg Cd per individual. As the amount of food consumed by an adult person is about 10 kg per week the mean content of cadmium in food should not exceed 0.04–0.05 mg/kg [289]. The provisional guidelines set by the Japanese Ministry of Health and Welfare, whereby detailed investigations of cadmium environmental pollution are initiated, are 0.4 mg/kg in rice or concentrations in drinking water up to 10 µg/l [10]. Reference values for Cd concentrations in various food items have been published by the German Ministry of Health in 1974, i.e. meat – 0.008 mg Cd/kg; vegetables – 0.1 mg/kg and fruit – 0.05 mg Cd/kg. Among the most important sources of contamination by food are pork (1.7 µg Cd/person/day), milk (3 µg), potatoes (11.6 µg), vegetables (5.6 µg) and beer (3 µg [237]). Kidneys from animals destined for slaughter and subsequent human consumption as well as uncultivated mushrooms may be rich in cadmium [290]. In Britain, the Toxicity Sub-Committee [278] found that a persistently high dietary consumption of shellfish or brown crab meat with a high cadmium content may conceivably constitute a risk for selected individual consumers.

Food Containers. The national regulations concerning the amount of cadmium released from ceramics (for a compilation see [289]) are mostly based on tests involving treatment with 3% or 4% acetic acid (at room temperature) and subsequent determination of dissolved Cd. Limits range from 0.1 mg Cd/l in Sweden, 0.2 mg/l in the UK (hollow ware, kitchen utensils), 0.5 mg Cd/l in Italy and the USA, 0.7 mg/l in the UK (ceramic ware) to 1.0 mg Cd/l in Denmark (90 min at 100 °C) and South Africa (all types of food containers; [10]). Proposed EEC directives differentiate according to the surface area and use (tableware, child's plate, cookingware, hollow ware).

Water. A tentative upper limit for Cd in *drinking water* was initially set at 10 µg/l by the US Public Health Service (1962), USSR Government (1970), the World Health Organization (European: 1970; International: 1971), USA National Academy of Sciences (1972), Australian Government (1973) and the US Environmental Protection Agency (1975) [291]. The WHO Regional Office of Europe recommended 5 µg/l in 1973. The Council of Ministers of the EEC fixed the maximum concentration of Cd in "*surface water* used after purifivation as drinking water" at 5 µg/l and the goal is less than 1 µg/l [292]. The "DVGW W 151" (FRG) limits the use of surface water at 5 µg/l (purification by bank filtration and artificial recharge) and 10 µg/l (physico-chemical treatment). Cd contents in *irrigation water* usually are limited at 5 µg/l [293]. *Effluent standards* are 0.1 mg Cd/l in Japan (assuming minimal in-river dilution of factor 10 [294]; US standards permit 40 µg Cd/l when the receiving streams' low flow exceeds ten times the waste flow [295].

Air. In Japan the provisional guideline for cadmium in ambient air is 0.88–2.92 µg/m^3. The industrial maximum allowable concentrations are significantly higher: in the USA the time weighted average value (TLV or TWA) for cadmium dust is 0.05 mg/m^3, and in the USSR the TLV for cadmium fumes is 0.1 mg/m^3. In Finland the TLV for cadmium dust and fumes is 10 µg/m^3 and 20 µg/m^3 in Sweden. In Germany the MAK for cadmium oxide is 0.1 mg/m^3, and the TLVs proposed by the Health and Safety Executive in the UK are 0.2 mg/m^3 for cadmium as metal dust or soluble salt and 0.05 mg/m^3 for cadmium oxide fumes. The TLV for cadmium dust in Switzerland is 0.2 mg/m^3 [10].

Sewage Sludge. Unlike zinc, copper, nickel, or boron, cadmium accumulations in vegetation can reach levels that are toxic to animals before the vegetation itself shows any signs of damage. In that respect, guidelines limiting cadmium application, e.g. sludge in agriculture, tend to be very conservative [296]. U.S. Department of Agriculture data [297] suggest a guideline of 0.1 kg Cd/ha/y, assuming a 20-year lifespan, for use of municipal sludges (103 metric tons/hectar/life) containing up to 2000 mg Zn/kg, 1000 mg Cu/kg, 200 mg Ni/kg, and 20 mg Cd/kg (1% of Zn). A draft copy of the Illinois Environmental Protection Agency [298] guideline specifies that cadmium application are to be limited as follows: (i) a maximum application rate of 0.33 kg/ha/y; (ii) a maximum of 6.7 kg/ha/y if the Zn:Cd ratio of the sludge exceeds 100; (iii) a maximum application of 3.4 kg/ha/life if the Zn:Cd ratio of the sludge is less than 100. The provisional Ontario guidelines [299] limit the application to a maximum of 1.6 kg Cd/ha/life. Regulations presently do not consider soil characteristics, such as cation exchange capacity (C.E.C., [300]), pH [301, 302], and carbonate content [199, 303].

Acknowledgements

Thanks are due to Dr. F. Prosi, who contributed substantial material for the chapter on biological uptake and accumulation, and to Mr. D. Godfrey for his aid in preparing the English version of the text.

References

1. Bryan, G.W.: Proc. Royal Soc. B (London) *177*, 389 (1971)
2. Förstner, U., Wittmann, G.T.W.: Metal Pollution in the Aquatic Environment. Berlin-Heidelberg-New York: Springer-Verlag 1979
3. Friberg, L.: In: Proc. 9th Int. Congr. Ind. Med. *9*, 641 (1948)
4. Kobayashi, J.: In: Proc. 5th Int. Conf. Adv. Water Pollut. Res. *I–25* (1971)
5. Hagino, N., Yoshioka, K.: J. Japan Orthoped. Assoc. *35*, 812 (1961)
6. Friberg, L. et al.: Cadmium in the Environment. Cleveland: CRC-Press 1974, 248p.
7. Fleischer, M. et al.: Environ. Health Perspect., 253 (1974)
8. Task Group Environmental Health Effects of Cadmium. Geneva: World Health Organization 1975
9. Nordberg, G.F. (ed.): Effects and Dose-Response Relationships of Toxic Metals. Amsterdam: Elsevier 1976
10. Criteria (Dose/Effect Relationships) for Cadmium. Commission Europ. Communities. Pergamon, Oxford 1978
11. Proc. First Int. Cadmium Conf., San Francisco, 31. Jan.–2. Feb. 1977. Metal Bulletin Ltd., Park House 3, Park Terrace, Worcester Park, Surrey, KT4 7HY (1978)
12. Nriagu, J.O. (ed.): Cadmium in the Environment. New York: Wiley-Interscience 1980
13. Anonym: Cadmium. Mining Ann. Rev. *1978*, 101
14. Rauhut, A.: Metall *32*, 947 (1978)
15. Versar Inc. Microeconomic Analysis of the Usage of Various Toxic Substances. Progr. Rep. EPA No. 68-01-2926, Springfield 1974
16. Kloke, A.: Nachrichtenbl. Dtsch. Pflanzenschutzd. *22*, 164 (1971)
17. Leland, W.T., Nier, A.O.: Physic Rev. *73*, 1206 (1948)
18. Brehler, B.: In: Handbook of Geochemistry (Wedepohl, K.H., ed.) *48-A*. Berlin-Heidelberg-New York: Springer-Verlag 1972
19. Krauskopf, K.B.: Introduction to Geochemistry. New York: McGraw-Hill 1967, 721p.
20. Wakita, H., Schmitt, R.A.: In: Handbook of Geochemistry *48-D*. Berlin-Heidelberg-New York: Springer-Verlag 1970
21. Marowsky, G., Wedepohl, K.H.: Geochim. Cosmochim. Acta *35*, 1255 (1971)
22. Gong, H., Rose, A.W., Suhr, N.H.: Geochim. Cosmochim. Acta *41*, 1687 (1977)
23. Mullin, J.B., Riley, J.P.: J. Marine Res. *15*, 103 (1956)
24. Damiani, V. et al.: In: Interactions between Sediments and Fresh Water (H.L. Golterman, Ed.), The Hague/Wageningen: Junk and Pudoc 1977, p. 83
25. Turekian, K.K., Wedepohl, K.H.: Bull. Geol. Soc. Amer. *72*, 175 (1961)
26. Horn, M.K., Adams, J.A.S.: Geochim.Cosmochim.Acta *30*, 279 (1966)
27. Altschuler, Z.S.: In: Abstr. 10th Int.Congr. Sedimentology, Jerusalem, p. 16 (1978)
28. Jonasson, I.R.: Geol. Surv. Canada Pap. *76–13*, p.37 (1976)
29. Tobschall, H.J., Göpel, C., Rast, U.: Abstr. 10th Int. Congr. Sedimentology, Jerusalem, p. 682 (1978)
30. Abernathy, R.F., Peterson, M.J., Gibson, F.H.: U.S. Bur. Mines Rep. Invest. No. 7281, 1 (1969)
31. Wixson, B.G.: In: Proc. 1st. Int. Cadmium Conf. San Francisco, 31. Jan.–2. Feb. 1977, p. 97 (1978)
32. Ruch, R.R., Gluskoter, H.J., Shimp, N.F.: Illinois Geol. Surv. Environ. Geol. Notes *61*, 43 p. (1973)
33. Förstner, U.: Cadmium in Sediments. In [12]
34. Boyle, E.A., Sclater, F.R., Edmond, J.M.: Nature *263*, 42 (1976)
35. Trefry, J.H., Presley, B.J.: In: Marine Pollutant Transfer (H.L. Windom, R.A. Duce, eds.). Lexington: Heath & Co. 1976, p. 39
36. Hem, J.D.: Water Resources Res. *8*, 661 (1972)
37. Morel, F., McDuff, R.E., Morgan, J.J.: In: Trace Metals and Metal-Organic Interactions in Natural Waters (P.C. Singer, ed.). Ann Arbor: Ann Arbor Sci. Publ. 1973, p. 157
38. Lu, C.S.J., Chen, K.Y.: Environ. Sci. Technol. *11*, 174 (1977)
39. Dyrssen, D., Wedborg, M.: In: The Sea (E.D. Goldberg, ed.). New York: Wiley-Interscience 1974, p. 181

40. Zirino, A., Yamamato, S.: Limnol.Oceanogr. *17*, 661 (1972)
41. Lerman, A., Childs, C.W.: In: Trace Metals and Metal-Organic Interactions in Natural Waters (P.C. Singer, ed.). Ann Arbor: Ann Arbor Sci. Publ. 1973, p. 201
42. Florence, T.M., Batley, G.E.: Talanta *24*, 151 (1977)
43. Hart, B.T., Davies, S.H.R.: Caulfield Inst.Technol. Water Studies Centre, Techn. Rep. 6, 188 p. (1978)
44. Pierce, J.O., O'Laughlin, J.W., Hemphill, D.D.: [11], p. 88 (1978)
45. McLellan, J.S. et al.: Phys. Med. Biol. *20*, 88 (1975). ES&T Jan. 1980, p. 23
46. Norval, E., Butler, L.R.P.: CSIR Rept., Pretoria S.A. (1974)
47. Anderson, J.: Atom. Absorpt. Newsl. *13*, 31 (1974)
48. Symeonides, C., McRae, S.G.: J. Environ. Qual. *6*, 120 (1977)
49. Farrah, H., Pickering, W.F.: Water Air Soil Pollut. *9*, 491 (1978)
50. Alloway, B.J., Davies, B.E.: J. Agric. Sci. (Cambridge) *76*, 321 (1971)
51. John, M.K., Van Laerhoven, C.J., Chuah, H.H.: Environ. Sci. Technol. *6*, 1005 (1972)
52. Jones, R.L., Hinesly, T.D., Ziegler, E.L.: J. Environ. Qual. *2*, 351 (1973)
53. Anderson, A., Nilsson, K.O.: Ambio *3*, 198 (1974)
54. Williams, C.H., David, D.J.: Austr. J. Soil Res. *11*, 43 (1973)
55. Navrot, J., Singer, A., Banin, A.: J. Soil Sci. *29*, 505 (1978)
56. Davis, W.E. et al.: Nat. Inventory of Sources and Emissions, 1968. Cadmium: Rep. Nat. Air Pollution Control Administration, Febr. 1970, pp. 1–44 (Cit. [7]).
57. Chizhikov, D.M.: Cadmium. New York-London: Pergamon Press 1966
58. Bertine, K.K., Goldberg, E.D.: Science *178*, 233 (1971)
59. De Groot, A.J., Allersma, E., Van Driel, W.: Symp. Water in Dienst van Industrie en Milieu. Publ. No. 110 N., Sekt 5, 24 p. (1973)
60. Klein, L.A. et al.: J. Water Poll. Contr. Fed. *46*, 2653 (1974)
61. Field, R.A., Lager, T.A.: J. Environ. Eng. Div. ASCE *101*, 107 (1975)
62. Schleichert, U., Hellmann, H.: Auftreten u. Herk. v. Zink in Gewässern. Literaturber. Bundesanstalt Gewässerkunde, Koblenz 1973
63. Koppe, P., Dietz, F.: Hydrochem. Hydrogeol. Mitt. *3*, 181 (1978)
64. Miettinen, J.K.: In: 16th Session Joint FAO/WHO Committee of Expert on Food Additives, Geneva, Apr. 4–12, 1972. Cit. [10]
65. Yost, K.J.: In: Proc. First Intern. Cadmium Conf. San Francisco, 31. Jan.–2. Feb. 1977, p. 147 (1978)
66. Fulkerson, W., Goehler, H.E.: Cadmium, the Dissipated Element. Oak Ridge National Laboratory 1973
67. Zoetman, B.C.J., Brinkmann, F.J.J.: In: Europ. Colloq. "Hardness of Drinking Water a. Public Health". Luxembourg 21–23 May 1975
68. Schroeder, H.A. et al.: J. Chronic Dis. *20*, 179 (1967)
69. Proctor, P.D. et al.: In: Trace Substances in Environmental Health, Univ. Missouri, Columbia *7*, 63 (1973)
70. Arbeitsgruppe "Metalle" im Schwerpunktprogramm "Schadstoffe im Wasser" der Dtsch. Forschungsgemeinschaft, DFG, Bad Godesberg 1975
71. Bouquiaux, J.: In: Problems of the Contamination of Man and his Environment by Mercury and Cadmium, Luxembourg, July 1973
72. Förstner, U., Müller, G.: Ruperto Carola (Univers. Heidelberg) *51*, 67 (1973)
73. Mink, L.L., Williams, R.E., Wallace, A.T.: Ground Water *10*, 310 (1972)
74. Schleichert, U.: Dtsch. Gewässerkdl. Mitt. *19*, 150 (1975)
75. Bergren, B., Odén, S.: Inst. Markvetenskap Lantbrukshögskolan. Uppsala, Sweden 1972
76. Jannson, S.L.: Landwirtschaftl. Forsch. *27*, 61 (1972)
77. Page, A.L.: Fate and Effects of Trace Elements in Sewage Sludge when Applied to Agricultural Lands. U.S.EPA Natl. Environ. Res. Center, Cincinnati/Ohio 1974, 98 pp.
78. Blakesley, P.A.: In: US EPA Workshop, Champaign/Urbana, Ill. 1973
79. Preuss, E., Kollmann, H.: Naturwissenschaften *61*, 270 (1974)
80. Bucher, E.: Z. Weinbau/Keller *22*, 102 (1975)
81. Fulkerson, W.: Cadmium – The Dissipated Element Revisited. Oak Ridge National Laboratory, Tenn. 1975
82. Nightingale, H.I.: J. Amer. Water Works Ars. *67*, 443 (1975)

83. Lund, L.J., Page, A.L., Nelson, C.O.: J.Environ.Qual. *5*, 330 (1976)
84. Balke, K.D., Kussmaul, H., Siebert, G.: Z. Dtsch. Geol. Ges. *124*, 447 (1973)
85. Qasim, S.R., Burchinal, J.C.: J. Water. Poll. Contr. Fed. *42*, 371 (1970)
86. Walker, W.: Ground Water *11*, 11 (1973)
87. Mueller, J.A., Anderson, A.R., Jeris, J.S.: *48*, 2309 (1976)
88. Schafer, H.A.: In: South. Calif. Coastal Water Res. Proj., Ann. Rep. 1976, p. 57
89. Martin, J.H., Bruland, K.W., Broenkow, W.W.: In: Marine Pollutant Transfer (Windom, H.L., Duce, R.A., eds.). Lexington: D.C. Heath, p. 159 (1976)
90. Hartung, R.: Proc. Intern. Conf. Pers. Chem. Aquatic Ecosystems, Ottawa I–93 (1974)
91. Bowen, H.J.M.: Proc. Symp. Heavy Metals in the Environment, Toronto, 1975, I/1–9 (1977)
92. Barnes, R.S., Schell, W.R.: Proc.Environ. Resources Conf., US EPA, Cincinnati/Ohio, p. 45 (1973)
93. Garrels, R.M., MacKenzie, F.T., Hunt, C.: Chemical Cycles and the Global Environment. Los Altos/Calif.: William Kaufmann Inc. 1975
94. Heindl, R.A.: US Bur. Mines Bull. *650*, 805 (1970)
95. Lagerwerff, J.V., Specht, A.W.: In: Proc. 4th Ann. Conf. on Trace Substances in Environmental Health. Univ. Missouri, Columbia Mo. 1971, p. 85
96. Klein, D.H.: Environ. Sci. Technol. *6*, 560 (1972)
97. Miesch, A.T., Huffman, C.Jr.: In: Area Pollution Study. US EPA, Office of Air Programs, Publ. No. AP-91, p. 65 (1972)
98. Chau, Y.K., Weiler, R.R., Chawla, V.K.: Proc. 13th Conf. Great Lakes Res., p. 213 (1970)
99. Klein, D.H.: Water, Air, Soil Pollut. *4*, 3 (1975)
100. Abdullah, M.I., Royle, L.G.: Nature *238*, 329 (1972)
101. Abdullah, M.I., Royle, L.G., Morris, A.W.: Nature *235*, 158 (1972)
102. Förstner, U.: In: Chemistry and Biochemistry of Estuaries (E. Olausson, I. Cato, eds.). London: Wiley 1980, in press
103. Preston, A. et al.: Environ. Pollut. *3*, 69 (1972)
104. Kremling, K., Petersen, H.: Meteor Forsch.-Ergebn. A *19*, 10 (1977)
105. Dutton, J.W.R. et al.: Mar. Pollut. Bull. *4*, 135 (1973)
106. Halcrow, W., MacKay, D.W., Thornton, I.: J. Mar. Biol. Assoc. U. K. *53*, 721 (1973)
107. Elderfield, H., Thornton, I., Webb, J.S.: Mar. Pollut. Bull. *2*, 44 (1971)
108. Abdullah, M.I., Royle, L.G.: J. Mar. Biol. Assoc. U. K. *54*, 581 (1974)
109. Förstner, U.: Metallanreichergg. in rezenten See-Sedimenten – geochem. background u. zivilisator. Einflüsse. Mitt. Nationalkomitee FRG, Internat. Hydrolog. Progr. UNESCO, Heft 2, Koblenz, 66 p (1978)
110. Murakami, T. et al.: Eisei Kaguku (Japan) *21*, 275 (1975)
111. Kneip, T.J., Re, G., Hernadez, T.: D.D. Hemphill (ed.) Trace Substances in Environmental Health, Vol. VIII. Univ. Missouri, Columbia/Mo., p. 173 (1974)
112. Wentsel, R.S., Berry, J.W.: Proc. Indiana Acad. Sci. *84*, 481 (1974)
113. Bloom, H., Ayling, G.M.: Environ. Geol. *2*, 3 (1977)
114. Chen, K.Y., Lockwood, R.A.: J. Environ. Eng. Div. ASCE *102*, 347 (1976)
115. Skei, J.M., Price, N.B., Calvert, S.E.: Water Air Soil Pollut. *1*, 452 (1972)
116. Asami, T.: Ibaraki Daigaku Nogakubu Gakujutsu Hokaku (Japan) *22*, 19 (1974)
117. Vivian, C.M.G., Massie, K.S.: Environ. Pollut. *14*, 47
118. Förstner, U., Müller, G.: Geoforum *14*, 53 (1973)
119. Kiba, T., et al.: Bunseki Kagaku (Japan) *24*, 18 (1975)
120. Perhac, R.M.: J. Hydrol. *15*, 177 (1972)
121. Tyler, P.A., Buckney, R.T.: Intern. Rev. ges. Hydrobiol. *58*, 873 (1973)
122. Schleichert, U., Hellmann, H.: Bilanzier. u. Herk. toxischer Schwermetalle für d. Rhein-Einzugsgebiet. 1976. Interim Rep. Bundesanstalt f. Gewässerk. Koblenz, pp. 53
123. Fitchko, J., Hutchinson, T.C.: J. Great Lakes Res. *1*, 46 (1975)
124. Stoffers, P. et al.: Environ. Sci. Technol. *11*, 819 (1977)
125. Holmes, C.W., Slade, E.A., McLerran, C.J.: Environ. Sci. Technol. *8*, 255 (1974)
126. Azumi, T., Yoneda, A.: Himeji Kogyo Daigaku Kenkyu (Japan) *28 A*, 110 (1975)
127. Kronfeld, J., Navrot, J.: Water Air Soil Pollut. *4*, 127 (1975)
128. Pasternak, K.: Acta Hydrobiol. *16*, 51 (1974)
129. Thornton, I., Watling, H., Darracott, A.: Sci. Total Environ. *4*, 325 (1975)

130. Laskowski, N. et al.: Naturwissenschaften *62*, 136 (1975)
131. Maxfield, D. et al.: Environ. Pollut. *7*, (1974)
132. Ribordy, E.: In: J.-P. Vernet, G. Scolari (eds.) Etude de la pollution des sédiments du Léman et du bassin du Rhône. Rapports Comm. Intern. Protect. Eaux Léman, campagne 1977 (1978)
133. Štern, J., Förstner, U.: Geologija (Ljubljana) *19*, 259 (1976)
134. Schafer, H.A., Bascom, W.: Calif. Cstl. Water Res. Proj., Ann. Rep., p. 77 (1976)
135. Suess, E., Erlenkeuser, H.: Meyniana (Kiel) *27*, 63 (1975)
136. Salomons, W., De Groot, A.J.: In: W.E. Krumbein (ed.) Environmental Biogeochemistry, Vol. 1. Ann Arbor: Ann Arbor Sci. Publ. 1978, p. 149
137. Shacklette, H.T.: Cadmium in Plants, US Geol. Surv. Bull. *1314–G* (1972)
138. Cannon, H.L.: In: Proc. 3rd Ann. Conf. on Trace Substances in Environmental Health. Univ. Missouri, Columbia/Mo., p. 21 (1971)
139. Connor, J.J., Shacklette, T.H., Erdmann, J.A.: US Geol. Surv. Prof. Paper *750-B*, 151 (1971)
140. Rühling, A.: Adv.Ekol.Botanik Lands Univ. 1969
141. Windom, H.L. et al.: In: Marine Pollutant Transfer (H.L. Windom, R.A. Duce, eds.). Lexington: D.C. Heath & Co. 1976, p. 135
142. Windom, H.L. In: Estuarine Research, (L.E. Cronin, ed.) Vol. 1, p. 137. New York: Academic Press 1975
143. Webber, M.D., Corneau, D.G.M.: In: Proc. Int. Conf. Heavy Metals in the Environment, Toronto 1975, Vol. I, 205 (1977)
144. Chumbley, C.G.: Permissible Levels of Toxic Metals in Sewage Used on Agricultural Land. ADAS Adv. Paper No. 10, Min. Agr. Fish Food, London (1971)
145. Webber, J.: Conf. Proc. Res. Prog. Abatement of Municipal Pollution under Provisions of the Canada-Ontario Agreement of Great Lakes Water Quality, No. 2, p. 10 (1974)
146. Davies, B.E.: Proc. Int. Seminar Soil Environment a. Fertility Management in Intensive Agricult. Tokyo/Japan, 1977, p. 394
147. Davies, B.E.: Sci.Total Environ. *9*, 243 (1978)
148. Lindsay, W.L., Norvell, W.A.: Agron. Abstracts, Amer.Soc.Agron., Madison, Wisc., p. 84 (1969)
149. John, M.K.: Soil Sci. *113*, 222 (1972)
150. Bingham, F.T. et al.: J. Environ. Qual. *4*, 207 (1975)
151. Keeney, D.R., Walsh, L.M.: Proc. Int. Conf. Heavy Metals in the Environment, Toronto 1975, II/1, 379 (1977)
152. McLean, A.J.: Can. J. Soil Sci. *49*, 327 (1969)
153. Keeley, J.W., Engler, R.M.: Dredged Material Res. Progr., US Army Corps of Engineers, Vicksburg, Miss. Rept. D-74-14, p. 13 (1974)
154. Gambrell, R.P. et al.: Dredged Material Res. Progr. US Army Corps of Engineers, Vicksburg, Miss. Rept. D-77-4 p. 309 (1977)
155. Jackson, M.L.: Soil Chem. Analysis. Englewood Cliffs/New Jersey: Prentice Hall Inc. 1958
156. Patchineelam, S.R.: Unpubl. Thesis Univ. Heidelberg, p. 136 (1975)
157. Gupta, S.K., Chen, K.Y.: Environ. Lett. *10*, 129 (1975)
158. Engler, R.M., Brannon, J.M., Rose, J.: 168th Meet. Amer.Chem. Soc. Atlantic City, N.Y., p. 17 (1974)
159. Malo, B.A.: Environ. Sci. Technol. *11*, 277 (1977)
160. Deurer, R., Förstner, U., Schmoll, G.: Geochim. Cosmochim. Acta *42*, 425 (1978)
161. Chao, L.L.: Soil Sci. Soc. Amer. Proc. *36*, 764 (1972)
162. Schwertmann, U.: Z. Pflanzenernähr. Düng. Bodenkde. *105*, 194 (1964)
163. Holmgren, G.S.: Soil Sci. Soc. Amer. Proc. *31*, 210 (1967)
164. Chester, R., Hughes, M.J.: Chem.Geol. *2*, 249 (1967)
165. Rashid, M.A.: Soil Sci. *111*, 298 (1971)
166. Volkov, I.I., Fomina, L.S.: In: (Degens, E.T., Ross, D.A., eds.) The Black Sea, Geology, Chemistry, and Biology. Am. Assoc. Petrol. Geol., Mem. *20*, 456 (1974)
167. Cooper, B.S., Harris, R.C.: Mar. Pollut. Bull. *5*, 24 (1974)
168. Gibbs, R.J.: Science *180*, 71 (1973)
169. Förstner, U.: unpublished

170. Lee, C.R., Engler, R.M., Mahloch, J.L.: DMRP, US Army Corps of Engineers, Vicksburg, Miss. Misc. Pap. D-76-5 (1976)
171. Förstner, U., Patchineelam, S.R.: Chemiker-Z. *100*, 49 (1976)
172. Salomons, W., Mook, W.G.: Abstr. 10th Int. Congr. Sedimentology, Jerusalem 1978, p. 569
173. Müller, G., Förstner, U.: Environ. Geol. *1*, 33 (1975)
174. Duinker, J.C., Nolting, R.F.: Mar. Pollut. Bull. *8*, 56 (1977)
175. Van der Weijden, C.H., Arnoldus, M.J.H.L., Neurs, C.T.: Neth. J. Sea Res. *11*, 130 (1977)
176. Rohatgi, N.K., Chen, K.Y.: J. Environ. Eng. Div. Amer. Soc. Chem. Engnrs. *102*, 675 (1976)
177. Beamish, R.J.: Water Air Soil Pollut. *6*, 501 (1976)
178. Pasternak, K.: Acta Hydrobiol. *16*, 273 (1974)
179. Rohde, G.: ANS Mitt. p. 24 (1974)
180. Banat, K., Förstner, U., Müller, G.: Chem. Geol. *14*, 199 (1974)
181. Chau, Y.K., Shiomi, M.T.: Water Air Soil Pollut. *1*, 149 (1972)
182. Schöttler, U.: Vom Wasser *49*, 295 (1977)
183. Gardiner, J.: Water Resour. Res. *8*, 157 (1974)
184. Pacific Northwest Laboratories, Contract No. DACWO 7-73-C-0080, see Khalid, R.A. et al. DMRP, US Army Engineer Waterways Experiment Station, Rep. D-77-4, p. 221 (1977)
185. Jackson, T.A.: Environ. Geol. *2*, 173 (1978)
186. Gilbert, T.R., Clay, A.M., Leighty, D.A.: Environmental Chemistry Div., Tyndall Air Force Base, Fla. Rept. AFCED-TR-6-22, 89 p. (1976)
187. Windom, H.L.: Rep. US Army Corps of Engineers, Mobile Distr. 46 p. (1973)
188. Lee, G.F.: In: Interactions between Sediment and Fresh Water (Golterman, H.L., ed.). The Hague: Junk Publ. 1977, p. 444
189. Sly, P.G.: J. Environ. Sci. Health *A-12*, 475 (1977)
190. May, E.B.: Alabama Marine Resour. Bull. *10*, 1 (1974)
191. Patrick, W.H.Jr. Cambrell, R.P., Khalid, R.A.: J. Environ. Sci. Health *A-12*, 475 (1977)
192. Rosmanith, J.: ISU-Mitt. Univers. Aachen *2*, 115 (1972)
193. Liebscher, L., Smith, H.: Arch. Environ. Health *17*, 881 (1968)
194. Koeppe, D.E.: Sci. Total Environ. *7*, 197 (1977)
195. Haghiri, F.: J. Environ. Qual. *3*, 180 (1974)
196. Miller, J.E., Hassett, J.J., Koeppe, D.E.: J. Environ. Qual. *5*, 157 (1976)
197. Schroeder, H.A., Balassa, J.J.: Science *140*, 819 (1963)
198. Root, R.A., Miller, R.J., Koeppe, D.E.: J. Environ. Qual. *4*, 474 (1975)
199. Hyde, H.C. et al.: J. Water Poll. Control Fed. *51*, 2475 (1979)
200. Bryan, G.W.: In: (Johnston, R., ed.) Marine Pollution. London: Academic Press 1976, p. 185
201. Martin, J.H.: Limnol. Oceanogr. *15*, 756 (1970)
202. Lowman, F.G., Rice, T.R., Richards, F.A.: In: Radioactivity in the Marine Environment. Nat. Acad. Sci., Washington, p. 161 (1971)
203. Lloyd, R.M.: In: Biol. Problems in Water Pollution, 3rd Seminar 1962. US Dept. of Health, Education and Welfare, p. 181 (1965)
204. Tabata, K.: Bull. Tokai Reg. Fish. Res. Lab. *58*, 203 (1969)
205. Knauer, G.A., Martin, J.H.: Limnol. Oceanogr. *18*, 597 (1973)
206. Zitko, V., Brown, V.M., Andrew, R.W.: In: (Andrew, R.W., Hodson, P.V., Konasewich, D.E., eds.) Toxicity to Biota of Metal Forms in Natural Water, Int. Joint. Comission, Windsor, Ontario, 1976, pp. 9, 59 and 127
207. Poldoski, J.E.: Environ. Sci. Technol. *13*, 701 (1979)
208. Sunda, W.G., Engel, D.W., Thuotte, R.W.: Environ.Sci.Technol. *12*, 409 (1978)
209. Jackson, G.A., Morgan, J.J.: Limnol. Oceanogr. *23*, 268 (1978)
210. Ramamoorthy, S., Kushner, D.J.: Microbial Ecol. *2*, 162 (1975)
211. Bryan, G.W., Hummerstone, L.G.: J. Mar. Biol. Assoc. U. K. *53*, 839 (1973)
212. Peden, J.D. et al.: Mar. Pollut. Bull. *4*, 7 (1973)
213. Förstner, U., Prosi, F.: In: Pollution Aspects of Fresh Waters (Ravera, O. (ed.). Oxford: Pergamon Press 1979, p. 129
214. Reay, P.F.: J. Appl. Ecol. *9*, 557 (1972)
215. Prosi, F.: Dissert. Univ. Heidelberg, 210 p. (1977)

216. Butterworth, J., Lester, P., Nickless, G.: Mar. Pollut. Bull. *3*, 72 (1972)
217. Brooks, R.R., Rumsby, M.G.: Limnol. Oceanogr. *10*, 521 (1965)
218. Topping, G.: Aquaculture *1*, 379 (1973)
219. Leatherland, T.M., et al.: Deep-Sea Res. *20*, 679 (1973)
220. Jaakkola, T., Takahashi, H., Miettinen, J.K.: WHO-Meeting, Geneva, Feb. 1974, p. 8
221. Phillips, D.J.H.: Environ. Pollut. *13*, 281 (1977)
222. Goldberg, E.D.: Mar. Pollut. Bull. *6*, 111 (1975); Environ. Conserv. *5*, 101 (1978)
223. Ayling, G.M.: Water Res. *8*, 729 (1974)
224. Fowler, S.W., Oregioni, B.: Mar. Pollut. Bull. *7*, 26 (1976)
225. Stenner, R.D., Nickless, G.: Mar. Pollut. Bull. *6*, 89 (1975)
226. Lande, E.: Environ. Pollut. *12*, 187 (1977)
227. Segar, D.A., Collins, J.D., Riley, J.P.: J. Mar. Biol. Assoc. U. K. *51*, 131 (1971)
228. Nickless, G., Stenner, R., Terrile, N.: Mar. Pollut. Bull. *3*, 188 (1972)
229. Talbot, V.W., Magee, R.J., Hussain, M.: Mar. Pollut. Bull. *7*, 84 (1976)
230. Sims, R.R., Presley, B.J.: Bull. Environ. Contam. Toxicol. *16*, 520 (1976)
231. Watling, H.R., Watling, R.J.: Mar. Pollut. Bull. *7*, 45 (1976)
232. Thornton, I., Watling, H., Darracott, A.: Sci. Total Environ. *4*, 325 (1975)
233. Boyden, C.R., Romeril, M.G.: Mar. Pollut. Bull. *5*, 74 (1974)
234. Ratkowsky, D.A. et al.: J. Fish. Res. Board Canada *31*, 1165 (1974)
235. Zook, E.G., Greene, F.E., Morris, E.R.: Cereal Chem. *47*, 720 (1970)
236. Guthrie, B.E.: New Zealand Med. J. *82*, 418 (1975)
237. Essing, H.B. et al.: Arch. Hyg. Bakt. *153*, 490 (1969)
238. Pfeilsticker, K., Markard, C.: Z. Lebensmittel Unters.Forsch. *158*, 129 (1975)
239. Ministerium für Arbeit, Gesundheit u. Soziales Nordrhein-Westfalen: Umweltprobleme durch Schwermetalle im Raum Stolberg, Düsseldorf 1975
240. Goodman, G., Robert, T.M.: Nature (London) *231*, 287 (1971)
241. Pringle, B.H. et al.: J. Sanit. Eng. Div. *94*, 155 (1968)
242. Sansoni, B. et al.: In: Probl. Contamination of Man and Environment by Mercury and Cadmium. Comm. Eur.Comm., Luxembourg 1974, p. 149
243. Kjellström, T. et al.: Arch. Environ. Health *31*, 321 (1975)
244. Tolan, A., Elton, G.A.H.: In: Probl. Contamination of Man and Environment by Mercury and Cadmium. Comm. Eur. Comm., Luxembourg 1974, p. 335
245. Duggan, R.E., Corneliussen, P.E.: Pestic. Monit. J. *5*, 331 (1972) Cit. [244]
246. Maranger, J.C., Smith, D.C.: Can. J. Publ. Health *63*, 53 (1972) Cit.: [244]
247. Rautu, R., Sporn, A.: Nahrung *14*, 25 (1970). Cit. [244]
248. Lener, J., Bibr., B.: Vitalstoffe *15*, 139 (1970)
249. Yamagata, N., Shigematu, I.: Bull. Inst. Publ. Health, Tokyo *19*, 1 (1970)
250. Yamagata, N. et al.: Bull. Inst. Publ. Health, Tokyo *20*, 170 (1971)
251. Yamagata, N., Iwashima, S.: Japan/Assoc.Publ. Health, Sept. 1973 Cit. [6]
252. Guthrie, B.E.: Proc. Univ. Otago Med. School *51*, 47 (1973)
253. Festy: Personal communication [10]
254. Wester, P.O.: Artherosclerosis *20*, 226 (1974)
255. Kjellström, T.: Thesis, Dept. Environ. Hyg. Karolinska Inst., Stockholm 1977
256. Princi, F., Geever, E.F.: Arch. Ind. Hyg. Occup. Med. *1*, 651 (1950)
257. Potts, C.L.: Ann. Occup. Hyg. *8*, 55 (1965)
258. Lewis, et al.: J. Chronic Dis. *25*, 717 (1972)
259. Rahola, T., Aaran, R.K., Miettinen, J.K. In: Assess. Radioact. Contamination in Man, Intern. Atom. Energ. Author., Vienna 1972
260. Einbrodt, H.J., Rosmanith, J., Projnar, D.: Naturwissenschaften *63*, 148 (1976)
261. Lauwerys, R.R., Buchet, J.P., Roels, H.A.: Int. Arch. Occup. Environ. Health *36*, 275 (1976)
262. Kagi, J.H.R., Vallee, B.L.: J. Biol. Chem. *236*, 2435 (1961)
263. Sabbioni, E., Marafante, E.: Environ. Physiol. Biochem. *5*, 132 (1975)
264. Goyer, R.A., Cherian, G.M., Richardson, L.D.: In: Proc. 1st. Intern. Cadmium Conf. San Francisco, 31. Jan.–2.Feb. 1977, p. 183 (1978)
265. Lewis, G.P. et al.: Lancet *1*, 291 (1972)
266. Shuman, M.S., Voors, A.W., Gallagher, P.N.: Bull. Environ. Contam. Toxicol. *12*, 570 (1974)

267. Roels, H.A. et al.: Comm. Europ. Comm./Envir. Protect.Agency / WHO-Intern. Symp., Paris, June 1974
268. Ishizaki, A., Fukushima, M., Sakamoto, M.: Japan J. Hyg. *26*, 268 (1971)
269. Sudo, Y., Nomiyama, L.: Ind. Med. (Japan) *14*, 117 (1972)
270. Tsuchiya, K.: Kankyo Hoken Rep. No. 11, Japan Ass. Publ. Health, Apr. 1972, 72 p.
271. Hammer, D.I. et al.: In: Trace Substances in Environmental Health, Univ. Missouri, Columbia 1972, Vol. V., p. 269
272 Tsuchiya, K., Sugita, M.: Nord. Hyg. Tidskr. *53*, 105 (1971)
273. Flick, D.F., Kraybill, H.F., Dimitroff, J.M.: Environ.Res. *4*, 71 (1971)
274. Ohnesorge, F.K.: In: Probl. Contamination of Man and Environment by Mercury and Cadmium. Comm. Europ.Comm., Luxembourg, July 1973, p. 409 (1974)
275. Godfrain, J.C. et al.: Rev. Médicine Vét. *128*, 441 (1977)
276. Vallee, B.L., Ulmer, D.D.: Ann. Rev. Biochem. *41*, 91 (1972)
277. Karbe, L.: Mar. Biol. *12*, 316 (1972)
278. Cohen, M.: In: Probl. Contamination of Man and Environment by Mercury and Cadmium. Comm. Eur. Comm., Luxembourg, July 1973, p. 639 (1974)
279. Kanisawa, M., Schroeder, H.A.: Exp. Mol. Pathol. *10*, 81 (1969)
280. Shigematsu, I., Hagesawa, Y.: In: US-Japan Coop. Sci. Progr., Conf. Effects Environ. Trace Metals Human Health, Honolulu, Hawaii 1971
281. Heath, J.C. et al.: Nature *193*, 592 (1962)
282. Gunn, S.A., Gould, T.C., Anderson, W.A.D.: Proc. Soc. Exp. Biol. Med. *115*, 653 (1964)
283. Roe, F.J.C. et al.: Brit. J. Cancer *18*, 674 (1964)
284. Knorre, v.D.: Arch. Geschwulstforsch. *38*, 257 (1971)
285. Lucis, O.J., Lucis, R., Aterman, L.: Oncology *26*, 53 (1972)
286. Schroeder, H.A., Balassa, J.J., Vinton, W.H.Jr.: J. Nutr. *83*, 239 (1964)
287. Bauchinger, M. et al.: Mutation Res. *40*, 57 (1976)
288. Malik, R.K.: In: Probl. Contamination of Man and Environment by Mercury and Cadmium, Comm. Europ. Comm., Luxembourg, July 1973, p. 647 (1974)
289. Engberg, Å., Bro-Rasmussen, F.: Comm. Europ. Comm. Doc. No. V/F/3799/74e, Luxembourg 1975
290. Chemisches Untersuchungsamt für das Saarland: Schwermetallgehalt in Lebensmitteln. Saarbrücken 1976. Zitiert in Umweltbundesamt, Berichte 4/77, Luftqualitätskriterien für Cd, p. 99
291. Hattingh, W.H.J.: Water South Africa *3*, 104 (1977)
292. Van Wambeke, L.: In: Proc. 1st. Intern. Cadmium Conf. San Francisco, 31. Jan.–2. Feb. 1977, p. 77 (1978)
293. Haberer, L., Normann, S.: Vom Wasser *38*, 157 (1971)
294. OECD, Cadmium and the Environment, Rep. by C.L. Nobbs, Paris 1975
295. Hannah, S.A., Jelus, M., Cohen, J.M.: J. Water Poll. Contr. Fed. *49*, 2297 (1977)
296. Garrigan, G.A.: ibid. *49*, 2380 (1977)
297. Chaney, R.L.: In: Factors involved in Land Application of Agricultural and Municipal Wastes. Natl. Progr. Staff, Agric. Res. Serv., Soil Water Air Serv., US Dept. Agric. Beltsville, Md., p. 97 (1974)
298. Illinois Environmental Protection Agency, Design Criteria for Municipal Utilization on Agricultural Land, Draft Copy, Techn. Pol. WPC-3 (1976)
299. Frank, F., et al.: Prov. Guidelines for Sewage Sludge Utilization on Agricultural Lands. Ad hoc Comm. of Ontario Ministry of Agricult. a. Food, Ont. Min. Environ., March 1976
300. Chaney, R.L., Hornick, S.B.: In: Proc. 1st. Intern. Cadmium Conf. San Francisco, 31. Jan.–2. Feb. 1977, p. 125 (1978)
301. Pahren, H.R. et al.: J. Water Poll. Control Fed. *51*, 2588 (1979)
302. Tirsch, F.S., Baker, J.H., DiGiano, F.A.: ibid. *51*, 2649 (1979)
303. Leeper, G.W.: Managing the Heavy Metals on the Land. New York: Marcel Dekker 1978

Polycyclic Aromatic and Heteroaromatic Hydrocarbons

M. Zander

Rütgerswerke AG
D-4620 Castrop-Rauxel, Federal Republic of Germany

Origin and Formation

Only a limited number of polycyclic aromatic hydrocarbons (PAH) and structurally related hetero-aromatic systems such as anthracene, pyrene or carbazole are industrially produced in pure form. They normally serve as a starting material for the synthesis of dyestuffs, herbicides and pesticides or pharmaceuticals. None of these commercially available pure PAH and hetero-aromatic systems show acute toxic or carcinogenic effects. Coal tar, which is a by-product in the manufacture of metallurgical coke, is the main source for the industrial production of these chemicals. The quantity of coal tar which is co-produced in the coal carbonization process amounts presently to about 16 M tonnes p.a.; three-quarters of this raw material is processed in the existing 127 coal tar refineries throughout the world [1]. The exposure of the environment with these chemicals occuring in the up-grading of coal tar can, however, be neglected in comparison with the exposure with PAH and hetero-aromatic systems from other sources.

PAH are always formed when organic material containing carbon and hydrogen is subjected to temperatures exceeding 700 °C, i.e., in pyrolytic processes and with incomplete combustion. If the starting material also contains hetero-atoms, e.g., oxygen, nitrogen, and sulphur, then hetero-aromatics are formed in addition to PAH. Since pyrolysis or incomplete combustion are processes that take place everywhere in our ecological system, it is not surprising that PAH's and related hetero-aromatics occur everywhere in the environment. However, no reliable data on the total emission of polycyclic organic matter (POM) are available, but Table 1 gives a summary of the estimated benzo[a]pyrene emission in the United States [2]. Global emission of benzo[a]pyrene has been estimated as 5,000 tons/yr [3]. The question as to whether the omnipresence of POM is related solely to civilization or has, in addition, a biogenic origin [4, 5] ist still under discussion. However,

Table 1. Estimated benzo [a] pyrene emission in the United States [2]. The data characterize the situation in the late sixties

Source	Tons/year		
Transportation sources			
Gasoline-powered			
Automobiles	10		
Trucks	12		
Diesel-fuel-powered			
Trucks and buses	0.4		
		22	≈2%
Heat and power generation sources			
Coal			
Hand-stoked residential furnaces	420		
Intermediate units	10		
Coal-fired steam power plants	1		
Oil			
Low-pressure air-atomized and others	2		
Gas	2		
Wood	40		
		475	≈ 38%
Refuse burning			
Enclosed incineration	33		
Open burning			
Forest and agricultural	140		
Vehicle disposal	50		
Coal refuse fires	340		
		563	≈ 45%
Industrial plants			
Cracking units	6		
Asphalt air-blowing	<1		
Coke production	192		
		198	≈16%
		1,260 tons/year	

arguments pointing to an origin conditioned solely by civilization are on the increase [6]. The POM contents of fossil materials, and in particular of mineral oil [7], are not given consideration in this respect.

Although the mechanism of POM formation in combustion and pyrolysis processes is very complex and variable, a relatively clear picture of the overall reaction has outlined [8–10]. POM formation proceeds by free radical mechanism. Radical species containing one, two or many carbon atoms can combine rapidly at the high temperatures (500–800 °C) prevailing in the flame front or under pyrolytic conditions. Highly reactive transient species formed in the first steps of the reaction are stabilized by ring closure, condensation, dehydrogenation, Diels-Alder reactions, ring expansions and other path ways yielding a manifold of polycyclic systems. Pyrosynthesis is obviously a function of many variables, not the least of which is the presence of a chemically reducing atmosphere, common in the center of flames, where radical chain propagation is enhanced, allowing the build-up of complex PAH molecules. Although methane itself can lead to PAH [11] the formation of these large molecules is favoured by the presence of higher-molecular-weight radicals. It has also been shown that specific aromatic systems can serve as precursors for higher-molecular PAH [9, 10]. Figures 1 and 2 give examples for the formation of PAH under pyrolytic conditions according to the work by Badger and Lang et al. [8–10] respectively.

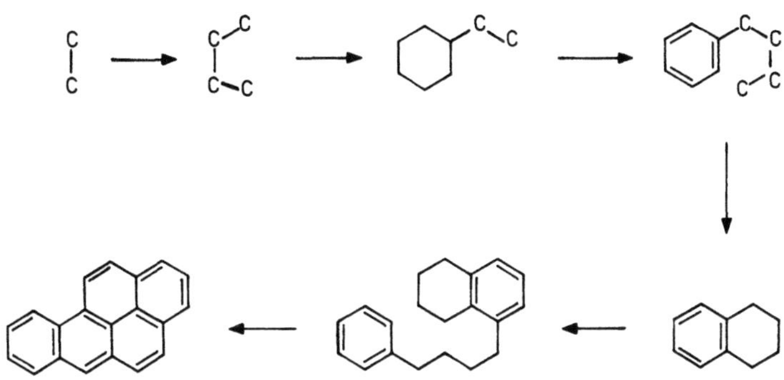

Fig. 1. Mechanism of benzo[a]pyrene formation under pyrolytic conditions [8]

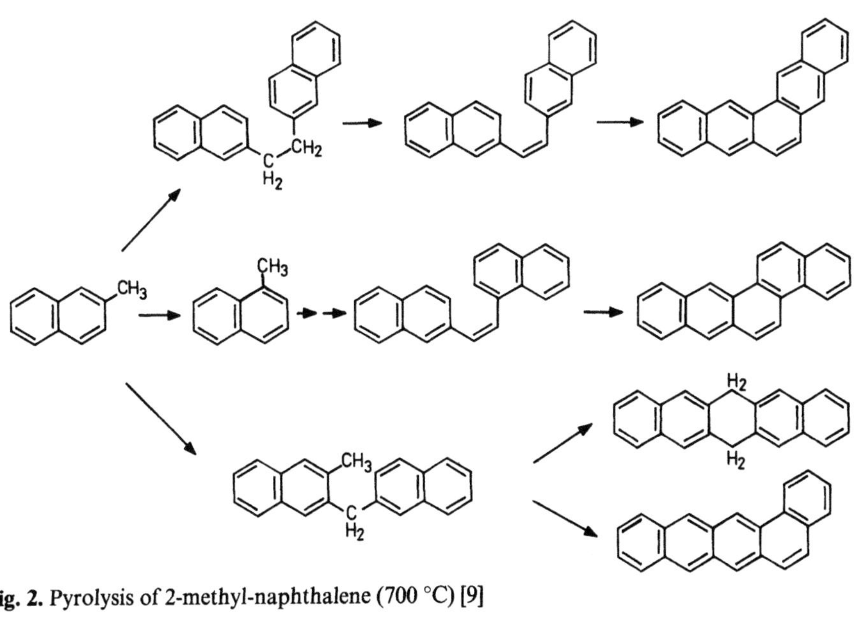

Fig. 2. Pyrolysis of 2-methyl-naphthalene (700 °C) [9]

Chemistry

Nomenclature

Although severals systems of nomenclature for PAH have been applied in the past [12, 13] the IUPAC system [14] has become generally accepted now. Common names for the basic ring systems, however, are still in use. For the construction of more highly condensed PAH the basic system is lettered (a, b, c, etc.) and the added rings numbered in the manner shown in Fig. 3. The IUPAC names for many known PAH are given in the Ring Index [15].

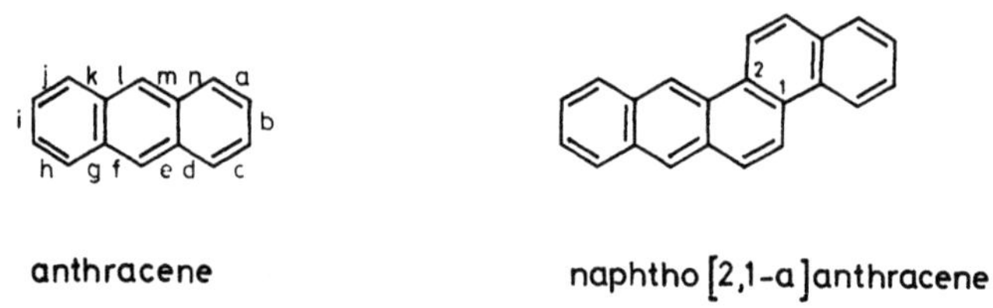

Fig. 3. IUPAC mode of PAH nomenclature

Building Principles

The number of isomeric PAH increases tremendously with the enlargement of molecular size. In the case of systems limited to benzene units, the number of isomers can be obtained very easily on a graph theoretical basis [16]. Table 2 shows several relevant examples. As to the 5-ring systems, all theoretically possible isomers have been described [17]. Of the theoretically possible 6-ring systems, only approx. 45% have been described in the literature to date. The ratio of the number of systems known today to the number of the theoretically possible systems decreases rapidly with increasing number of rings. However, in connection with environmental pollution very high-molecular PAH are less relevant due to their low volatility and solubility.

The host of PAH can be subdivided according to different building prin-

Table 2. Molecular size and number of isomers of PAH [16][a]

Number of benzene units	Number of isomers, PAH
3	3[b]
4	7
5	22
6	82
7	333
10	30,490
12	683,101

[a] Including phenalenyl radical structures
[b] As an example: phenanthrene, anthracene, phenalenyl radical

ciples. – Extremely useful, in particular from the theoretical point of view, is the classification of PAH according to the alternance principle [18]. In alternant PAH the carbon centres can be subdivided in two sets g (starred) and u (unstarred) in such a way, that each g-carbon is directly linked to u-carbons only and vice versa, while in non-alternant PAH C-C-bonds between carbons of the same parity occur. For alternant PAH the pairing theorem is valid while this is not the case with non-alternant PAH [18]. Thus, many differences concerning the physical and chemical properties of alternant and non-alternant PAH can be correlated with the different HOMO-LUMO situation in these two classes of π-electronic systems as given in Fig. 4.

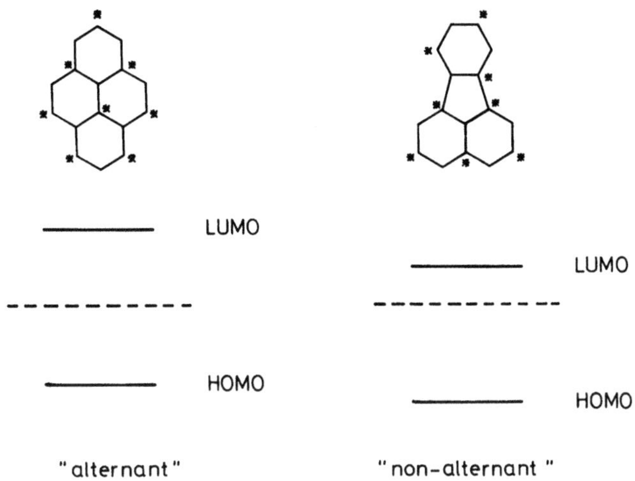

Fig. 4. HOMO/LUMO representation of alternant and non-alternant PAH

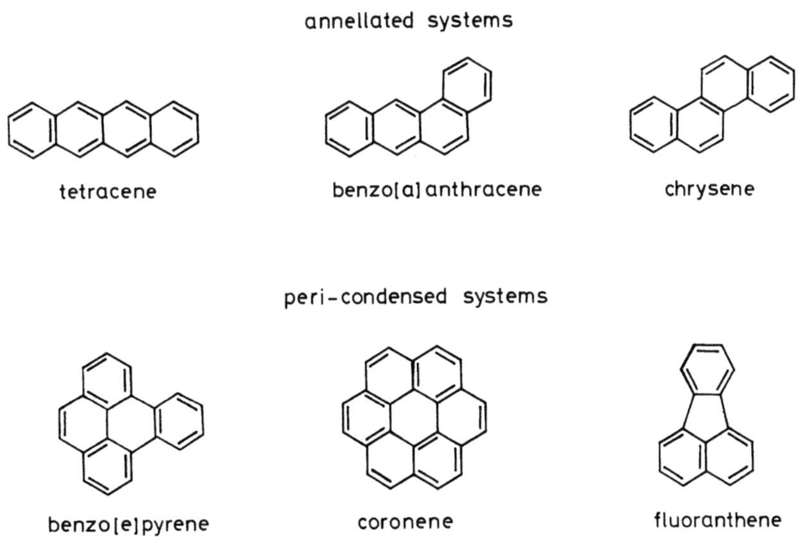

Fig. 5. Examples for annellated and peri-condensed PAH

Most common is the subdivision of the whole amount of PAH into annellated and peri-condensed systems. In annellated systems the tertiary carbons are centers of two inter-linked rings, whereas in peri-condensed systems some of the tertiary carbons are centers of three inter-linked rings. Examples of these two classes of PAH are given in Fig. 5.

Relationships Between Topology, Stability, and Reactivity of PAH

The resonance energy per π-electron ("specific resonance energy") of benzene, naphthalene, anthracene, and tetracene (acenes) is plotted versus the total number of π-electrons of these systems in Fig. 6. The diagram shows a stability

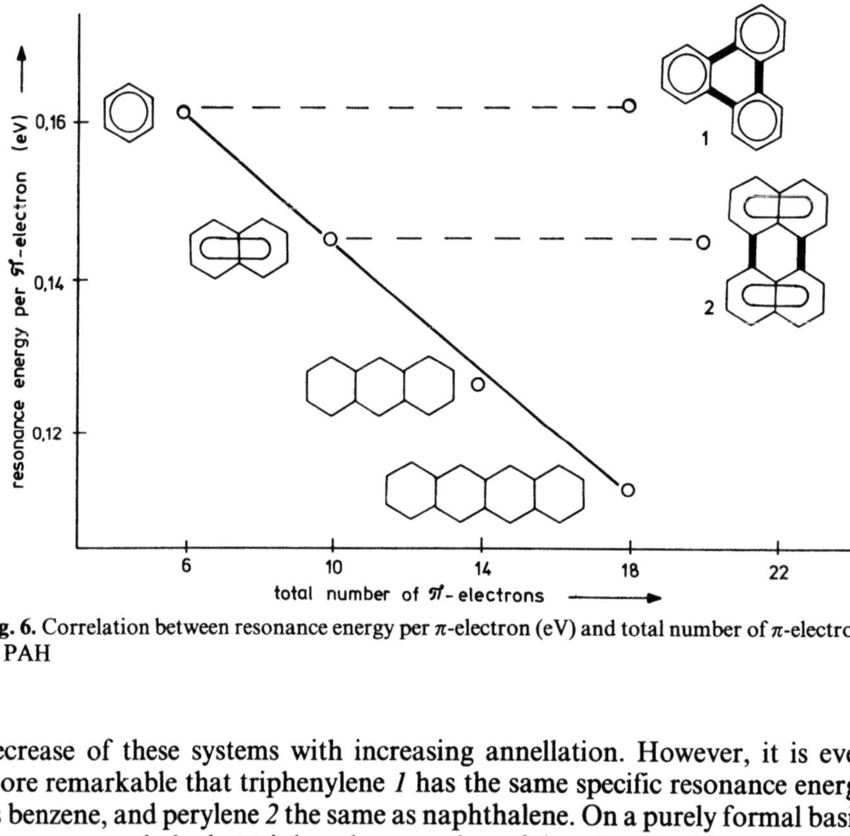

Fig. 6. Correlation between resonance energy per π-electron (eV) and total number of π-electrons of PAH

decrease of these systems with increasing annellation. However, it is even more remarkable that triphenylene *1* has the same specific resonance energy as benzene, and perylene *2* the same as naphthalene. On a purely formal basis, one must conclude that triphenylene consists of three localized benzene units, and perylene, on the other hand, of two localized naphthalene units. This is shown in the formulae, in which the thicker lines denote "quasi-single bonds". The existence of localized π-electron ranges in PAH was first postulated by Clar [19, 20], who developed this concept on the basis of numerous experimental results to build up a comprehensive structural model of PAH.

Polansky and Derflinger [21] developed the quantum chemical verification of this model by using the MO's of benzene and other partial structures recognizeable in PAH (butadienoids, allyloids, and ethylenoids) as bases in the Hückel approximation instead of the usual atom orbitals. In this way, they obtained the "character orders" of these partial structures, the character orders being a measure for the contribution of the bonding orbitals of a partial structure to the bonding orbitals of the entire molecule. According to the analogy principle, the bonding properties of a partial structure equate the more with those of the reference compound (benzene, butadiene etc.) the greater the character order of the partial structure under cconsideration. By way of example, the formula of dibenzo[b,n]perylene (Fig. 7) includes the

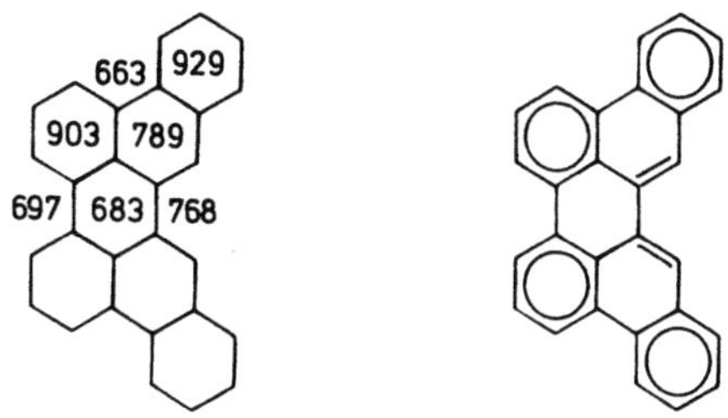

Fig. 7. Dibenzo[b,n]perylene – Representation of bonding properties [19, 21]

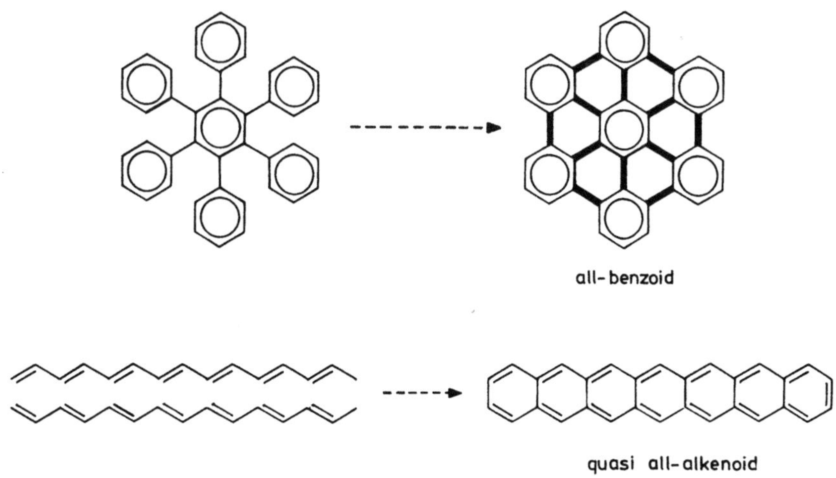

Fig. 8. The "All-benzoid" and quasi "all-alkenoid" building principle of PAH

relevant character orders. The figures in the hexagons are the benzoid character orders, the remaining figures the butadienoid character orders of the cisoid C_4 partial structures. The formula on the right gives the qualitative bonding properties in dibenzo[b,n]perylene according to Clar [19, 20]. The Polansky character orders correlate well with the experimental data such as NMR coupling constants, magnetic susceptibilities, half-wave potentials, and reaction rates [22].

The model indicates that in a series of isomeric PAH stability increases with the number of benzoid partial systems. In fact, PAH such as triphenylene *1* (see Fig. 6), which can also be thought of as purely benzoid partial systems linked by "quasi single bonds", are the most stable PAH known [23]. Such "allbenzoid" PAH can be thought as having been formed by innermolecular ring closures of polyphenyls (Fig. 8). The acenes are found at the other end of the stability scale of PAH (Fig. 8). An infinitely long acene is formally created by linking two polyene chains; it could be described as being an "all-alkenoid". The extremely unstable heptacene, with 7 linear annellated rings, is the most high molecular representative of this structural principle known so far. Heptacene could be designated as being a "quasi all-alkenoid". Between these two extreme cases – the all-benzoid and the quasi all-alkenoid structural principle – numerous different structural principles of differing stability exist; the simple relationship between the number of benzoid partial systems and stability is to be noted in all cases.

The localization energy concept [24] has proved useful to describe reactivity of PAH's quantitatively. The localization energy L_u is the energy – mostly given in units of resonance integral β-required to isolate a π-electron at the centre u from the remaining π-system. The smaller L_u, the greater the relative reaction rate constant of an addition step under consideration. Of the various known reactivity indices, which are a measure for the localization energy [25], Dewar's reactivity number N_u [26] has two striking advantages: the N_u values correlate extremely well with experimental data, and the N_u values can also be calculated extremely easily for systems having a very large number of centres. One disadvantage of Dewar's method lies in the fact that in its simple form it is applicable only to even-alternant π-electron systems.

In principle, the relevant N_u value can be calculated in respect of each carbon centre of a PAH. Accordingly, there is a N_u pattern for each PAH. Initially, no relationships are to be recognized between the N_u pattern and the topology of the systems. Recently, however, it has been shown, that N_u values and Polanskys character orders [21] correspond to some extent in a significative way; since Polansky's approach gives a good description of the bonding properties of PAH an understanding of the relationship between the N_u pattern and the topology of the systems has also been derived [27, 28].

Synthetic Methods

In most cases the synthesis of a definite PAH consists in adding new rings to an easier available starting PAH. If the starting system shall be enlarged by one benzene unit and (6-m) carbons of this benzene unit are already present in the starting system then m carbons have to be added by the synthesis. The most important cases are those for which $m = 0$, 2 or 4 respectively. The case $m = 0$ corresponds to an innermolecular ring closure. The case $m = 2$ can be advantageously verified by Diels-Alder reactions. Several suitable methods exist for the case $m = 4$ one of which is cyclialkylation. The respective examples for the synthesis of PAH by innermolecular ring closure [29, 30], Diels-Alder reaction [31], and cyclialkylation [32] are depicted in Fig. 9–11.

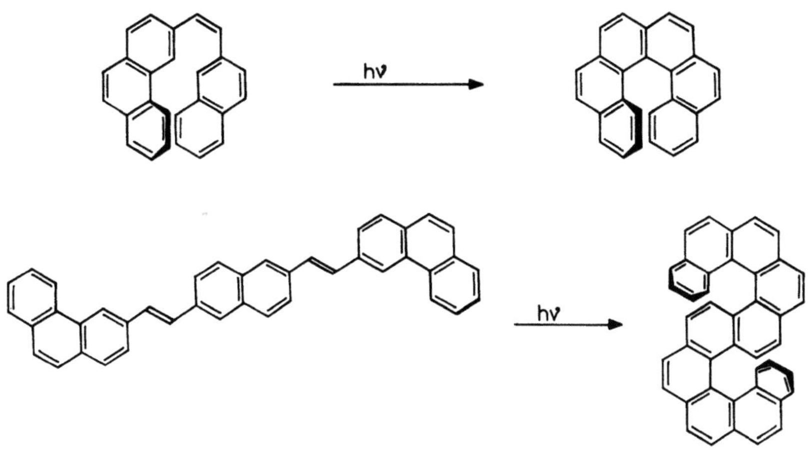

Fig. 9. PAH synthesis by (photochemical) intramolecular ring closure [29, 30]

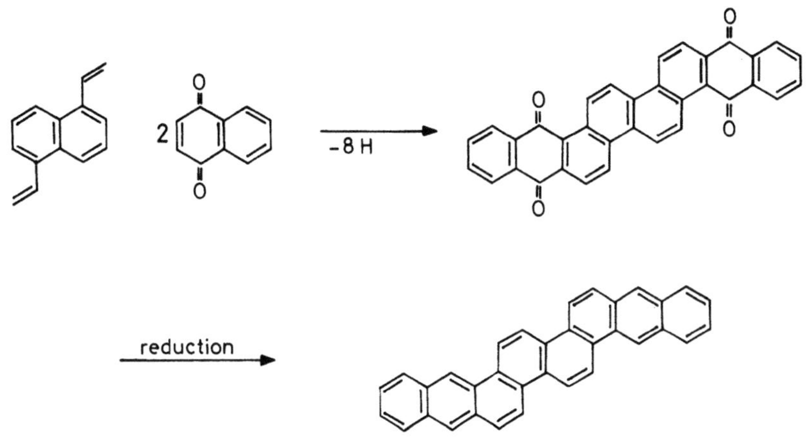

Fig. 10. PAH synthesis by Diels-Alder reaction [31]

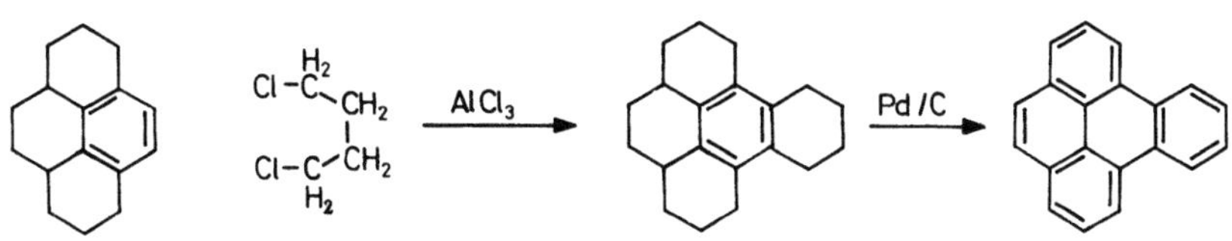

Fig. 11. PAH synthesis by cyclialkylation [32]

The most important methods for the purification of PAHs obtained by synthesis are liquid chromatography, high vacuum sublimation, complexation using electron acceptor compounds like picric acid following decomposition of the adducts and crystallization [33].

The standard reference books on the synthesis of PAHs are still Clar's two volumes [17].

Analytical Methods

Since the environmental analytical chemist is interested, generally, in the analytical determination of numerous compounds present only in low concentrations in complex mixtures separation techniques with high sensitivity have to be employed to gain the required information. Thus, gas liquid chromatography (GLC) and high pressure liquid chromatography (HPLC) are the most widely used methods in PAH analysis.

What is possibly the most complete GLC separation so far of PAHs up to mole masses of approximately 300, was obtained by using 92 m glas capillary columns with polyphenylether sulfones as the stationary phase [34]. Capillary GLC mostly in combination with mass spectrometry has been widely used for the determination of PAHs and related heterocyclic systems in environmental samples. Thus, Lao et al. [35] detected 150 different PAH in city air using the capillary GLC/mass spectrometry combination and Grimmer et al. [36] succeeded in characterizing 150 components of vehicle exhaust gas as PAH, 73 of these could be positively identified by comparison with test substances.

Stationary phases with high selectivity are of continuing interest, in particular for high-temperature GLC. Among thermally stable liquid crystal phases that can be used in combination with mass spectrometry without bleeding, N,N′-Bis-(p-phenylbenzylidene)*a,a′*-bi-p-toluidine can be employed at working temperatures up to 290 °C. Excellent separations of PAH with 4–7 rings have been achieved with this phase [37]. Inorganic salts such as LiCl or $CaCl_2$ comprise another group of selective stationary phases that have been used in high temperature GLC of PAHs [38]. Snowdon [39] reported on the application of column packings consisting of eutectic salt mixtures (KNO_3/$LiNO_3$/$NaNO_3$) on Chromosorb for the separation of PAH and PAH homologues.

The application of HPLC to the analysis of PAHs has been reviewed recently [40]. Since, compared with HPLC, GLC is much easier to apply in quantitative analysis and has an excellent separation capacity, GLC should always be used where this is possible as regards the volatility of the PAH and their thermal stability. On the other hand, HPLC should be employed in the range of very high molecular and/or thermally unstable PAH which cannot, at present, be analysed satisfactorily by GLC. Soluble PAH with mole masses of up to approximately 600 have been separated by HPLC [41, 42]. The advantages of HPLC have also been exploited for the analysis of thermally unstable PAH metabolites [43, 44]. HPLC can be performed with high selectivity not only by using suitable stationary phases [45], but also by the application of selective detection methods, for example fluorescence quenching [46]. On-line coupling of HPLC with UV spectroscopy has proved useful for the identification of PAH in complex mixtures [41].

Although GLC and HPLC are superior to thin layer and paper chromatography regarding the separation capacity these methods are useful for obtaining rapid information on the concentration of distinct PAHs in complex samples and are therefore still widely applied in water analysis [47].

Compared with the chromatographic analytical methods spectroscopic methods are less important in environmental PAH analysis, although fluorimetry [48] and phosphorimetry [49] because of their excellent sensitivity and selectivity proved useful in special cases.

Transport Behaviour in the Environment

Air

PAH in the atmosphere are predominantly associated with particulate matter, especially soot [50]. It has been speculated that the PAH appear to be adsorbed primarily on the surface of soot by hydrogen bonding [51]. The benzene soluble portion of this material is approximately 10% by weight but the PAH component is much smaller than that.

Particle size is the physical property with the greatest influence on the behaviour of PAH-containing aerosols. Generally, the particle size spectrum of atmospheric aerosols extends from less than 0.01 μm to greater than 10 μm, but PAHs appear to be associated largely with particles less than 5 μm in diameter. There is considerable variation in the size-concentration distribution of particles with location in space and time, but the large-particle portion of the spectrum (greater than 0.1 μm) often tends, on the average, to follow a power law form:

$$\frac{dN}{dD_\rho} = \text{const.}\ \varnothing\ D_p^{-4}$$

where N is the number of particles, D_p the particle diameter, $\varnothing$ the volume fraction and the constant is approximately 0.40. Although there is no reliable theoretical explanation, as yet, for the apparent regularity observed, some speculation has been reported [52, 53]. Particle shape [54] and density are important factors which determine the rate of aerosol deposition. For urban aerosols a density range from 1.8 to 2.1 g/cm^3 has been reported [55].

Particulate PAH released in the atmosphere in one location may be transported to very distant areas [56], whereby the local meteorology as well as the wind fields must be taken into account.

In connection with the removal of particulate PAH from the atmosphere deposition of large particles by gravitational settling is important. For some time, it is possible to estimate roughly the deposition rate of particles on an obstacle if the air-flow field near it is known [57, 58] and some limited data have been reported for deposition rates on vegetation [59–61]. The development of rain clouds, on the other hand, influences aerosols containing PAH significantly, whereby the size distribution and the chemical composition as a function of size may be modified. When precipitation begins to fall from clouds, smaller particles will be deposited as in the case of the dry scavenging mechanism. This washout is believed to be significant in removing PAH from the atmosphere [50].

Water

Although the solubility of pure PAH in water is extremely low (for example benzo[a]pyrene $4 \cdot 10^{-3}$ mg/l, dibenzo[a,h]anthracene $5 \cdot 10^{-4}$ mg/l) these compounds can be solubilized by other organic substances in particular detergents [62]. Besides that PAH are capable to form associates with colloids present in water and in this form can be transported through natural occuring water. Thus, PAH have been detected in tissues of organisms from marine habitats far removed from intensive human activity [63].

Chemical and Photochemical Reactions

PAH can undergo various types of ground state reactions such as electrophilic and nucleophilic substitution, 1,2- and 1,4-cycloaddition reactions, oxydation, hydrogenation and intra- as well as intermolecular condensation reactions. The reactivity behaviour of PAH has been comprehensively reviewed by Clar [17]. Most of these reactions are known for a long time, but only recently it has been well documented that various PAH and structurally related hetero-aromatic systems can also undergo Lewis acid catalyzed innermolecular skeleton rearrangements [64] this being a type of PAH transformation that can complicate the synthesis of polycyclics.

Since most of the reactions of PAH are addition reactions in the rate determining step or true addition reactions the reactivity behaviour can be quantitatively described by using the localization energy concept (see section on p. 116). In this context Dewar's pertubational molecular orbital method [26] and Fukui's frontier orbital method [65] are extremely useful to provide the organic chemist with the relevant data. Since the Bell-Evans-Polanyi principle is valid for various reaction types of π-electronic systems free-energy relationships play a dominant role in PAH chemistry.

For the environmental chemist reactions of electronically excited PAH (photochemical reactions) are of particular interest because the fate of PAH under environmental conditions is determined to a high degree by its photochemical behaviour. Tricyclic or larger PAH and related heterocyclic systems have strong UV absorption at wavelenghts longer than 300 nm (present in solar radiation) and most are readily photooxidized. Photooxidation is probably one of the most important proccesses in the removal of polycyclics from the environment.

Photooxidation of PAH in solution involves energy transfer from the triplet state of the aromatic system, producing singlet oxygen, which reacts with the compound, yielding its peroxide. Normally, for endoperoxide formation two anthracene 9,10-like positions are required (Fig. 12). Photolysis or pyrolysis of PAH endoperoxides yields a variety of reaction products via dealkylation and ring cleavage [66] as shown by the possible pyrolysis products of 9,10-dimethyl-anthracene peroxide in Fig. 12. It is important to note that quinones can also be produced when no endoperoxide can be formed for

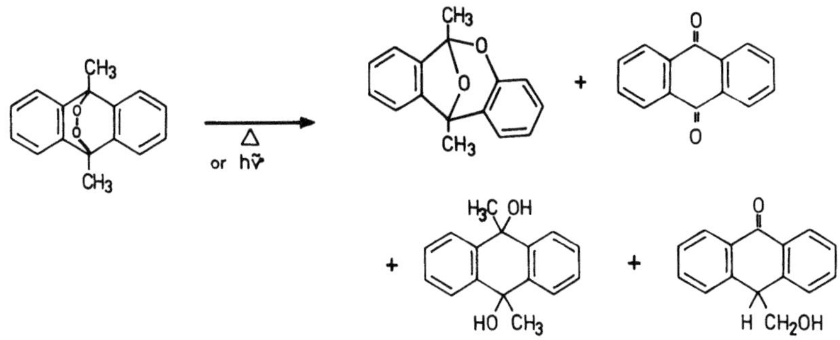

Fig. 12. Pyrolysis (photolysis) of 9,10-dimethyl-anthracene peroxide [66]

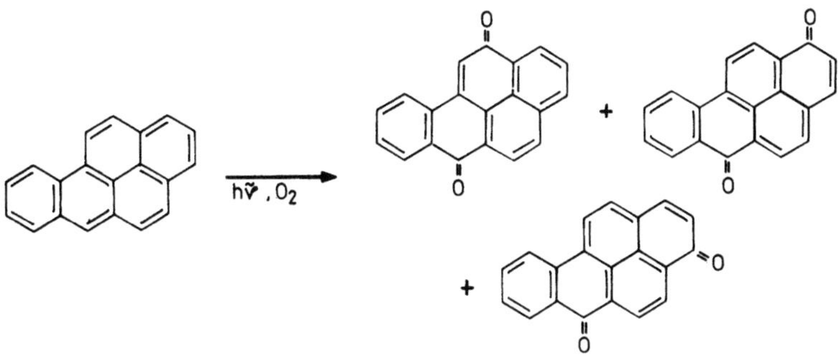

Fig. 13. Quinone formation from benzo[a]pyrene by irridation

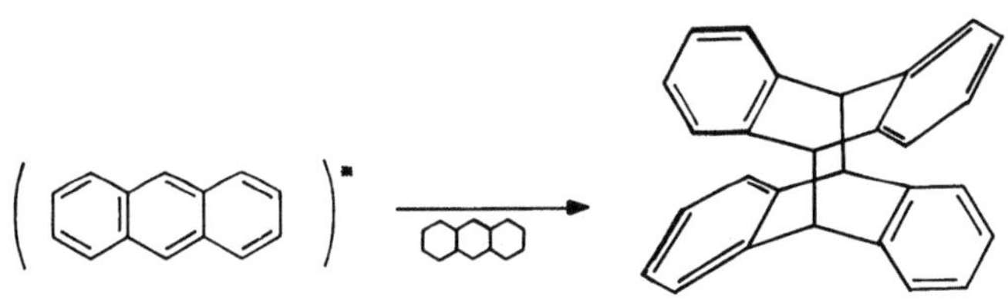

Fig. 14. Anthracene photodimer formation

steric reasons. Thus, benzo[a]pyrene yields a mixture of three quinones by irridation in solution (Fig. 13).

It has been speculated that photooxidation of PAH in the adsorbed state does not proceed *via* endoperoxides [67]. Although only few reports of photooxidation of adsorbed PAH have appeared it is quite evident that PAH are photoxidized with higher rates in the adsorbed state than in solution [50]. Since POM in the environment is mostly associated with particulate matter photooxidation studies on PAH in the adsorbed state [68] have gained particular relevance for the environmental chemist. Half-lives for photooxidation of PAH in particular of benzo[a]pyrene under various conditions have been estimated and on average are less than one day.

Non-substituted acenes form easily photodimers by reaction of one PAH molecule excited to its singlet state with another PAH molecule in its ground state (Fig. 14). The primary step of this reaction is the formation of an excimer; since the reaction proceeds via the singlet state excited molecule dimerization does not take place under conditions that enhance intersystem crossing into the triplet manifold.

Polansky [69] demonstrated how within the pars orbital concept the character orders of partial structures of electronically excited molecules can be derived. Here, the character orders are so defined that they describe the analogy between the reference compound in the ground state and the relevant partial structure in the electronically excited molecule. The pars orbital concept can be of use in the interpretation and prediction of the photochemical behaviour of PAH. However, the method has so far been applied only in the simple HMO fashion as it is restricted to PAH whose first absorption transition is the ^1L_a transition (Clar's para band).

For the environmental chemist the behaviour of PAH towards agents commonly applied for the purification of drinking water is of interest. While PAH dissolved in water will be oxidized by ozone, chlorinating agents mostly yield chlorine substituted PAH besides oxidation products. Benzo[a]pyrene, for example, during chlorination is transformed into a variety of products, none of which is carcinogenic [70, 71].

Metabolism

The metabolism of PAH in terrestrial mammals has been extensively studied, above all to achieve an understanding how PAH act as carcinogens. Although arene epoxides as the primary metabolites of PAH have been postulated already in 1950 [72] this was not definetively proven before 1968 [73, 74]. The mechanism of the carcinogenic activity of benzo[a]pyrene according to our present knowledge is summarized in Fig. 15. With the participation of cytochrome P 450 which is present in the endoplasmic reticulum of the cell benzo[a]pyrene is oxidized yielding the arene oxide *1*. Thus, in contrast with older theories [75] not the so-called "K-region" (K for Krebs) is attacked but a ring belonging to the "bay-region" of the PAH. During the next step the enzyme epoxide-hydratase [76] transforms *1* into the trans-dihydro diol *2*, which then again undergoes an epoxidation with the participation of cytochrome P 450 yielding the "ultimate carcinogen" *3*. The trans-diol epoxide *3* exists in two stereo isomers *4* and *5* and each of them can be separated into two pairs of enantio isomers (+)−*4*, (−)−*4*, (+)−*5*, (−)−*5*. It was found only recently that from these four molecules (+)−*5* exhibits a high carcinogenic activity in experiments with mice while the other three compounds and benzo[a]pyrene itself have minor or no activity at all [77]. The electrophile *3* ([+]−*5*) reacts with nucleophilic bases of the DNA, guanine being the favoured base for the attack of the diol epoxide.

By using the simple PMO method [26] it could be shown that the formation of carbonium ions is energetically favoured in the bay-region of PAHs [78].

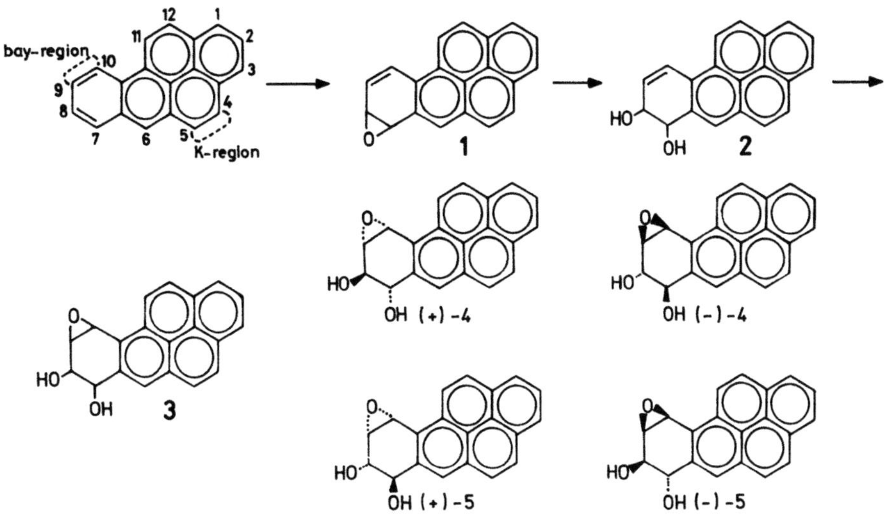

Fig. 15. Metabolic activation of benzo[a]pyrene to the "ultimate carcinogen"

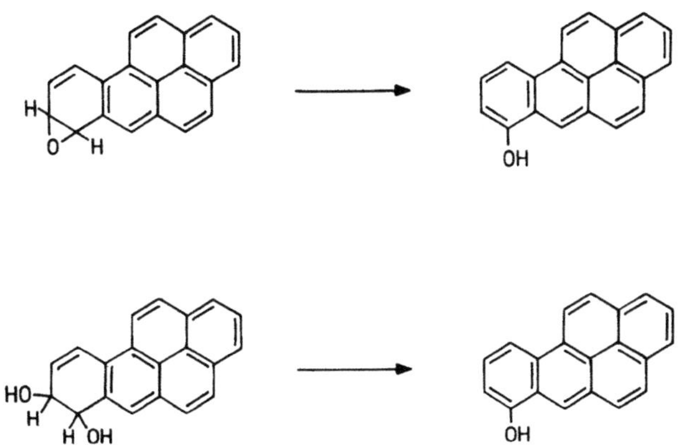

Fig. 16. Metabolic formation of benzo[a]pyrene phenols

Consequently, an epoxide ring in the bay-region of a PAH should be the critical structure element of the ultimate carcinogen. This has been definetively proven with several modell substances in mutagenecity screening tests and animal experiments [79–81].

Urinary metabolites of PAH are usually phenols that arise by isomerization of arene oxides (NIH-shift) [73, 74] or by elimination of water from trans-dihydro diols (Fig. 16). But none of these mechanisms has been definitely proven. Moreover only one of the two possible isomeric phenols which in principle can be formed from the arene epoxide or trans-dihydro diol has been detected. Frequently transdihydro diols in vivo will be conjugated to sulphuric acid and then excreted as the monosulphates [82]. Also the excretion of phenols and dihydro diols conjugated to glucoronic acid was observed [83, 84].

Quinones are formed in small amounts as metabolites of PAH. Benzo[a]-anthracene-7,12-quinone and the 1,6-, 3,6-, and 6,12-quinone of benzo[a]pyrene have been observed [85], but the mechanism of quinone formation needs additional clarification.

The formation of 6-hydroxymethyl-benzo[a]pyrene from benzo[a]pyrene in rat liver microsomes was described by several authors [86–89]. However, neither the origin of the C_1-fragment nor the enzyme that catalyzes the C_1-transfer are known so far (Fig. 17).

Fig. 17. C_1-transfer during benzo[a]pyrene metabolism

With methylated PAH oxidation of the methyl groups occurs independently of the ring oxidation. In vivo oxidation of 7,12-dimethyl-benzo[a]anthracene to 7-hydroxymethyl-12-methyl-benzo[a]anthracene and also of 3-methyl-cholanthrene to 3-hydroxymethyl-cholanthrene was observed [90]. Normally the oxidation proceeds by formation of carboxyclic acids that occur as urinary metabolites (Fig. 18).

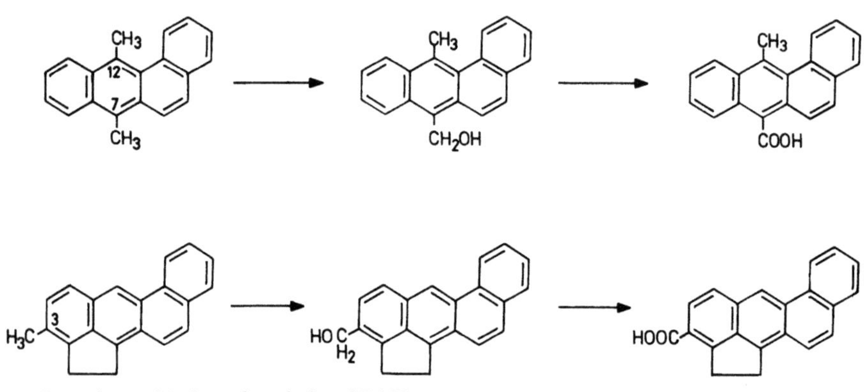

Fig. 18. In vivo oxidation of methylated PAH

The conjugation of arene oxides with glutathione is considered to be a detoxification reaction of PAH. This reaction is catalyzed by the glutathione-S-epoxide transferase system present in the cytoplasma (Fig. 19) and consisting of several isoenzymes [91–94]. K region epoxides are favourably conjugated with glutathione [95].

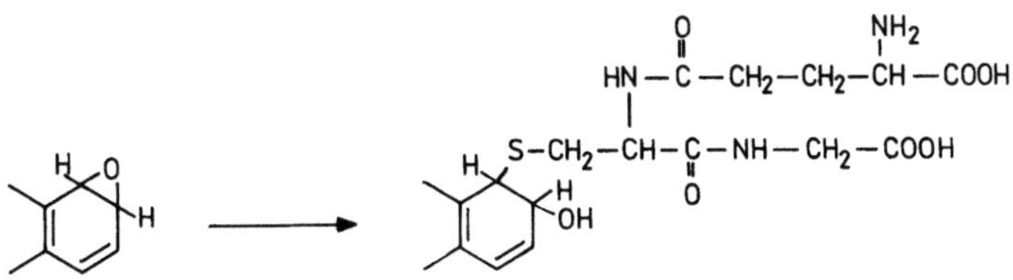

Fig. 19. In vivo conjugation of arene oxides with gluthathione

The conjugation of PAH with proteines has been well documented. The rate of conjugation decreases in the order epoxides, phenols, dihydro diols. There is some evidence that K region epoxides are favourably conjugated with proteines [96] and this obviously is a parallel to the reaction with glutathione. It is well established that the reaction of PAH metabolites with proteines predominates the conjugation to nucleic acids [97] this latter reaction being the ultimate step in the malignant cell transformation.

Biodegradation

Bacteria can oxidize PAH that range in size from benzene to benzo[a]pyrene but for more highly condensed PAH this is not clear [98]. Benzo[a]pyrene is oxidized by microorganisms, for example Beijerinckia sp., to form 7,8-dihydroxy-7,8-dihydro-benzo[a]pyrene and 9,10-dihydroxy-9,10-dihydrobenzo[a]pyrene with further more extensive degradation [99] (Fig. 20). Microbial degradation is a major mechanism for compound removal in sediments and bacteria capable of degrading hetero-aromatic systems have also been isolated. Generally, benzo[a]pyrene metabolizing cultures are microorganisms with long exposure to benzo[a]pyrene [100]. Growth rates of bacteria on PAH are directly related to the solubilities of the PAH.

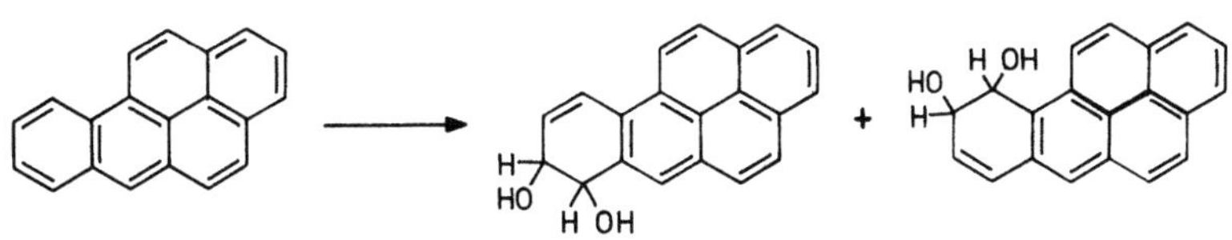

Fig. 20. Oxidation of benzo[a]pyrene by microorganisms [99]

The Overall Environmental Fate of PAH

The overall environmental fate of PAH depends on several factors. – By way of example, Table 3 gives half-lives of dissolved benzo[a]pyrene for individual transformation or removal processes. Half-life for photolysis is two orders of magnitude smaller than the half-lifes of the other transformation pathways.

Table 3. Half-lives of dissolved benzo [a] pyrene for individual transformation and removal processes [100]

Process	River	Pond (eutrophic)	Lake (eutrophic)	Lake (oligotrophic) (h)
Photolysis[a]	3.0	7.5	7.5	1.5
Oxidation	>340	>340	>340	>340
Volatilization	140	350	700	700
All processes, except dilution	2.9	7.3	7.4	1.5

[a] Summer sunlight

Sorption half-lives have not been measured, but they are probably at least 100 times smaller than the half-lives for photolysis. The degradation of PAH in the atmosphere by photooxidation has been discussed in an earlier section (see section on chemical and photochemical reactions).

The rate of the various transformation and removal processes which lead to a reduction of PAH and hetero-aromatic systems in the environment is considered to depend on the physical and chemical properties of the individual compounds. Solubility and adsorbtivity are the most important physical properties in this context, while amongst the chemical properties photochemical reactivity is particularly relevant. These properties depend on size and topology of the systems and it is by no means surprising, that some PAH are readily destroyed under environmental conditions while others are remarkably stable. Moreover, it is interesting to note, that special types of PAH are suspected to occur in interstellar matter [101]. – Several mechanisms involving PAH under environmental conditions may cause reactive species to be delivered to genetic and other biological material, but this aspect needs further clarification.

Since benzo[a]pyrene and various other PAH are lipophilic these compounds can be bioaccumulated to high levels. For example, high accumulation of benzo-[a]pyrene in clams was reported (24 h exposure to 0.0305 ppm ^{14}C-benzo[a]pyrene in seawater) the tissue concentration of 7.2 ppm indicating a 236 fold bioaccumulation over exposure water concentration. This result is consistent with the observed slow release of benzo[a]pyrene : 29% remained after 10 days [63]. Lower bioaccumulation factors were observed with other species.

Concentrations of benzo[a]pyrene in air, water, soil, sediments, aquatic organisms and food are listed in Table 4 and give a rough idea of the overall contamination of the environment with PAH.

Toxicology

In general, the acute toxicity of PAH and related heteroaromatic systems is low and is restricted to necrotisation of the glandulae suprarenales.

Table 4. Concentrations of benzo [a] pyrene in the environment and biota

Occurrence		Concentration	
Air (ng/m^3)			
Sydney (Australia)	year: 1962/63	winter: 8	summer: 0.8
Kopenhagen (Denmark)	1956	17	5
Salford (Great Britain)	1953		210
Paris (France)	1958	300–500	
Toronto (Canada)	1961/62	5.4	6.4
Taschkent (Soviet Union)	1965	110	
Los Angeles (USA)	1971/72	ca. 1,3	
Berlin (Germany)	1970	18	
Water (ng/1)			
tap-water		2.5–9	
ground water		1.0–10	
rain water		2.2–7.3	
Surface water		130–350 (Thames river)	
Soil (μg/kg)			
beach (Baltic sea)		$8 \cdot 10^{-4}$	
forest (Baltic sea)		$3.5 \cdot 10^{-2}$	
near high-way (Germany)		3.0	
Sediments (μg/kg)			
Greenland (depth 0.2 m)		5	
Italy (highly industrialized area, depth 15–45 m)		1,000–3,000	
French mediterranean coast (depth 14 m)		400	
Bodensee (Germany)		max. 1,600	
Aquatic organisms (μg/kg)			
Oysters (French Coast)		1–70	
Mussels (France)		2–30	
Codfish and shellfish (Greenland)		16–60	
Plankton (Greenland)		5	
Algae (Italy)		2	
Food (μg/kg)			
Smoked meat and sausages		0.2–0.9	
Smoked fish		0.1–9.8	
Kale		4 – 16	

Various PAH, heterocyclics and derivatives have been examined as for carcinogenicity in short-term screening tests such as the Ames test [102–105], cell transformation test [106] and sebaceous gland supression test [107, 108]. Although a 90–95% correlation between the results from mutagenicity screening tests and carcinogenicity observed in animal experiments was claimed [109] it becomes increasingly evident that these in vitro tests cannot replace in vivo experiments with animals if reliable information on carcinogenic effects is required.

It is well established that certain PAH and related heterocyclics produce skin cancer in test animals and there are various hints that the same might be true for lung cancer [110]. Some proven polycyclic carcinogens are listed in Fig. 21. Although cocarcinogens such as long chain alkanes [111] play an

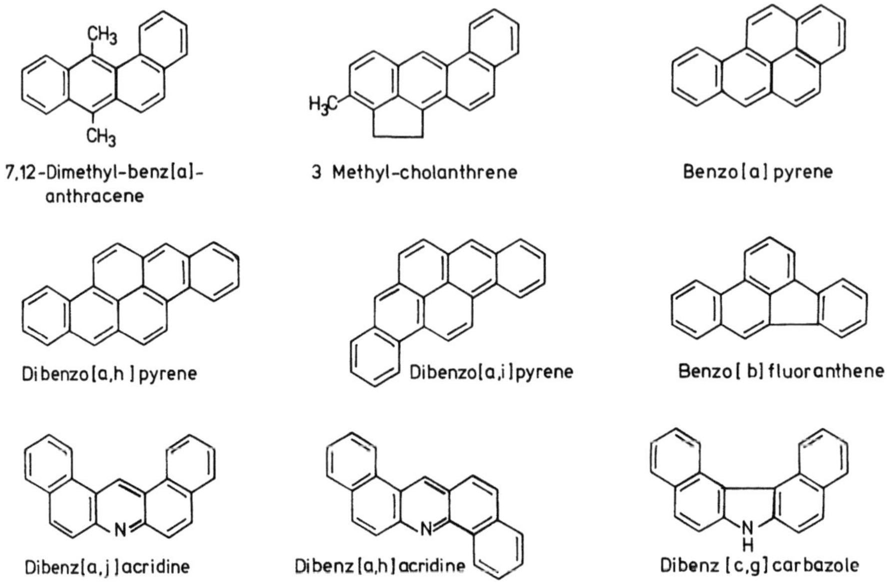

Fig. 21. Some proven carcinogenic PAH and related heterocyclic systems

important role in PAH carcinogenesis the effects of the PAH's itself probably exhibit an additive behaviour [2].

A carcinogenic risk to man from chemicals in the environment can in principle evaluated only from epidemiological studies [2, 50] or under certain circumstances from case studies. Since man is exposed to a host of chemicals in the environment unambigous cause/effect relations cannot be deduced from epidemiological studies for specific compounds and even not for a class of compounds. Similarities of metabolism of benzo[a]pyrene in human and mouse cells cultured in vitro have been reported. However, the relevance of this finding for evaluating the risk to man cannot yet be assessed [110].

References

1. Collin, G., Mildenberg, R.: Chem. and Ind. *1978*, 567
2. Grimmer, G.: in: Luftqualitätskriterien für ausgewählte polycyclische aromatische Kohlenwasserstoffe, ed. Umweltbundesamt (FRG), 1979
3. Suess, M.J.: Sci. Total Environ. *6*, 239 (1976)
4. Gräf, W.: Med. Klinik, *15*, 561 (1965)
5. Borneff, J., Selenka, F., Kunte, H., Maximos, A.: Environ. Res. *2*, 22 (1968)

6. Blumer, M.: Sci. American 1976, März, p. 34
7. Arcos I.C., Argus, M.F.: Chemical Induction of Cancer, vol. IIA. Academic Press, New York – London 1974
8. Badger, G.M.: Proc. Roy. Soc. *1956,* 87
9. Lang, K.F., Buffleb, H., Zander, M.: *Erdöl/Kohle 16,* 944 (1963)
10. Zander, M.: Intern. J. Environ. Anal. Chem. *4,* 109 (1975)
11. Burrows, I.E., Lindsey, A.J.: Chem. Ind. *1961* 1395
12. Stelzner, R., Kuh, A.: Lit. Reg. Org. Chem. *3,* 21 (1921)
13. Patterson, A.M.: J. Amer. Chem. Soc. *47,* 543 (1925)
14. IUPAC-Rules 1957
15. The Ring Index. Amer. Chem. Soc., Washington 1960
16. Lunnon, W.F.: in: Graph Theory and Computing. Academic Press, New York – London 1972 p. 87
17. Clar, E.: Polycyclic Hydrocarbons. Academic Press New York-London 1964
18. Coulson, C.A., Rushbrooke, G.S.: Proc. Cambridge Phil. Soc. *36,* 193 (1940)
19. Clar, E., Zander, M.: J. Chem. Soc. (London) *1958,* 1861
20. Clar, E.: The Aromatic Sextet. John Wiley and Sons, New York 1972
21. Polansky, O.E., Derflinger, G.: Int. J. Quantum Chem. *1,* 379 (1967)
22. Fratev, F., Polansky, O.E., Zander, M.: Z. Naturforsch. *30 a,* 1704 (1975)
23. Zander, M.: Angew. Chem. *72,* 513 (1960)
24. Brown, R.D.: Quart. Revs. *6,* 63 (1952)
25. Streitwieser, A.: Molecular Orbital Theory for Organic Chemists. John Wiley and Sons, New York 1961
26. Dewar, M.S.J., Dougherty, R.C.: The PMO Theory of Organic Chemistry. Plenum Press, New York 1975
27. Zander, M.: Z. Naturforsch. *33 a,* 1395 (1978)
28. Zander, M.: Z. Naturforsch. *34 a,* 521 (1979)
29. Moradpour, A., Nicoud, J.F., Balavoine, G., Kagan, H., Tsouraris, G.: J. Amer. Chem. Soc. *93,* 2353 (1971)
30. Laarhoven, W.H., Cuppen, Th.J.M.: Tetrahedron Lett. *1971,* 163
31. Zander, M., Franke, W.H.: Chem. Ber. *107,* 727 (1974)
32. Hausigk, D.: Chem. Ber. *101,* 473 (1968)
33. Zander, M.: in: Houben-Weyl (Meth. d. org. Chem.), G. Thieme Verlag, Stuttgart, in press
34. Borwitzky, H., Schomburg, G.: J. Chromatogr. *170,* 99 (1979)
35. Lao, R.C., Oja, H., Thomas, R.S., Monkman, J.L.: Symp. Rec. Adv. Analyt. Chem. Pollutants, Halifax, N.S., Aug. 23–25, 1972
36. Grimmer, G., Böhnke, H., Glaser, A.: Erdöl/Kohle *30,* 411 (1977)
37. Janini, G.M., Muschik, G.M., Schroer, I.A., Zielinski, W.L.: Analyt. Chem. *48,* 1879 (1976)
38. Sauerland, H.D., Zander, M.: Erdöl/Kohle *19,* 502 (1966)
39. Snowdon, L.R.: Analytic. Chem. *50,* 379 (1978)
40. Thoms, R., Zander, M.: Erdöl/Kohle *30,* 403 (1977)
41. Thoms, R., Zander, M.: Fresenius Z. Analyt. Chem. *282,* 443 (1976)
42. Blümer, G.-P., Zander, M.: DGMK-Compendium (Suppl. to Erdöl/Kohle) 1978/79, p. 1472
43. Soedigdo, S., Angus, W.W., Flesher, J.W.: Analyt. Biochem. *67,* 684 (1975)
44. Dark, W.A.: Internat. Laboratory *1975,* 44
45. Blümer, G.-P., Zander, M.: Fresenius Z. Analyt. Chem. *288,* 277 (1977)
46. Blümer, G.-P., Zander, M.: ibid., *296,* 409 (1979)
47. Borneff, J., Kunte, H.: Arch. Hyg. *153,* 220 (1969)
48. Bartle, K.D., Collin, G., Stadelhofer, J.W., Zander, M.: J. Chem. Techn. Biotechn. *29,* 531 (1979)
49. Zander, M.: Phosphorimetry – The Application of Phosphorescence to the Analysis of Organic Compounds. Academic Press, New York 1968
50. Particulate Polycyclic Organic Matter, National Academy of Sciences, Washington, D.C. 1972
51. Went, F.W.: Proc. Nat. Acad. Sci. USA *46,* 212 (1960)
52. Hidy, G.M.: Proc. 3rd Univers. Rochester Conf., Environm. Toxicol., June, 1970

53. Junge, C.E.: J. Atmos. Sci. *26,* 603 (1969)
54. Kotrappa, P.: Proc. 3rd Univers. Rochester Conf. Environm. Toxicol. June, 1970
55. Corn, M., Montgomery, T.L., Esmen, N.A.: Environ. Sci. Techn. *5,* 155 (1971)
56. Krstulovic, A.M., Rosie, D.M., Brown, P.R.: Internat. Laboratory *1977,* 11
57. Fuks, N.A.: The Mechanics of Aerosols. Macmillan, New York 1964
58. Hidy, G.M., Brock, J.R.: The Dynamics of Aerocolloidal Systems. Pergamon Press, New York 1970
59. Chamberlain, A.C.: Atmos. Environ. *4,* 57 (1970)
60. Neuberger, H., Hosier, C.L., Kocmond: in: Tromp, S.W. and Weihe, W.H. (Eds.) Biometeorology, Vol. 2. Pergamon Press, Oxford 1967, p. 693
61. Rosinski, J., Nagamoto, C.T.: Kolloid-Z. *204,* 78 (1966)
62. Andelmann, J.B., Suess, J.N.: Bull. World Health Org. *43,* 479 (1970)
63. Neff, J.M., Cox, B.A., Dixit, D., Anderson, J.W.: Marine Biol. *38,* 279 (1976)
64. Blümer, G.-P., Gundermann, K.-D., Zander, M.: Chem. Ber. 110, 2005 (1977)
65. Fukui, K., Yonezawa, T., Shingu, H.: J. Chem. Phys. *20,* 722 (1952)
66. Southern, P.F., Waters, W.A.: J. Chem. Soc. (London) *1960,* 4340
67. Kortüm, G., Braun, W.: Liebigs Ann. Chem. *632,* 104 (1960)
68. Inscoe, M.N.: Analytic. Chem. *36,* 2505 (1964)
69. Polansky, O.E.: Z. Naturforsch. *29 a,* 529 (1974)
70. Gräf, W., Nothoff, G.: Arch. Hyg. *147,* 135 (1965)
71. Reichert, J.K.: Arch. Hyg. *152,* 277 (1968)
72. Boyland, E.: Biochem. Soc. Symp. *5,* 40 (1950)
73. Jerina, D.M., Daly, J.W., Witkop, B., Zaltzman-Nirenberg, B., Udenfriend, S.: J. Amer. Chem. Soc. *90,* 6525 (1968)
74. Daly, J.W., Jerina, D.M., Witkop, B.: Experientia *28,* 1129 (1972)
75. Pullmann, A., Pullmann, B.: Adv. Cancer Res. *3,* 117 (1955)
76. Oesch, F.: Xenobiotica *3,* 305 (1973)
77. Buening, M.K.: Proc. Nat. Acad. Sci. USA *75,* 5358 (1978)
78. Jerina, D.M. et al.: in: de Serres, F.J. (Ed.) In Vitro Metabolic Activation in Mutagenesis Testing. Elsevier, Amsterdam 1976
79. Wood, A.W. et al.: Proc. Nat. Acad. Sci. USA *74,* 3176 (1977)
80. Wislocki, P.G. et al.: Cancer Res. *38,* 693 (1978)
81. Slaga, T.J. et al.: Cancer Res. *38,* 1699 (1978)
82. Boyland, E., Sims, P.: Biochem. J. *84,* 583 (1962)
83. Berenbom, M., Young, L.: Biochem. J. *49,* 165 (1951)
84. Sims, P.: Biochem. J. *73,* 389 (1959)
85. Freundenthal, R.I., Hundley, S.G., Cattaneo, S.M.: in: Jones, P.W. and Freundenthal, R.I. (Eds.) Carcinogenesis, Vol. 3. Raven Press, New York 1978
86. Flesher, J.W., Sydnor, K.L.: Int. J. Cancer *11,* 433 (1973)
87. Sloane, N.H.: Biochim. Biophys. Acta *107,* 599 (1965)
88. Sloane, N.H., Davis, T.K.: Arch. Biochem. Biophys. *163,* 46 (1974)
89. Sloane, N.H., Uzgiris, V.I., Seeley, D.: in: Jones, P.W. and Freundenthal, R.I. (Eds.) Carcinogenesis, Vol. 3. Raven Press, New York 1978
90. Sims, P.: Biochem. J. *98,* 215 (1966)
91. Habig, W.H., Pabst, M.J., Jacoby, W.B.: J. Biol. Chem. *249,* 7130 (1974)
92. Gelboin, H.V. et al.: in: Screening tests in chemical carcinogenesis, IARC Sci. Publ. No. 12, Lyon 1976
93. Boyland, E., Williams, K.: Biochem. J. *94,* 190 (1965)
94. Pabst, M.J., Habig, W.H., Jacoby, W.B.: Biochem. Biophys. Res. Commun. *52,* 1123 (1973)
95. Booth, J., Sims, P.: Biochem. Pharmacol. *23,* 2547 (1974)
96. Raha, C.R., Gallagher, C.H., Shubik, P.: Proc. Soc. Exp. Biol. Med. *143,* 531 (1973)
97. Sims, P., Grover, P.L.: Adv. Canc. Res. *20,* 165 (1974)
98. Gibson, D.T.: Science *161,* 1093 (1968); Groenewegen, Ir.D., Stolp, H.: Erdöl/Kohle *28,* 206 (1975)
99. Shabad, L.M.: Z. Krebsforsch. *70,* 204 (1968)
100. Stanford Res. Internat., part II, Environm. Protection Agency, Washington 1978
101. Donn, B.: Astrophys. J. *152,* 129 (1968)

102. Ames, B.N.: in: Hollaender, A. (Ed.) Chemical mutagenesis, principles and methods for their detection. Plenum Press, New York 1971
103. Ames, B.N., Durston, E.W., Yamasaki, E., Lee, F.D.: Proc. Nat. Acad. Sci. *70,* 2281 (1973)
104. Ames, B.N., McCann, J., Yamasaki, E.: Mutation Res. *31,* 347 (1975)
105. Ames, B.N., McCann, J.: IARC Sci. Publ. *12,* 493 (1976)
106. Berwald, Y., Sachs, L.: J. Nat. Cancer Inst. *35,* 641 (1965)
107. Smith, W.E., Cooper, N.S., Wynder, E.L.: Proc. Amer. Assoc. Cancer Res. *1,* 45 (1954)
108. Kracht, J., Klein, U.E., Baghirzade, M.: Verhandl. Dtsch. Ges. Pathologie, 45. Tagung, Münster 1961
109. Purchase, I.F.H. et al.: Nature *264,* 624 (1976)
110. IARC Monographs, Vol. 3, Lyon 1973
111. Bingham, E., Falk, H.L.: Arch. Environm. Health *19,* 779 (1969)

Fluorocarbons

J. Russow

Hoechst Aktiengesellschaft
D-6230 Frankfurt am Main, Federal Republic of Germany

Introduction

"Fluorocarbons" is a term used to designate partially or completely halogenated alkanes containing fluorine, and with one or more, but primarily 1 and 2, C atoms. The general formula is

$$C_nH_{2n+2-x-y}\,Cl_xF_y, \text{ where } x + y \leqslant 2n + 2.$$

Table 1. Physical data of fluorcarbons

Formula	Short symbol	bp [°C] at 1 bar	Vapor pressure (bar) [20 °C]
CCl_3F	F 11	23.8	0.9
CCl_2F_2	F 12	-29.8	5.8
$CClF_3$	F 13	-81.4	32.4
$CHClF_2$	F 22	-40.8	9.4
$CCl_2F \cdot CClF_2$	F 113	47.6	0.4
$CClF_2 \cdot CClF_2$	F 114	3.6	1.9

The most important members of this group are listed in Table 1 along with their physical characteristics [1]. Because of their technical importance and wide-spread use the compounds are also indicated by the short designation used here, which consists of a series of three digits.

1st digit from right = number of F atoms
2nd digit from right -1 = number of H atoms
3rd digit from right $+1$ = number of C atoms (omitted where number of C-atoms = 1).

Cl atoms contained in a fluorocarbon molecule are not indicated in this system. The compounds which are by far the most important are the compounds F 11 and F 12; they are used principally as aerosol propellants, refrigerants and as blowing agents for plastic foams. Furthermore, F 22 has assumed greater importance as a refrigerant and an intermediate product for the manufacture of the technically extremely important polymeric perfluorinated hydrocarbons polytetrafluoroethylene (PTFE) and its various copolymers.

The fluorocarbons first attained environmental significance when they were detected in extremely low concentrations, in the ppt range, in the atmosphere [2, 3]. Owing to their stability they can be used as inert tracers in studies of air flow and dispersion conditions in the lower atmosphere [4].

In recent years fluorocarbons have attained greater topical importance through the hypothesis of Rowland and Molina, according to which the fluorocarbons are responsible for depleting the stratospheric ozone layer, which may have adverse effects on the life inhabiting the earth's surface [5–7]. The scientific discussion of this set of problems, which has led to intensive research on the chemical, photochemical, and dynamic transport processes in the stratosphere, has by no means come to an end, there are still very serious discrepancies between theory and measurement, so that final judgment cannot yet be made [8–11].

Production and Use

Very recently there have been indications that fluorocarbons may be produced in nature, although the amounts detected are negligibly small [12, 13, 69]. Large-scale industrial production began in the 1930s. Tables 2 and 3 summarize world production figures for F 11 and F 12. These figures are based on a survey made by the Manufacturing Chemists Association (MCA), Washing-

Table 2. Fluorocarbon 11 production and release (world total) in million kilograms

Year	Production		Release	
	Per year	Cumulative	Per year	Cumulative
1940	0.2	0.7	0.1	0.4
1950	6.6	18.6	5.4	14.4
1960	49.7	286.9	39.7	24.79
1970	241.1	1699.1	205.1	1435.2
1971	266.6	1965.7	225.4	1660.6
1972	310.5	2276.2	253.8	1914.5
1973	354.3	2630.5	290.6	2205.0
1974	377.6	3008.1	320.9	2526.0
1975	322.5	3330.6	312.4	2838.3
1976	349.9	3680.5	303.6	3141.9
1977	330.7	4011.2	305.6	3447.5

Table 3. Fluorocarbon 12 production and release (world total) in million kilograms

Year	Production		Release	
	Per year	Cumulative	Per year	Cumulative
1940	4.5	18.8	1.7	13.3
1950	34.6	198.3	27.1	129.5
1960	99.4	828.3	83.2	631.9
1970	336.9	2957.5	296.2	2475.2
1971	360.5	3318.1	319.1	2494.3
1972	401.7	3719.7	348.3	3142.6
1973	447.5	4167.2	386.2	3528.8
1974	473.6	4640.8	420.3	3949.1
1975	419.7	5060.5	412.6	4361.7
1976	449.8	5510.4	396.3	4758.0
1977	424.4	5934.7	376.5	5134.5

ton, D.C., (USA), among 20 manufacturers within the countries outside the Eastern bloc, which account for 95% of the total world production [14–16]. The tables also show the emissions into the atmosphere which are accounted for by the use of these products. F 11 and F 12 enter the atmosphere unaltered after a longer (e.g., refrigerants) or shorter (e.g., aerosols) delay. According to a careful analysis of their use 85% of current production will be emitted into the atmosphere within a year. Of the total cumulative production today more than 95% has been released into the atmosphere [15, 16].

The applications of the fluorocarbons are a consequence of their special chemical and physical properties: they are chemically and thermally very stable and thus practically inert, and they predominantly occur as gases which can be easily liquefied under pressure. The main applications are:

a) propellant for aerosols (inert, non-flammable, practically non-toxic)
b) refrigerant for air conditioning and refrigeration systems (favourable thermodynamic properties)
c) blowing agent for the manufacture of plastic foams (good heat insulating properties of the entrapped gas)

Table 4. Fluorocarbon 11 and fluorocarbon 12 sales and uses for year 1977; 20 reporting companies outside of the eastern block representing 95% of the total world production and sales in million kilograms

Application field	F 11	F 12
Refrigeration	24.7	154.3
Blowing agent	106.9	20.6
Aerosol propellant	164.6	174.5
Other uses	19.1	33.5
Total sales	315.3	382.9

d) solvent for dry cleaning and for cleaning of high-grade electronic components (non-corrosive, high dissolving power).

Table 4 shows the 1977 sales for the Western world broken down by areas of application [14, 15].

Chemistry

The starting materials for the manufacture of the fluorocarbons are generally the corresponding chloroalkanes, which are produced by the chlorination of alkanes, e.g.

$$CH_4 + 3\,Cl_2 \longrightarrow CHCl_3 + 3\,HCl \qquad (1)$$

$$CH_4 + 4\,Cl_2 \longrightarrow CCl_4 + 4\,HCl. \qquad (2)$$

In the presence of a catalyst fluorine from hydrogen fluoride is substituted for chlorine in the chloroalkanes, forming fluorocarbons and hydrogen chloride, e.g.

$$CHCl_3 + 2\,HF \xrightarrow{(SbX_5)} \underset{\text{F 22}}{CHClF_2} + 2\,HCl \qquad (3)$$

$$CCl_4 + 2\,HF \xrightarrow{(AlF_3)} \underset{\text{F 12}}{CCl_2F_2} + 2\,HCl \qquad (4)$$

$$CCl_4 + HF \xrightarrow{(CrOF)} \underset{\text{F 11}}{CCl_3F} + HCl. \qquad (5)$$

The catalysts used are a) antimony(V) halides in a homogenous liquid-phase reaction and b) solid fluorides like AlF_3 or CrOF in a heterogeneous gas-phase reaction [1].

For the technically important products the conversion rates and yields are practically 100%. The mixture of reactants must be carefully prepared. Apart from the catalyst, which must be renewed occasionally, there are practically no unusable by-products. The losses of this type of gas, which is liquefiable under pressure, are also very low; at the present state of the art total losses during production, storage, filling and transport are expected to be about ½%.

There is little to be said concerning the chemical reactivity of these compounds because the commercially most important compounds are manufactured and used just because of their excellent thermal and chemical stability, which increases with increasing fluorine content. The compounds are non-flammable and practically non-toxic provided they are, like those mentioned in the introduction, fully halogenated. In view of their applications these compounds are usually handled in a very pure form (absolutely acid-free, water content <0.001%, evaporation residue <0.005 Vol.%), so that their handling creates no problems, such as corrosion due to impurities. They are not known to react chemically in any way.

Analytical Methods

With two exceptions the fluorocarbons are in gaseous form at room temperature and normal pressure, but they are stored and transported in liquefied form under pressure. In practice the only quantitative determination of fluorocarbons that is of any importance is for the purpose of determining the purity and the concentration in gas mixtures, and in both cases the fluorocarbons are present in relatively high concentrations. In recent years quantitative determination in the trace range has become of major interest, as, for example, for the determination of concentrations in the atmosphere. 1971 Lovelock succeeded in quantitatively determining F 11 and F 12 in the range from 50 to 100 ppt (1 ppt corresponds to $1{:}10^{12}$) by gas chromatography employing an electron capture detector [2, 3].

Several analytical methods have been developed which are similar in principle, but are distinguished from one another by the difference in the enrichment steps taken before GC analysis. An important problem of quantitative flurocarbons determination is the preparation of suitable standard mixtures, and different techniques are proposed for doing this. The present state of the art, which only a few project groups have mastered, permits a high precision in analytical results with standard deviations of a few percent. In contrast the accuracy is much lower, with a margin of error of probably more than $\pm 10\%$, which above all makes comparison difficult among the data of the various work groups [17]. A paper by Sze and Wu [18] summarizes the older determinations of concentration in the atmosphere which were available at about mid 1976; Table 5 was taken from this paper. A cautious approach to these data, made necessary by the level of analytical accuracy and comparability and the degree to which samples are free of contamination, allows us to draw three conclusions:

in the observation period from 1971 until 1974 the concentrations of F 11 and F 12 in the air near the ground increased by a factor of about 1.5;

the concentration exhibits a distinct decline from the northern to the southern hemisphere, which is related to the higher rate of emission in the northern hemisphere (see below);

the concentration is constant within the troposphere and decreases with increasing altitude in the region of the tropopause (8–13 km altitude) and in the stratosphere.

A more recent compilation by Sandalls [29] contains some additional concentration data.

Transport Behaviour in the Environment

The fluorocarbons are emitted into the lower atmosphere (troposphere) during or after their use. In accordance with their areas of application and the given economic conditions, the fluorocarbons are released largely in the northern hemisphere. Sales surveys have yielded the emission figures for F 11 and F 12, the two most important types, as shown in Tables 2 and 3 [14–16].

Table 5. Measurements of fluorocarbon concentrations units = ppt in vol (from [18])

Author	Altitude (km)	F-11	F-12	Time and location	Remarks
Heidt et al. [19]	40–45	<0.2	<5	Sept. 1973	By balloon-borne cryogenic sampling system
	34.0	3	35	June 1975	
	31.0	9	48		
	28.6	11	-	TX (32 °N)	
	24.5	18	-		
	23.0	45	86		
	16.9	95	-		
	13.0	83	133		
	12.0	94	78	Sept. 1973	
Hester et al. [20]	18.3	60 ± 4	98 ± 18	23 May 1974 36.15–39.30 Lat 106.17–106.45 Long	Average value from 2 flights
	12.2	75	140	23 May 1974 33.10–34.34 Lat 104.30–105.10 Long	
	6.4	80 ± 3	125 ± 7	May 1974 34.45–33.50 Lat 106.20–105.00 Long	Average value from 2 flights
Krey and Lagomarsino [21]	13.7	59[a]		} April 1974 (60 °N–37 °S)	Compressed air sample, data analyzed by gas chromatography
	15.2	70			
	16.8	65			
	18.3	57			
	19.2	47			
	12.2	69		} October 1974 (75 °N–10 °S)	
	13.7	77			
	15.3	77			
	16.8	66			
	18.3	67			
	19.2	41			

Author	Altitude (km)	F-11	F-12	Time and location	Remarks
Lovelock [22]	Surface	79.8	101.7	June/July 1974 W. Ireland	By gas chromatography
		88.6	115.2	October 1973 N. Atlantic	
		101–119		June 1974 Central England	
		57		Sept. 1974 Capetown, S. Africa	
Lovelock et al. (1973) [2]		49[b]		Nov.–Dec. 1971 50 °N–60 °S	Oceanographic cruise of Shackleton Latitude
Schmeltekopf et al. [23]	26.2 ± 1	<20	75 ± 5	June 1975	Stratospheric measurements
	22.3 ± 0.7	30 ± 3	135 ± 10		
	17.7 ± 0.5	80 ± 10	210 ± 10		
Goldan et al. [24]	Surface	48 ± 5	90 ± 10	August 1975	
	1–4 km	150	330	Sept. 1975	
	10 km	160 ± 15	230 ± 30	August 1975	
	14 km	120 ± 15	225 ± 30	August 1975	
	18.5 km	$1.3 \pm ^{0.7}_{0.3}$	25 ± 3	August 1975	
	22 km	10 ± 1	84 ± 8	August 1975	
	25.5 km	7.6 ± 1	100 ± 10	August 1975	
Williams et al. [25]	21	23	120	26 Sept. 1975	Preliminary results
	18.5	49	160	26 Sept. 1975	
	18.5	20	50–60	12 Aug. 1968	
	15	110	140	26 Sept. 1975	
Rasmussen [26]	Surface	120–130	210–230	May 1975	
Wilkniss et al. [27, 28]	at surface	43		Nov. 1971 (10 °S)	By gas chromatography
	at surface	53		Nov.–Dec. 1972 (10 °S)	
	at surface	72		Mar.–Apr. 1974 (10 °S)	
	at surface	61		Dec. 1972 (20 °S–80 °N)	
	at surface	80		Mar.–Apr. 1974 (20 °S–20 °N)	

[a] Mean concentration averaged over latitudes

[b] Mean aerial concentration averaged over 50° N–60° S. Concentration ranges from 70 ppt at 50° N to 38 ppt at 60° S

These tables reveal the increase in F 11 and F 12 emissions up until 1974. Since that time emissions have remained constant or declined slightly; the reason for this is the decline in the use of fluorocarbons in aerosols for economical (substitution by propane/butane) and environmental (fluorocarbons-ozone hypothesis) reasons.

The emission of fluorocarbons is also reflected in the concentrations measured in the atmosphere. Whereas the first measurements from the year 1971 yielded concentrations of 40–50 ppt F 11 [3, 27, 28] (cf. Table 5), those made in mid 1978 show a concentration of about 160 ppt F 11 and about 280 ppt F 12 [30]. For the two-year period from November 1975 to November 1977 concentration measurements between 35° and 65° north latitude determined an average growth rate of 12.9 ppt/year or 12%/year for F 11 and 18.5 ppt/yr or 10%/yr for F 12 [31]. With regard to sampling it is always difficult to know whether the sample contains the true background concentration or a very much higher concentration because, owing to meteorological conditions, the sample originated in an area with high industrial and population densities, so that it exhibits excessive F 11 and F 12 levels.

For example, Pack et al. [4] have shown that occasionally greatly increased fluorocarbon concentrations at a very remote sampling point are due to these air masses originating in industrial areas. Figure 1, taken from Fraser [32], is

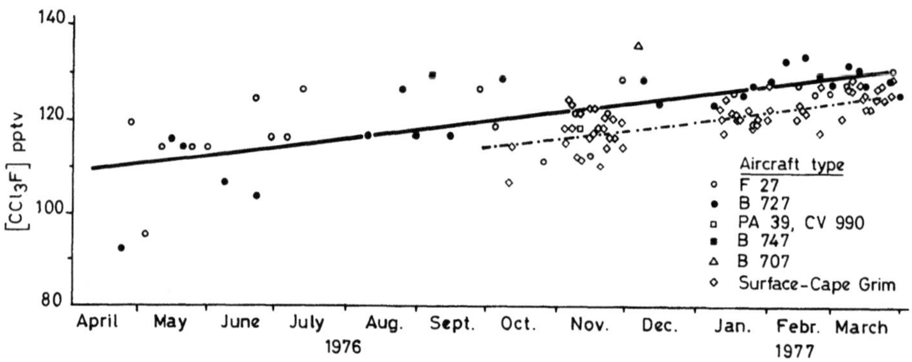

Fig. 1. Variation with time of observed CCl_3F concentration in surface air at Cape Grim and in aircraft air samples. Best fit curves: —— aircraft data. ln $[CCl_3F] = 4.65\ (\pm 0.02) + 5.0\ (\pm 0.7) \times 10^{-4}\ t$, r^2 (coefficient of determination) $= 0.51$; –·–·– Cape Grim data. ln $[CCl_3F] = 4.60\ (\pm 0.03) + 5.1\ (\pm 0.7) \times 10^{-4}\ t$, $r^2 = 0.49$; t = days since 31/12/75. Data points are averages (from [32])

a typical example for an increase in F 11 concentrations over time. At the same time it also shows the scatter of the empirical data. Within the troposphere the fluorocarbons disperse very quickly, so that the concentration within the troposphere is uniform (cf. Fig. 2 from Fraser [32]). This diagram also clearly reveals the emission, and hence the increased concentration, over heavily populated and industrialized areas. Above the troposphere, however, the concentration decreases with altitude, which is due to stratospheric decomposition mechanisms. Figure 3 is given as a typical example of concentration

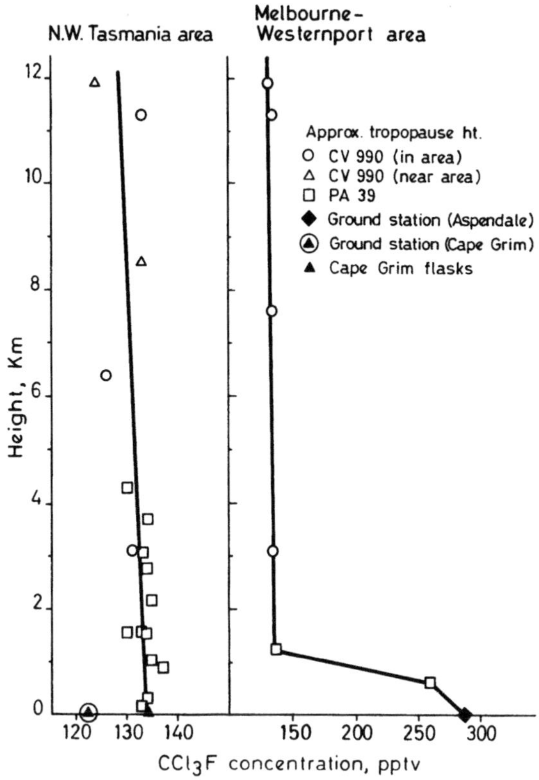

Fig. 2. Vertical profiles of CCl_3F observed over NW Tasmania and Melbourne on 11/11/76; air samples from Cv990 by co-operation with NASA-Ames (from [32])

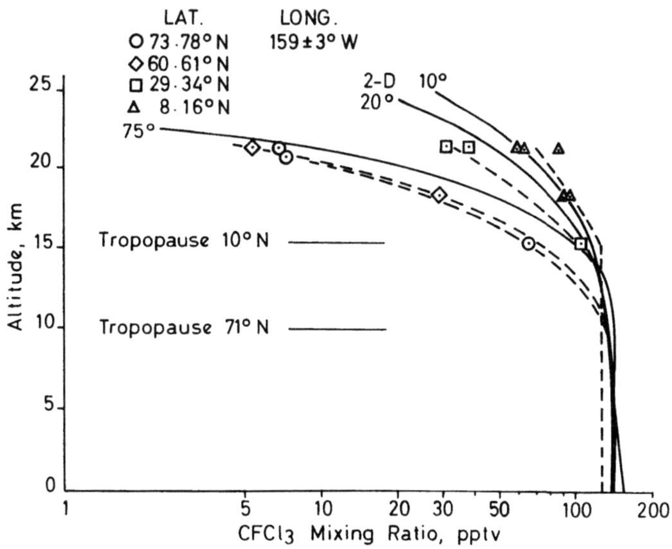

Fig. 3. Lower stratosphere mixing ratios of $CFCl_3$ at various latitudes in the Northern Hemisphere. Tropopause altitudes estimated from National Meteorological Center analysis. Tropospheric mixing ratios from airborne (CV-990) measurements. 2-D curves from Borucki [68] (from [33])

profiles for F 11; it was taken from Vedder et al. [33]. Further profiles can be found in [18, 19, 34].

Since fluorocarbons are emitted largely in the northern hemisphere and the period of atmospheric exchange between the two hemispheres is about 1 to 2 years, the concentration measurements also exhibit a north-south gradient. Figure 4 is such a north-south profile, taken from Lovelock [3], and is presented as a characteristic example. Very recent measurements yield flatter gradients [31], because emission has ceased its rise and is now declining (cf. Tables 2 and 3).

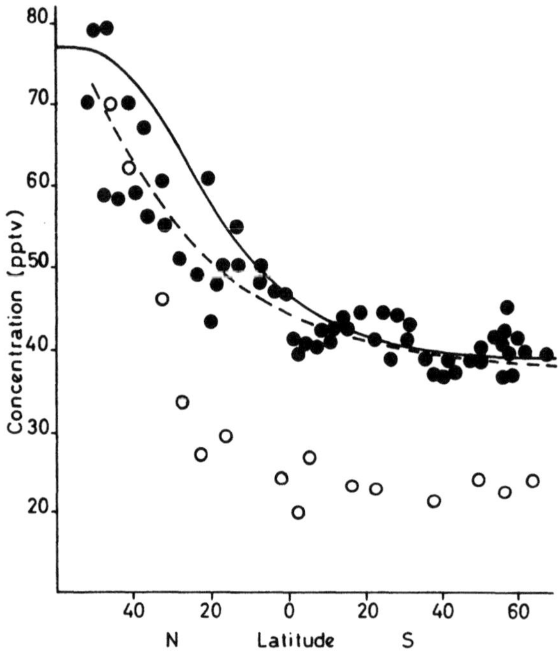

Fig. 4. Distribution of CCl_3F in and over the North and South Atlantic Ocean. ● Aerial concentrations ($\times 10^{-12}$) by volume. ○ Seawater concentration ($\times 10^{-12}$); as aerial concentrations in equilibrium with water. —— Theoretical prediction. - - - Best fit third degree polynomial (from [3])

Chemical and Photochemical Reactions

The compounds being considered here are characterized by high chemical stability. For this reason the fluorocarbons are not involved, for example, in the chemical processes of smog formation in the lower troposphere. In connection with the ozone depletion hypothesis, which maintains that after the fluorocarbons diffuse into the stratosphere they are photochemically degraded there, and the reaction products formed react with the stratospheric ozone through a chain mechanism, an intensive search has been conducted for degradation mechanisms in the *lower atmosphere (troposphere)*.

Any decomposition of fluorocarbons within the troposphere even if this is only at a rate of a few percent per year diminishes the flux of chlorine containing compounds into stratosphere and reduces the predicted ozone depletion attributed to fluorocarbons drastically because of non linear relations between ozone depletion and tropospheric residence time.

Thus, for example, hydrolysis is just barely measurable in aqueous solution ($\sim 10^{-7}$ mol/l.yr), and it can be catalytically accelerated by traces of heavy metals [35]. Solubility in water at 20 °C amounts to 1,500 ppm for F 11 and 600 ppm for F 12. Solubility in seawater is somewhat lower [36]. Because of the low partial pressure in the atmosphere, the concentration in seawater, at $0.1–0.4 \times 10^{-9}$ g/l, is very low [31, 37], so that the oceans do not play any appreciable role as a sink [31, 36, 38].

Various studies have led to the conclusion that F 11 or F 12 is degraded in a heterogeneously catalyzed photolytic process on solid surfaces like those of silica gel, sand or desert dust under exposure to ultraviolet light [39–42]. According to more recent studies decomposition also occurs without light (dark reaction) if the substrates are very dry [43, 44]. Decomposition is accelerated by the presence of oxygen. Carbon dioxide has been definitely detected as a decomposition product [40, 41]. It is entirely unknown whether such decomposition mechanisms also play a role under natural conditions, as in desert areas, in determining the residence time in the troposphere [45, 46], as long as no quantitative evaluation of the experiments is available for atmospheric conditions. Comparison of the amounts of F 11 and F 12 emitted into the troposphere, the amounts which diffuse into the stratosphere and the amounts just analytically detectable in the troposphere, which, however, are very unreliable because of the low accuracy, leads to the conclusion that no decomposition occurs in the troposphere [31]. As a result the atmospheric residence time is calculated to be 40–45 yr for F 11 and 65–70 yr for F 12. On the other hand, evaluation of the north-south gradient in the troposphere, the uncertainty in the profile used for diffusion in the stratosphere and an overall stratospheric chlorine balance support the view that the possibility of shorter atmospheric residence times and thus also of decomposition in the troposphere cannot be excluded at this time [18, 47–49]. In an experiment designed on a grand scale (*A*tmospheric *L*ifetime *E*xperiment, ALE) the concentration is monitored constantly at four measuring stations scattered around the globe in order to obtain information on the atmospheric residence time [50]. On the basis of the facts that are known now it must be assumed that a major part of the fluorocarbons emitted so far has accumulated in the atmosphere, even if the major part is 95% this still has a very significant effect on the predictions concerning stratospheric ozone depletion.

Whereas the fluorocarbons in the lower atmosphere (troposphere) are relatively stable, in the *upper atmosphere (stratosphere)* they can be decomposed photochemically by the more energetic (shorter wavelength) UV-radiation [5, 6]. In the first step

$$CF_2Cl_2 + h\nu \longrightarrow {}^{\cdot}CF_2Cl + {}^{\cdot}Cl \quad (\lambda < 230\ \text{nm}) \tag{6}$$

$$CFCl_3 + h\nu \longrightarrow {}^{\cdot}CFCl_2 + {}^{\cdot}Cl \quad (\lambda < 230\ \text{nm}) \tag{7}$$

Cl radicals are split off. It is highly probable that the reaction with oxygen molecules follows this step:

$$^{\cdot}CF_2Cl + O_2 \longrightarrow CF_2O + {}^{\cdot}ClO \quad (8)$$

$$^{\cdot}CFCl_2 + O_2 \longrightarrow CFClO + {}^{\cdot}ClO \quad (9)$$

At present little is known about the subsequent fate of the compounds CF_2O and CFClO [51]. From these reactions the following average photodissociation lifetimes of F 11 and F 12 in the atmosphere have been calculated [6]:

	F 11	F 12	Altitude
Average photo-dissociation lifetimes	3×10^5 yr	4×10^6 yr	10 km
	6.6 yr	66 yr	20 km
	1.3 months	11 months	30 km
	4.7 days	1.3 months	40 km
	2.1 days	16 days	50 km

These figures lead to average atmospheric residence times of 50 and 100 yr, respectively.

Photochemical decomposition influences the shape of the concentration profile in the stratosphere, as Fig. 3 shows. Above the tropopause at an altitude of 8–13 km the concentration declines relatively rapidly with increasing altitude. The ˙Cl radicals formed by photolytic decomposition may – as stated by the fluorocarbon ozon hypothesis – interfere with the dynamic equilibrium, which is determined essentially by the ozone formation reaction (10, 11), the decomposition reaction (12) and the catalytic decomposition cycles (13+14=15), in which the catalytically effective radicals can be ˙X=˙NO, ˙Cl, and ˙HO.

Ozone formation:

$$O_2 + h\nu \longrightarrow 2\,O \ (\lambda < 240\ \text{nm}) \quad (10)$$

$$O + O_2 + M \longrightarrow O_3 + M \quad (11)$$

Ozone decomposition:

$$O_3 + O \longrightarrow 2\,O_2 \quad (12)$$

$$O_3 + X \longrightarrow O_2 + {}^{\cdot}XO \quad (13)$$

$$^{\cdot}XO + O \longrightarrow O_2 + {}^{\cdot}X \quad (14)$$

$$O_3 + O \longrightarrow 2\,O_2 \quad (15)$$

There can be no doubt that fluorocarbons diffusing into the stratosphere and their subsequent photolysis increase the total amount of chlorine in the stratosphere, but whether the fluorocarbons cause depletion of the ozone or not remains an open question. As scientific research on the atmosphere has progressed, the margin of error of the calculated hypothetical ozone depletion by fluorocarbons has increased: for the equilibrium condition in 50–100 yr at the present emission rate of 0.33 Mt F 12/yr and 0.27 Mt F 11/yr the ozone depletion is currently calculated to be 15–20% with a probability range of 0–40% [10]. Furthermore, there are obviously several irreconcilable discrepancies between theoretical calculations and measurements and they have a bearing on factors which regulate the equilibrium system, which suggests that it is not yet possible to completely describe the chemistry of the stratosphere. On the other hand the ozone measurements and the statistical evaluation of the data, which are subject to large natural fluctuations, have not yet supplied proof that ozone depletion has occurred, although according to the calculations this should already be the case [52–56].

Metabolism

Only a few studies of fluorocarbon metabolism are known. Such studies are probably of little interest because, for one, we know that if these compounds have been inhaled in high concentrations and preferentially absorbed by the blood, they will be practically completely eliminated within an hour if fresh air is supplied and do not accumulate [57–61]. Although experiments with C^{14}-labeled F 11 and F 12 have demonstrated a low residual activity in the organism, it is unknown whether metabolites are actually involved in any way [62].

Biodegradation

Studies made so far of decomposition on soil samples under natural conditions (microorganisms and flora) have not yielded any clear-cut results. If there is any degradation at all, it must be very little at normal atmospheric concentrations [63].

Accumulation

There are no known indications that the fluorocarbons accumulate in living organisms. It is also hardly to be expected because of their high volatility.

Biological Effects and Toxicity

The compounds exhibit only very low toxicity. The toxicological data are well known because relatively high local concentrations can occur when they are used. Their potential hazard when they are abused, for example, as in "sniffing", has been carefully studied.

Gulden [64] and Paulet [58] have summarized information about fluorocarbon toxicity. An acute two-hour inhalation test has yielded a tolerable inhalation concentration of 1.25 vol. % for F 11 and 10 vol. % for F 12 (rat, guinea pig). If these concentrations are exceeded, narcotic effects appear, which are tolerated by the experimental animals without any permanent damage [65]. Only very high concentrations can cause death: LC_{50} over 30 min for F 11: 10–25 vol. %, depending on the animal species; LC_{50} over 30 min for F 12: 76–80 vol. %, depending on the animal species [58]. In subacute inhalation trials with the compounds F 11, 12, 22, 113 and 114 in rats (1 vol. %) or dogs (0.5 vol. %) over 90 days with 6 h exposure daily no effects could be determined in either the external or histological findings [66]. According to these findings these compounds must be classified as practically non-toxic or relatively harmless under the Hodge and Sterner system [67]. The threshold limit value (TLV) has been set at 1,000 ppm for most fluorocarbons.

A number of studies have been concerned with intake via the respiratory route. A few minutes after exposure the fluorocarbons can already be detected in the blood in concentrations of a few µg/ml [60]. Arrhythmias appear only when the blood level exceeds 40 µg/ml, a value which can be attained only when extremely high concentrations are inhaled [58]. After cessation of inhalation exposure the fluorocarbons are eliminated within an hour via the expired air. At present it is not yet certain whether to some small extent these compounds are metabolized (cf. the section on Metabolism).

The LD_{50} value in rats for the two liquid fluorocarbons F 11 and F 113 lies above 15 g/kg, which means that these compounds fall into toxicity class 6 – relatively harmless [64]. The corresponding values for the other fluorocarbons, which are gases under normal conditions, cannot be determined.

References

1. Ullmanns, Encyklopädie d. Techn. Chem. 4th ed. (1976) Vol. *11*, 632, Verlag Chemie GmbH, Weinheim. Topics Curr. Chem. *14*, 129 (1970)
2. Lovelock, J.E.: Nature*230*, 379 (1971)
3. Lovelock, J.E., Maggs, R.J., Wade, R.J.: Nature, *241*, 194 (1973)
4. Pack, D.H., Lovelock, J.E., Cotton, G., Curthoys, Ch.: Atmospheric Environment *11*, 329 (1977)
5. Molina, M.J., Rowland, F.S.: Nature *249*, 810 (1974)
6. Rowland, F.S., Molina, M.J.: Atomic Energy Commission Rep. No. 1974-1, Irvine, California USA, Sept. 5, 1974
7. Crutzen, P.J.: Geophysical Res. Lett. *1*, 205 (1974)
8. US-National Acad. Sci.: Response to the Ozone Protection Sections of the Clean Air Act Amendments of 1977: An Interim Report. Washington 1977
9. Sze, N.D., McElroy, M.B., Wofsy, S.C., Kong, D., Daesen, R.: Atmospheric and Environmental Res., Inc., Cambridge, USA; Report Oct 1978
10. Ehhalt, D.H.: Chlorfluormethane und ihr Einfluß auf die stratosphärische Ozonschicht. Bericht im Auftrag des Umweltbundesamtes Jülich, Germany, Sept. 1978
11. Boville, B.W., Evans, W.F.J.: Current Status of the Stratospheric Pollution Problem; UNEP Co-ordinating Committee on the Ozone Layer, Second Session Bonn, Germany Nov 28-Dec 1, 1978

12. Prinn, R.G., Barshay, S.S.: Halocarbons and other minor species in volcanic emissions: Theoretical considerations. Rep. CAP Associates to the Manufacturing Chemists Assoc., Bedford, Mass. USA, May 18, 1978
13. Rasmussen, R.A.: Update on EX-GC and GC-MS Analyses of Trace Gases in Hawaiian Volcanic Emissions. Spec. Rep., Beaverton/Oregon USA, Dec 21, 1978
14. Alexander Grant & Company, Washington: 1977 World Production and Sales of Fluorocarbons FC-11 and FC-12, Manufacturing Chemists Assoc., June 26, 1978
15. Manufacturing Chemists Assoc., Fluorocarbons Technical Panel, D.C.: World Production and Release of Chlorfluorocarbons 11 and 12 Through 1977, July 17, 1978
16. McCarthy, R.L., Bower, F.A., Jesson, J.P.: Atmospheric Environment *11*, 491 (1977)
17. Rasmussen, R.A.: Interlaboratory Comparison of Fluorocarbon Measurements. Preprint, submitted to Atmospheric Environment (1978)
18. Sze, N.D., Wu, M.F.: Atmospheric Environment *10*, 1117 (1976)
19. Heidt, L.E., Lueb, R., Pollock, W., Ehhalt, D.H.: Geophys. Res. Lett. *2*, 445 (1975)
21. Krey, P.W., Lagomarsino, R.J. (1975): Health and Safety Lab., ERDA, Environmental Quarterly Rep., HASL-294, 97–123
22. Lovelock, J.E.: Nature *252*, 292 (1974)
23. Schmeltekopf, A.L., Goldan, P.O., Henderson, W.R., Harrop, W.J., Thomson, T.L., Fehsenfeld, F.S., Schiff, H.I., Crutzen, P.J., Isaksen, I.S.A., Ferguson, E.E.: Geophys. Res. Lett. *2*, 393 (1975)
24. Goldan, P.D. (1975): Private communication
25. Williams, W.J., Kosters, J.J., Goldman, A., Murcray, D.G.: Paper presented at AGU meeting, San Francisco, Dec. 1975
26. Rasmussen, R. (1975): Private communications
27. Wilkniss, P.E., Swinnerton, J.W., Bressan, D.J., Lamontagne, R.A., Larson, R.E.: J. Atmos. Sci. *22*, 158 (1975)
28. Wilkniss, R.A., Swinnerton, J.W., Lamontagne, R.A., Bressan, D.J.: Science *187*, 832 (1975)
29. Sandalls, F.J., Hatton, D.B.: Atmospheric Environment *11*, 321 (1977)
30. Krasnec, J.: Paper at the AGU meeting Dec. 1978, San Francisco
31. Singh, H.B., Salas, L.J., Shigeishi, H., Scribner, E.: Science *203*, 899 (1979)
32. Fraser, P.J.B., Pearman, G.I.: Atmospheric Environment *12*, 839 (1978)
33. Vedder, J.F., Tyson, B.J., Brewer, R.B., Boitnott, C.A., Inn, E.C.Y.: Geophys. Res. Lett. *5*, 33 (1978)
34. Seiler, W., Müller, F., Oeser, H.: Pure and Applied Geophys. *116*, 554 (1978)
35. Grewer, T.: Internal Rep. Hoechst AG, Frankfurt – Germany 1962, 1966
36. Junge, C.: Naturforsch. *31a*, 482 (1976)
37. Rasmussen, R.A., Pierotti, D., Krasnec, J., Halter, B.: Trip Report on the Cruise of the Alpha Helix Research Vessel from San Diego, California to San Martin, Peru March 5 to 20. 1976
38. Liss, P.S., Slater, P.G.: Nature *247*, 181 (1974)
39. Ausloos, P., Rebbert, R.E., Glasgow, L.: J. Res. Nat. Bur. Standards, Washington, *82*, 1 (1977)
40. Gäb, S., Schmitzer, J. Thamm, H.W., Korte, F.: Angew. Chem. *90*, 398 (1978)
41. Gäb, S., Schmitzer, J., Thamm, H.W., Parlar, H., Korte, F.: Nature *270*, 331 (1977)
42. Rasmussen, R.A.: Halocarbon Photolysis over Sand in Ambient Air. Special report, Nov. 1978
43. Gäb, S.: personal communication
44. Ausloos, P.: personal communication to MCA
45. Alyea, F.N., Cunnold, D.M., Prinn, R.G.: Atmospheric Environment *12*, 1009 (1978)
46. Pierotti, D., Rasmussen, L.E., Rasmussen, R.A.: Geophysical Research Lett. *5*, 1001 (1978)
47. Jesson, J.P., Meakin, P., Glasgow, L.C.: Atmospheric Environment *11*, 499 (1977)
48. Meakin, P., Gumerman, P.S., Glasgow, L.C., Jesson, J.P.: Atmospheric Environment *12*, 1271 (1978)
49. Glasgow, L.C., Gumerman, P.S., Meakin, P., Jesson, J.P., Atmospheric Environment *12*, 2159 (1978)
50. Cunnold, D., Alyea, F., Prinn, R.: J. Geophys. Res. *83*, 5493 (1978)
51. National Academy of Sciences: Halocarbons: Effects on Stratospheric Ozone. Washington, D.C., Sept. 1976

52. Hill, W.J., Sheldon, P.N.: Geophysical Research Lett. *2,* 541 (1975)
53. Hill, W.J., Sheldon, P.N.: Geophysical Research Lett. *4,* 21 (1977)
54. Parzen, E., Pagano, M., Newton, H.J.: Report to MCA: Statistical Time Series Analysis of Worldwide Total Ozone for Trends. March 1977
55. Tiede, J.J., Sheldon, P.N., Hill, W.J.: Preprint of a Paper at the WMO-Symposium June 26–30, 1978, Toronto
56. Angell, J.K., Korshover, J.: Paper at the WMO-Symposium June 26–30, 1978, Toronto
57. Paulet, G., Lanoë, J., Thos, A., Toulouse, P., Dasonville, J.: Toxicol. Appl. Pharmacol. *34,* 204 (1975)
58. Paulet, G.: Europ. J. Toxicology. Supplément *9,* 385 (1976)
59. Morgan, A., Black, A., Walsh, M., et al.: Int. J. Appl. Radiat. Isot. *23,* 285 (1972)
60. Williams, F.W., Draffan, G.H., Dollery, C.T.: Thorax *29,* 22 (1974)
61. Cox, P.J., King, L.J., Parke, D.V.: Biochem. J. *130,* Proc. Biochem. Soc. 13 p, 87 p. (1972)
62. Blake, D.A., Mergner, G.W.: Toxikol. Appl. Pharmakol. *30,* 396 (1974)
63. Seiler, W.: personal communication
64. Gulden, W.: Aerosol Rep. *12,* 248 (1973)
65. Scholz, J.: in: Fortschr. biolog. Aerosol-Forschg. 1957–61. Nückel, H. (ed.), Schattauer Verlag, Stuttgart 1961
66. Leuschner, F.: unpublished
67. Hodge, H.C., Sterner, J.H., cit. by Spector, W.S.: Handbook of Toxikology Vol. I, Acute Toxicities of Solids, Liquids and Gases to Laboratory Animals, W.B. Saunders Comp. 1958
68. Borucki et al.: Amer. Inst. of Aeronautics and Astronautics J. *14,* 1738 (1976)
69. Kranz, R.: Naturwissenschaften *53,* 593 (1966)

Chlorinated Paraffins

V. Zitko

Fisheries and Environmental Sciences,
Fisheries and Oceans, Biological Station
St. Andrews, N. B. (EOG 2XO), Canada

Chlorinated paraffins discussed in this chapter are compounds obtained by chlorination of C_{10}–C_{30} paraffins to a chlorine content between 10 and 70%. The most frequent types of chlorinated paraffins are based on C_{12}, C_{15}, and C_{24} feedstocks and are chlorinated to 40–70% chlorine. Depending on the chlorine content, chlorinated paraffins range from mobile through highly viscous liquids to solids.

Production and Applications

Although first uses of chlorinated paraffins were reported during World War I, a large-scale commercial production started only around 1930. Production figures for this period are not available. During World War II, the U.S. annual production was about 23×10^6 kg. After a decrease during the late 1940's, the annual production of the 1950's was about 18×10^6 kg, and increased to about 21×10^6 kg through the 1960's. Between 1970 and 1975, the production increased to about 45×10^6 kg/yr and is currently an estimated $(50–70) \times 10^6$ kg/yr. The world production of chlorinated paraffins is probably 3–4 times the U.S. production. According to Mills [18], the 1978 annual production of chlorinated paraffins was 105×10^6 kg in western Europe, 60×10^6 kg in North America, and 65×10^6 kg in other free-world countries.

Chlorinated paraffins are manufactured by a number of companies and marketed under a variety of tradenames. Some of these are given in Table 1. Additional manufacturers are listed in Table 2. It is understood that chlorinated paraffins are also produced in China, Czechoslovakia, East Germany, Poland, Romania, and the USSR [18].

In addition to tradenames, chlorinated paraffin preparations are further characterized by numbers. These are often related to the chlorine content of

Table 1. A partial list of tradenames and manufacturers of chlorinated paraffins

Tradename	Manufacturer
Arubren CP	Farbenfabriken Bayer (West Germany)
Cereclor	ICI Limited (UK, France, Italy, Spain, Australia, USA, Canada)
Chlorez	Dover (USA)
Chloroflo	Dover (USA)
Chlorowax	Diamond Shamrock (USA)
Chloroparaffine Hoechst	Farbwerke Hoechst (West Germany)
Chloroparaffine Hüls	Chemische Werke Hüls (West Germany)
Clorafin	Hercules (USA)
CPF	Pearsall (USA)
CW	Keil (USA)
FLX	Pearsall (USA)
Kloro	Keil (USA)
Paraoil	Dover (USA)
Unichlor	Neville (USA)
Witaclor	Deutsche Dynamit Nobel (West Germany)

Table 2. Additional manufacturers of chlorinated paraffins [18]

AECI (South Africa)	Meltur Chem. (India)
Ajinomoto (Japan)	Plasticlor (Mexico)
Asahi Denka (Japan)	Rhone Pulenc (France)
Bann Quimica (Brazil)	Rio Rodano (Spain)
Caffaro (Italy)	Sintesis Quimica (Argentina)
Ciclomeros (Mexico)	Toyo Soda (Japan)
Electrochlor (Argentina)	Ugimica (Spain)
Excel Ind. (India)	Ugine Kuhlmann (France)
FPQ (Brazil)	Wintershall (West Germany)
Kop (South Africa)	

the preparation, but, as a rule it is necessary to consult the manufacturer's technical data sheets to obtain the average formula of the preparation, either directly or by calculation from the average molecular weight and the chlorine content. Properties of some Cereclor chlorinated paraffins are given in Table 3.

Applications of chlorinated paraffins include plasticizers, additives to paints, adhesives, mastics and caulks, additives to lubricants, particularly cutting oils and heavy duty gear oils, and additives to printing inks. According to Howard et al. [11], lubricating oil additives, solvent and plasticizer uses, secondary vinyl plasticizers, and traffic paints represented 45, 27, 24, and 4% respectively, of the U.S. market in 1973. In 1976, the amount of chlorinated paraffins used in plastics in the U.S. (35×10^6 kg) exceeded the amount of non-halogenated phosphoric acid esters (21×10^6 kg). Chlorinated paraffins thus became the second largest group of flame retardant for plastics, exceeded only by aluminum hydroxide [12].

According to Mills [18], the major use of chlorinated paraffins in 1978 was as a plasticizer in flexible PVC (PVC cables, flooring, shoes, extrusions, etc.),

Table 3. Properties of some typical Cereclor (ICI) chlorinated paraffins [18]

Grade	Chlorine content %	Molecular weight of the CP	Appearance	Colour hazen units	Density at 25 °C g/mL	Viscosity at 25 °C poises
42	42	600	Clear very pale yellow liquid	250	1.16	22
48	48	700	Clear viscous yellow liquid	300	1.24	250
54	54	780	Clear viscous yellow liquid	450	1.32	5000
70	70	1100	White powder	100	1.63	Solid/ Softening point 90 °C
S45	45	390	Clear water white mobile liquid	80	1.16	2
S52	52	440	Clear water white liquid	100	1.25	16
S58	58	500	Clear viscous pale yellow liquid	150	1.35	350
50LV	49	320	Clear almost colourless liquid	100	1.19	0.8
56L	56	370	Clear very pale yellow liquid	125	1.30	8
60L	60	400	Clear very pale yellow liquid	125	1.36	35
63L	63	430	Clear very pale yellow liquid	125	1.42	150
65L	65	440	Clear viscous yellow liquid	150	1.44	300
70L	70	500	Clear very viscous liquid	200	1.54	3500

and the relative use of chlorinated paraffins was PVC (45%), lubricants (25%), paints (13%), flame retardants (10%), and others (7%).

Three reviews of the manufacture, properties, applications, analytical chemistry, etc. of chlorinated paraffins are available [11, 15, 27].

Chemistry

Chlorinated paraffins are produced by liquid-phase chlorination of paraffinic stocks at 50–150 °C, often in the presence of a solvent such as carbon tetrachloride. Both batch and continuous processes are used. In the latter, the chlorination unit may consist of 3–4 cylindrical reactors, connected in series, with chlorinated paraffins flowing counter currently to chlorine.

Different stocks, ranging from the kerosine-gas oil fraction of petroleum to very pure straight chain paraffins, may be used. As discussed below, the presence of branched paraffins in the feedstock decreases the thermal stability of the final product and also causes its yellow to dark brown color. Many chlorinated paraffins produced at present are based on very pure straight chain feedstocks, are very stable thermally and practically colorless [7].

Aromatic hydrocarbons in the feedstock may result in chlorinated aromatic hydrocarbons in the product. The content of aromatics in the usual C_{12} feedstock (range C_9–C_{14}) is typically about 0.5–1% and is reported by the supplier [5]. The content of aromatics in the C_{24} feedstock (normal range C_{20}–C_{30}) is less than 0.1% [5]. No information is available on the content of aromatics in the C_{15} feedstock. It is probably somewhat below that of the C_{12} feedstock. European feedstocks are treated to reduce aromatics to 50–100 μg/g [18].

The chlorination is a radical reaction and the reactivity of hydrogen atoms decreases with their increasing acidity. At 300 °C in the gas phase, the relative chlorination rates on tertiary, secondary, and primary carbon atoms are approximately 4:2:1 [6]. These reactivities and statistical factors determine the distribution of chlorine atoms along the chain of the paraffin [9, 20]. As a result, chlorinated paraffins are extremely complex mixtures of chloroparaffins with different chain lengths, degree of chlorination, and distribution of chlorine atoms along the chain. Consequently, it is difficult to elucidate experimentally their structural details. It is interesting to note that a hexakontane, containing 33% chlorine, still contains 1.8% of monochloro- and 4% of dichlorohexakontane [13].

The stability of chlorine in chlorinated paraffins is inversely related to the ease of chlorination. Thus the stability decreases in the order
primary > secondary > alicyclic > benzylic > allylic > tertiary chlorine.

Since the presence of other than primary and secondary chlorines is due to impurities, chlorinated paraffins of high thermal stability require very pure straight chain paraffins as feedstocks.

The loss of chlorine from chlorinated paraffins occurs primarily by dehydrochlorination which may start at temperatures above 250 °C [8]. Dehydrochlorination followed by a complete volatilization of products was observed on heating of C_{22} chlorinated paraffins containing up to 25% chlorine. Under similar conditions, appreciable residues were formed from chlorinated paraffins containing 31 and 38% chlorine. The mechanism of color formation and the degradation products have not been identified [21]. Ions C_4H_3, C_5H_5, C_6H_6, C_7H_7, C_5H_4Cl, $C_3H_3Cl_2$, C_9H_7, C_7H_6Cl, and $C_5H_6Cl_2$ were observed on direct pyrolysis of chlorinated paraffin (Cereclor 70) – $(BiO)_2CO_3$ or Sb_2O_3 mixtures in the ion source of a mass spectrometer at 260–350 °C [2].

Binding of hydrochloric acid, formed by dehydrochlorination, retards the process and stabilizers such as epoxides and organotin compounds are usually added to chlorinated paraffin preparations. Traces of iron also promote dehydrochlorination and complexing agents such as EDTA and NTA may be used as stabilizers.

The preparation of chlorinated paraffins containing butoxy, hydroxy, nitrile, quinolinyl or dialkylphosphoryl groups has been described. An acetoxylated chlorinated paraffin preparation had a slightly better plasticizing efficiency than chlorinated paraffin [3]. The modified chlorinated paraffin preparations do not appear to be widely used at the moment.

Determination

The determination of chlorine appears to be the only technique for the quantitation of trace levels of chlorinated paraffins. Since many other chlorine-containing compounds, both natural and anthropogenic, may be encountered, several cleanup procedures have been used.

The situation is somewhat less complex for technical products, containing chlorinated paraffins as additives. The levels of chlorinated paraffins are much higher than those likely present in environmental samples. A separation of chlorinated paraffins from polymers, based on gel permeation chromatography, has been described recently [17].

Extraction of Biological Samples. Solvent mixtures such as acetone-hexane [1] or cyclohexane-isopropanol [23] have been used. Routine extraction conditions, as used for the extraction of PCB's and common organochlorine pesticides, appear to be applicable as well [16, 25].

Cleanup of Extracts. Solvent partitioning, column and thin-layer chromatography are the basic steps used in the cleanup. Partitioning between hexane and dimethyl formamide [1], or between hexane and acetonitrile [16] is the usual preliminary step for the separation of chlorinated paraffins from the bulk of co-extracted lipids. The pair hexane-dimethyl sulfoxide may be possibly used to separate highly chlorinated (70% chlorine) from less chlorinated (40% chlorine) preparations [28].

Alumina, silica, and Florisil are the common adsorbents for the column chromatographic cleanup and separation of chlorinated paraffins not only from lipids but also from other organochlorine compounds. These procedures may be used in combination with solvent partitioning or independently.

On elution with hexane, chlorinated paraffins are only partially eluted from alumina, but the elution becomes quantitative if the lipid load on the column exceeds 10 mg/g alumina [24]. Chlorinated paraffins are eluted from silica by 10% ether in hexane [24] or by benzene, carbon tetrachloride or carbon disulfide [1], and from Florisil by 6% ether in hexane [16]. The chromatography on silica thus separates chlorinated paraffins from organochlorine compounds such as PCB's, DDE, and hexachlorobenzene. Since these experiments have been carried out with C_{24} chlorinated paraffins, there is some uncertainty about the behaviour of the C_{12} preparations under these chromatographic conditions.

Friedman et al. [10] used UV irradiation of the 6% ether fraction from Florisil to eliminate the interference by PCB's, DDE, and many other organochlorine pesticides in the subsequent quantitation. Svanberg et al. [23] cleaned up extracts by treatment with concentrated sulfuric acid.

Quantitation. C_{24} chlorinated paraffins are too non-volatile to be subjected to gas chromatography without decomposition and have been quantitated either by direct microcoulometry of cleaned-up extracts [24] or by sophisticated thin-layer chromatography, including forward and reverse solvent development of the plates, heat transfer from silica to alumina, and detection by silver nitrate [1].

Gas chromatography with microcoulometric [16] or mass spectrometric [14] detection can be used to quantitate C_{12} chlorinated paraffins. OV-1 or OV-101 columns, programmed from about 100 to 300 °C were used. Separation of chlorinated paraffins into distinct peaks was not accomplished. Chlorinated paraffins were eluted as a broad envelope with some indication of individual peaks. The data of Lahaniatis et al. [14] indicate that the retention time increases both with increasing carbon chain length and with increasing degree of chlorination. The components identified in a Witaclor 50 preparation ranged from $C_{12}H_{24}Cl_2$ to $C_{16}H_{27}Cl_7$.

Svanberg et al. [23] quantitated chlorinated paraffins by determining chlorine by neutron activation analysis.

Confirmation. Reductive dechlorination to the parent paraffinic stock can be used to confirm the presence of chlorinated paraffins. Zitko [26] and Panzel and Ballschmitter [19] used sodium bis (2-methoxyethoxy) aluminum hydride; Lahaniatis et al. [14] used sodium in ammonia.

Sosa [22] described the application of infrared spectrophotometry to detect potentially toxic chlorinated kerosine and gas-oil fractions in chlorinated paraffins. These fractions may be used to decrease the viscosity of C_{24} chlorinated paraffins, and may be present in amounts up to 15%. Smaller additions may not be detectable by this technique.

The possibility of such blending in the formulation of chlorinated paraffin preparations compounds the problems in the determination of these materials. Solvent partitioning may be used to detect differences between chlorinated paraffin preparations due to either different stocks of the same nominal carbon chain length or to blending [27].

Chlorinated Paraffins in the Environment

Because of the lack of selective and sensitive analytical techniques for chlorinated paraffins, very little is known about their occurrence and fate in the environment.

Many of the applications of chlorinated paraffins (additives to plastics, oils, and paints) are similar to the past "open-ended" applications of PCB's and are potentially likely to lead to environmental contamination. On the other hand, chlorinated paraffins are much less stable thermally than PCB's and, in all probability, are not released unchanged on incineration of solid waste, although some chlorinated fragments of low molecular weight might be released.

No data on the solubility of chlorinated paraffins in water are available. It can be expected that the solubility decreases with increasing carbon chain length and that it is generally extremely low (1 µg/L or less). Consequently, the tendency of chlorinated paraffins to be adsorbed on suspended solids in the aquatic environment should be very high. Many compounds with these properties tend to accumulate in aquatic fauna.

Although the data [25] are somewhat inconclusive, they certainly indicate that C_{24} chlorinated paraffins with 40 and 70% chlorine are not accumulated, or accumulated much less than PCB's by fish exposed to contaminated solids or food. On the other hand, C_{12} chlorinated paraffins with 50% chlorine were detected in fish following administration of contaminated food [16]. The accumulation coefficient was about 0.1. Chlorinated paraffins (Chloroparaffin Hüls 70C, C_{12}, 70% chlorine) were also taken up by fish (bleaks, *Alburnus alburnus*) from water [23]. The accumulation coefficient, based on nominal concentration in water, was about 460. Accumulation coefficients and excretion half-lives of chlorinated paraffins, estimated from the data of Bengtsson et al. [4], are given in Table 4.

Table 4. Accumulation coefficients and excretion half-lives of Witaclor chlorinated paraffins in bleaks *(Alburnus alburnus)*. Calculated from the data of Bengtsson et al. [4]

Formulation	Carbon chain length	% chlorine	Accumulation coefficient[a]	Half-life (days)
Witaclor 149	10–13	49	770	13
Witaclor 159	10–13	59	740	34
Witaclor 171P	10–13	71	140	7
Witaclor 350	14–17	50	40	30
Witaclor 549	18–26	49	10	7

[a] 14 days exposure

The data are insufficient to place much reliability on the estimated excretion half-lives. The accumulation coefficients indicate that the accumulation decreases with increasing carbon chain length and chlorine content.

The laboratory data are in agreement with the environmental survey for chlorinated paraffins presented by Baldwin and Bennett [1]. Out of 52 samples, representing eggs of 4 species of aquatic birds, 6 species of fish, and 2 species of shellfish, C_{24} chlorinated paraffins were detected in only one sample at about 0.06 μg/g, close to the detection limit of the method. C_{12} chlorinated paraffins were detected in 13 samples at levels around 0.05 μg/g.

Little is known about the persistence of chlorinated paraffins in the environment. Biodegradation data quoted by Howard et al. [11] indicate a limited biodegradation of chlorinated paraffins by an acclimated sewage seed, but the data are quite inconclusive. The levels of C_{24} chlorinated paraffins with 40 and 70% chlorine decreased in marine sediments to about 20% of the initial concentration after 28 days under anaerobic conditions, and the decrease was somewhat less pronounced under aerobic conditions [27]. The degradation products have not been identified and products containing polar groups would not have been extracted. Consequently, the degree of degradation might have been less than the data would indicate.

From the available information it appears that chlorinated paraffins have a lower potential than PCB's for contamination of the environment. However, the data are limited and a comprehensive assessment is impossible at the moment.

References

1. Baldwin, M.K., Bennett, D.: Analysis of Biological Samples for Chlorinated Straight-Chain Paraffins. Group Research Report TLGR.0058.74. Tunstall Laboratory 1974
2. Ballistreri, A., Foti, S., Montaudo, G., Pappalardo, S., Scamporrino, E.: Thermal Decomposition Products from Mixtures of Chlorinated Paraffin with Sb_2O_3 and $(BiO)_2CO_3$. Chem. Ind. (Milan), *60,* 501 (1978)
3. Bellorin, C., Sosa, J. M.: J. Appl. Polym. Sci. *22,* 851 (1978)
4. Bengtsson, B.-E., Svanberg, O., Lindén, E., Lunde, G., Bauman Ofstad, E.: Ambio *8,* 121 (1979)
5. Borror, J.A.: Personal Communication, Diamond Shamrock Organics Research & Development, Electro Chemicals Division, 1979
6. Bratolyubov, A.S.: Chem. Revs. (Russian) *30,* 602 (1961)
7. Caesar, H.J.: Chem. Ind. (London) *1978,* 615
8. Camino, G.C.O., Costa, L., Guaita, M.: J. Calorim. Anal. Therm. [Prepr.] *9A, B8,* 59 (1978)
9. Frensdorf, H.K., Ekiner, O.: J. Polym. Sci. *42,* 1157 (1967)
10. Friedman, D., Lombardo, P.: J. Assoc, Offic. Anal. Chem. *58,* 703 (1975)
11. Howard, P.H., Santodonato, J., Saxena, J.: Investigation of Selected Potential Environmental Contaminants: Chlorinated Paraffins. EPA-560/2-75-007, Office of Toxic Substances, U. S. Environmental Protection Agency, Washington, D.C. 20460, 1975
12. Ikeda, K.: Chem. Economy Engng. Rev. *9,* 31 (1977)
13. Koennecke, H.G., Hahn, P.: J. prakt. Chem. *16,* 37 (1972)
14. Lahaniatis, E.S., Parlar, H., Klein, W., Korte, F.: Chemosphere *1975,* 83
15. Lindén, E., Svanberg, O.: Chlorinated Paraffins in the Environment. A Literature Survey, Statens Naturvårdsverk, SNV PM 1035, NBL Rapp. 62, Nyköping 1978
16. Lombardo, P., Dennison, J.L., Johnson, W.W.: J. Assoc. Offic. Anal. Chem. *58,* 707 (1975)
17. Migliori, F.: The Application of Gel Permeation Chromatography to the Analysis of Binders in Paints Used for Road Marking. Rapp. Rech. LPC No. 77, Min. L'Equip. L'Amenag. Territoire, Laboratoire Central Des Ponts et Chaussees, Paris 1978
18. Mills, J.F.D.: Personal Communication, ICI Mond Division (1979)
19. Panzel, H., Ballschmitter, K.: Fresenius Z. Anal. Chem. *271,* 182 (1974)
20. Saito, T., Matsumura, Y.: Polymer J. *4,* 124 (1973)
21. Sosa, J.M.: J. Polym. Sci., Polym. Chem. Edit. 13, 2397 (1975)
22. Sosa, J.M.: Brit. Polym. J. *7,* 161 (1975)
23. Svanberg, O., Bengtsson, B.-E., Lindén, E., Lunde, G., Baumann, E.: Ambio *7,* 64 (1978)
24. Zitko, V.: J. Chromatogr. *81,* 152 (1973)
25. Zitko, V.: Bull. Environ. Contam. Toxicol. *12,* 406 (1974)
26. Zitko, V.: J. Assoc. Offic. Anal. Chem. *57,* 1253 (1974)
27. Zitko, V., Arsenault, E.: Chlorinated Paraffins: Properties, Uses, and Pollution Potential. Environment Canada Fisheries and Marine Service Technical Rep. 491, St. Andrews, N. B. (1974)
28. Zitko, V., Arsenault, E.: Adv. Environm. Sc. Technol., I.H. Suffet (Ed.), Vol. *8*.2, Wiley Interscience, New York 1977, p. 409

Chloroaromatic Compounds Containing Oxygen

Phenols, Diphenyl Ethers, Dibenzo-p-dioxins and Dibenzofurans

C. Rappe

Department of Organic Chemistry, University of Umeå
S-901 87 Umeå, Sweden

Chlorophenols

A monograph covering the chemistry, pharmacology and environmental toxicology of pentachlorophenol has recently been published [1].

Production, Use, Contaminants

Chlorinated phenols are manufactured in large amounts and the most widely used methods by which they are prepared, are direct chlorination and alkaline hydrolysis of the appropriate chlorobenzene, the particular method used is depending on the isomer desired. 2,4-Di-, 2,4,6-tri-, 2,3,4,6-tetra, and penta-chlorophenol (PCP) are prepared by direct chlorination, while 2,4,5-tri- and pentachlorophenol are manufactured via the hydrolysis of chlorobenzenes.

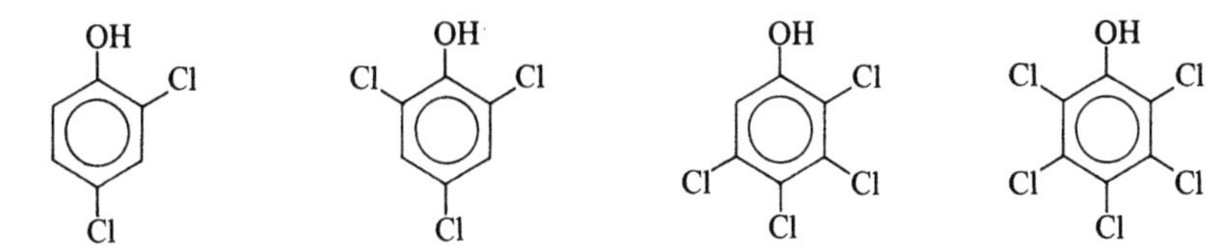

2, 4–Dichloro–phenol | 2, 4, 6–Trichloro–phenol | 2, 3, 4, 6–Tetra–chlorophenol | Pentachlorophenol

Isomers manufactured by direct chlorination

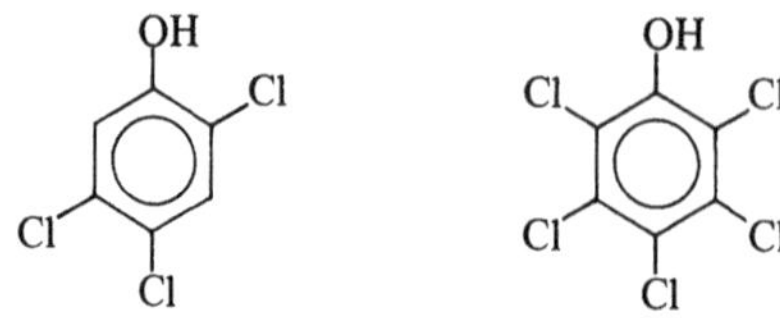

2, 4, 5–Trichlorophenol Pentachlorophenol

Isomers manufactured by hydrolysis of chlorobenzenes

Chlorinated phenols have found use in a great diversity of applications. The most important use of 2,4,6-tri, 2,3,4,6-tetra-, and pentachlorophenol is as wood perservatives, in addition they are used as bactericides, insecticides and herbicides; pentachlorophenol also being used as slimicide in pulp and paper mills and for curing hides. In the United States PCP is the second heaviest in use of all pesticides [2]. 2,4-Di- and 2,4,5-trichlorophenol are mainly used as starting materials for the phenoxy herbicides 2,4-D and 2,4,5-T. The US production of PCP is about 25,000 tons [2] and the total world production of all chlorophenols is estimated to be in the range of 150,000 tons [3]. The annual consumption of chlorophenols in Canada has been estimated to 1,500 tons [4]. In Sweden they are banned as wood perservatives from January 1, 1978, previously the annual consumption was estimated to 150 tons [5]. Most chlorophenol preparations are contaminated to a greater or lesser extent with a variety of products. In addition to other chlorophenols, they contain up to 8% of polyhalogenated phenoxyphenols, "predioxins" [6,8]. Polychlorinated diphenyl ethers, dioxins and dibenzofurans are often present in the range of ten to thousands of mg/kg, the levels and identity of these toxic contaminants depending on the route of synthesis [9–11] see also section "Occurrence of PCDDs and PCDFs in industrial chemicals". Lindström and Nordin [12] have recently identified 2,4,6-trichlorophenol, chloroguajacols and chlorocatecols in spent bleach liquors from sulphate kraft mills.

Analytical Methods

A variety of analytical techniques have been used to detect chlorophenols. These have included colorimetric methods, ultraviolet and infrared absorption, and paper, thin layer and gas chromatography. After suitable derivatization, gas chromatography seems to be the most sensitive, rapid and specific methods [13, 14].

Physical and Chemical Properties

All chlorophenols are solids at room temperature and they all have a pungent odor. The lower chlorinated chlorophenols are insoluble in water, ethanol, ether and acetone while the higher chlorinated ones are soluble in ethanol, ether and acetone. The volatility of the compounds generally decreases and the melting and boiling point generally increase as the number of chlorine atoms substituted on the benzene ring increases.

Transport Behaviour

Mass balances and flow diagrams of chlorophenol transport through the environment are not available because of the general lack of monitoring data. Their moderate volatility would suggest that atmospheric transport may be a significant route. However, in general they are considered as water and soil contaminants.

Chemical and Photochemical Reactions

The chlorophenols are weak acids and they will form ethers, esters and salts due to the phenol function. The lower chlorinated isomers easily undergo substitution reactions such as halogenation, nitration, alkylation and acylation.

When a dilute aqueous solution of pentachlorophenol was irradiated with sunlight or UV-light the main reaction was dechlorination to lower chlorinated phenols, but tetrachlorodihydroxyl benzenes and non-aromatic products like dichloromaleic acid could also be identified [15]. Irradiation of an aqueous solution of the sodium salt of PCP yielded octachlorodioxins [16]. Thermal degradation of salts of chlorinated phenols results in the formation of high yields of dioxins [17, 18] see also section "Formation of PCDDs and PCDFs".

Metabolism and Biodegradation

The microbial metabolism of PCP has been studied by Reiner et al. [19]. Three metabolites were isolated and identified as tetrachlorohydroquinone, tetrachlorobenzoquinone and trichlorohydroxybenzoquinone. PCP absorbed by goldfish and rainbow trout was quickly excreted into the surrounding water, mostly in a conjugated form accompanied with small amount of free PCP. The conjugate was identified as the PCP sulphate [20, 21]. In rats it has been shown that rapid dechlorination of PCP occurs. The dechlorination is mediated by liver microsomal enzymes and the dechlorination products formed are tetra- and trichloro-p-hydroquinone [5]. Autoradiographic studies after administration of ^{14}C PCP in rats and mice showed high levels in liver, kidney, blood and the gastrointestinal tract. These investigations also showed that PCP is fairly rapidly eliminated from the body. In the rat 90% of a single dose is eliminated within 3 days [5, 22].

Accumulation and Persistence

Pierce and others studied the fate of PCP in an aquatic ecosystem after an accidental spill [23]. PCP was found to persist in the water and in fish for over six months. Sediment samples retained high concentrations of PCP throughout the two-year period of investigation. Pentachloroanisol was found to be a major conversion product. 2,4,6-Trichlorophenol, tetra- and trichloroguaiacol were shown to bioaccumulate in liver fat of fish cought in the vicinity of a pulp mill producing full bleach sulphate pulp, the levels were in the range 0.4–11.5 mg/kg fat [24]. Dougherty and Piotrowska found an average of 20 μg PCP/kg urine in 60 university students in USA [25]. An human monitoring program, also in USA, has detected a range of 1–193 μg/kg of PCP in the urine of 34 occupationally nonexposed individuals [26]. Similar values was found for PCP in plasma and in fat [27, 28]. All these studies suggest that contamination of human population with PCP at a level of 10–20 μg/kg is quite general in the USA.

Biological Effects

Chlorinated phenols are acute toxic for mollusks, fish and mammals, the toxicity in general increased for the higher chlorinated compounds. Most sensitive are mollusks, a concentration of 15.8 μg/l caused marked reduction in the numbers of individuals [29]. The LC_{50} values for goldfish and fathead minnow is about 200 μg/l [30]. The chlorinated phenols have an effect of uncoupling the oxidative phosphorylation. Studies on the chronic toxicity of chlorophenols must take into consideration the degree of contamination by chlorinated dimers. Fahrig et al. [31] have shown that carefully purified 2,4,6- and pentachlorophenol have a weak mutagenic effect in a mammalian spot test.

Halogenated Diphenyl Ethers

The polychlorinated diphenyl ethers (PCDPEs) have physical and chemical properties similar to the PCBs, and they have been suggested as substitutes for PCBs as hydraulic fluids and as additives to pesticides. A 4-monochloro-4′-isobutyl-diphenyl ether is now in use as dielectric fluid in capacitors. The PCDPEs also occur as impurities in commercial chlorophenols in levels as high as 1,000 ppm [9, 10]. A number of the polybrominated analogoues (Br_4-Br_{10}) are now in use as flame retardants [32, 33].

For environmental purpose the most important chemical reactions for the PCDPEs is the thermal and photochemical conversion of these compounds to polychlorinated dibenzofurans and dioxins [34–38], see also section "Formation of PCDDs and PCDFs".

No report seems to be available on the environmental contamination with PCDPEs.

Tulp et al. [39] have recently studied the metabolism of the PCDPEs. In the rat they found two metabolic pathways, the predominant reaction bring aromatic hydroxylation, primarily in the *ortho* position. Scission of the ether bond was found to be a minor metabolic pathway. In two studies PCDPEs have shown to have the tendency to bioaccumulate [40, 41].

Chlorinated Dibenzo-p-dioxins and Dibenzofurans

Polychlorinated dibenzo-p-dioxins (PCDDs) and dibenzofurans (PCDFs) are two series of tricyclic compounds, which exhibit similar physical and chemical properties. Some of these compounds have extraordinary toxic properties and were the subject of much concern. They have been involved in accidents like the Yusho accident in Japan in 1968 [42], the intoxication at horse arenas in Missouri, USA, in 1971 [43], and the accident near Seveso, Italy, in 1976 [44]. Because they are both chemically stable and lipophilic in nature, they have a potential for accumulation in food chains and therefore they present a threat for man and the environment. There is no known technical use or production of PCDDs and PCDFs, and they do not occur naturally. Environmental contamination with hazardous PCDDs and PCDFs can result from industrial and agricultural chemicals containing these toxic impurities, from the accidental formation and release of these compounds into the environment, from burning or incineration of industrial chemicals acting as precursors and by generation from suitable precursors under environmental conditions.

Chemical and Physical Data

The structures the numbering of PCDDs and PCDFs are given below.

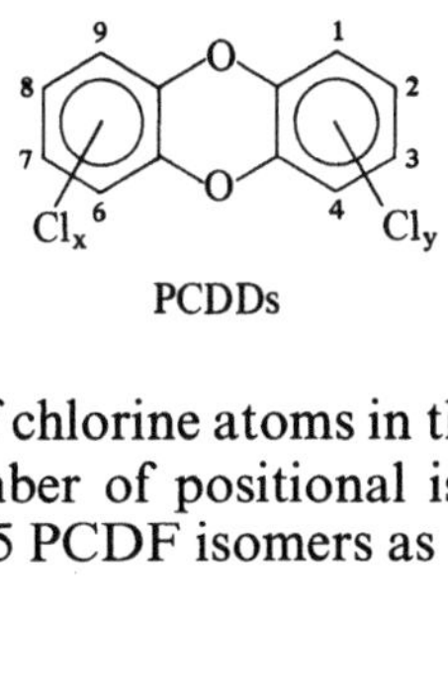

PCDDs

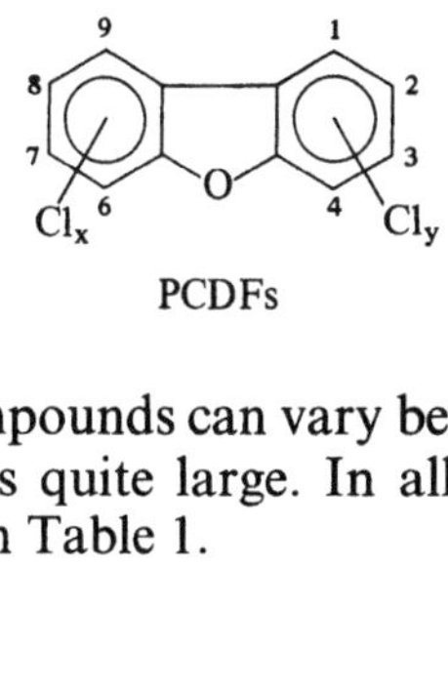

PCDFs

The number of chlorine atoms in these compounds can vary between one and eight, the number of positional isomers is quite large. In all, there are 75 PCDD and 135 PCDF isomers as shown in Table 1.

Table 1. Possible number of positional PCDD and PCDF isomers

Chlorine substitution	Number of isomers	
	PCDDs	PCDFs
mono-	2	4
di-	10	16
tri-	14	28
tetra-	22	38
penta-	14	28
hexa-	10	16
hepta-	2	4
octa	1	1
	75	135

A large number of the individual PCDDs and PCDFs have been synthesized by various methods and characterized mainly by gas chromatography/mass spectrometry [45, 46]. As a general trend in both series, solubility and volatility decreases with the incraesing number of chlorine atoms.

The most toxic and most extensively studied representative of these compounds is 2,3,7,8-tetra-CDD or TCDD.

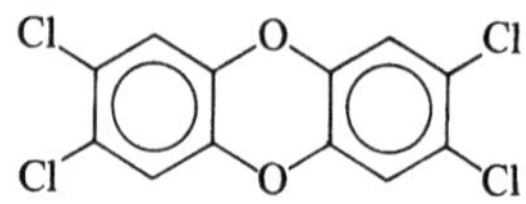

TCDD

Although TCDD is lipophilic, it is only slightly soluble in most organic solvents and very, very slightly soluble in water, see Table 2.

Table 2. Solubility of 2,3,7,8-tetrachlorobenzo-p-dioxin in various solvents at 25 °C [47]

Solvent	Solubility g/l
o-dichlorobenzene	1.4
chlorobenzene	0.72
benzene	0.57
chloroform	0.37
acetone	0.11
Herbicide Orange	0.058
lard oil	0.04
methanol	0.01
n-octanol	0.0048
water	0.0000002 (0.2 μg/kg)

The melting point of TCDD is 305–306 °C [48] but no boiling point has been given for this compound. The volatility must be quite low, although it can be analyzed by gas chromatography like the other PCDD and PCDF isomers.

The first synthesis of TCDD was reported by Sandermann et al., by the catalytic chlorination of the unchlorinated dioxin [49]. It has also been prepared in good yields by the pyrolytic dimerization of 2,4,5-trichlorophenol salts [50], and this method has been successfully used for the preparation of a large number of PCDD standards [17]. The main reactions for the synthesis of the PCDF standards are Pd (II) acetate catalyzed cyclization of chlorinated diphenyl ethers, Sandmeyer reactions [51] and pyrolytic conversion of individual PCB isomers [46].

Occurrence of PCDDs and PCDFs in Industrial Chemicals

2,4,5-T (Phenoxy Acids). The dioxin problem was first observed in connection with teratogenic effects found for the phenoxy acid 2,4,5-T. These effects were shown to be caused by 30 μg/g of 2,3,7,8-tetra-CDD present in this particular sample [52]. The levels of 2,3,7,8-tetra-CDD in drums of Herbicide Orange placed in storage in the USA and in the Pacific before 1970 have been found to vary between 0.1 and 47 μg/g [53]. Since Herbicide Orange was formulated as a 1:1 mixture of the butyl esters of 2,4-D and 2,4,5-T, the levels of 2,3,7,8-tetra-CDD in individual 2,4,5-T preparations used in the 1960's could be as high as 100 μg/g. As a result of governmental regulations, efforts were made during the 1970's to control and to minimize the formation of 2,3,7,8-tetra-CDD, and now all producers claim that their products contain less than 0.1 μg/g of 2,3,7,8-tetra-CDD.

In addition to 2,3,7,8-tetra-CDD it has also been reported that samples of Herbicide Orange as well as 2,4,5-T presently on the market may contain penta-CDDs, tetra- and penta-CDFs at similar levels [54]. However, the analytical technique used in this investigation (low-resolution GC/MS) does not allow the identification or quantification of the individual PCDD and PCDF isomers. In analyses using high-resolution GC/MS and MS confirmation, Rappe et al., have reported that in other samples of Herbicide Orange, as well as in European 2,4,5-T formulations from the 1960's, 2,3,7,8-tetra-CDD was the dominating compound of this group. Only minor amounts of other PCDDs and PCDFs could be found, primarily lower chlorinated PCDDs in samples of Herbicide Orange [55].

Up to now most of the interest from scientists, regulatory agencies and the public has been concentrated on 2,4,5-T as the only source of PCDDs and PCDFs. However, using the analytical techniques now available, it has been possible to identify other important sources of these hazardous products.

Chlorophenols. Chlorophenols have been found to contain a variety of contaminants, including PCDDs and PCDFs [3]. In female rats the hepatic effects of technical and pure grade pentachlorophenol have been investigated. The technical product, which was heavily contaminated, produced a number of hepatic effects that cannot be attributed to the pentachlorophenol itself, but were consistent with effects associated with PCDDs and PCDFs, present at the 100–1000 μg/g level [56].

It has been reported that several positional isomers of PCDDs and PCDFs can be found in chlorinated phenols. The presence of these isomers, as well as the relative ratio of PCDDs to PCDFs, seems to be dependent on the synthetic route, by which the chlorophenol is prepared. A series of pentachlorophenols from commercial sources in Switzerland, and possibly prepared by the alkaline hydrolysis of hexachlorobenzene, was found to have a ratio for PCDFs/PCDDs of about 1 and an almost identical pattern of hexa- and hepta-CDD isomers [10]. On the contrary, in commercial Scandinavian 2,4,6-tri- and 2,3,4,6-tetrachlorophenols, both prepared by chlorination of phenol, the ratio PCDFs/PCDDs was found to be greater than 20. The toxic 1,2,3,7,8,9-hexa-

CDD, which was the major hexa-CDD in the pentachlorophenol samples, was only a minor isomer in these lower chlorinated samples. A difference was also seen in the case of the hepta-CDDs [18, 57]. Concerning the PCDFs, all products contained the same major PCDF isomers (1,2,4,6,8-penta-, 1,2,3,4,6,8-, 1,2,4,6,7,8- and 1,2,4,6,8,9-hexa-, 1,2,3,4,6,7,8- and 1,2,3,4,6,8,9-hepta-CDF), but the ratio was different between the two groups of chlorophenols. In general the suspected most toxic isomers were present only as minor components. However, in the Scandinavian 2,4,6-trichlorophenol the toxic 2,3,7,8-tetra-CDF was found at a level of 0.5 mg/kg [11]. Obviously, different reaction conditions in the synthesis of these chlorophenols are the cause for the changed ratio between the amounts of PCDDs and PCDFs, and for the different PCDD and PCDF isomers being formed. These data are of importance in evaluating the risks associated with the production and use of these products.

Polychlorinated Biphenyls (PCBs). In 1970 Vos et al. identified PCDFs as toxic impurities in European PCBs at the μg/g level. The toxic effects of these PCBs were found to be parallel to the levels of PCDFs [58]. The same impurities have also been reported in American and Japanese PCBs [59]. Using packed column GC and MS Bowes et al. found that the most abundant PCDFs had the same retention times as 2,3,7,8-tetra- and 2,3,4,7,8-penta CDF [60]. Using high-resolution GC and MS it has been shown that commercial PCBs contain quite a complex mixture of PCDFs (up to 40 different isomers) [61]. A PCB used for two years in a heat exchange system was found to have a four-fold increase in PCDFs (15–20 μg/g). The dominating isomer was identified as 2,3,7,8-tetra-CDF at a level of 1.25 μg/g [62].

Hexachlorophene. The bactericide hexachlorophene is prepared from 2,4,5-trichlorophenol, the key intermediate in the production of 2,4,5-T. Due to additional purification, the level of 2,3,7,8-tetra-CDD in this product is usually <0.03 μg/kg. However, hexachlorophene also contains about 100 mg/kg of a hexachloroxanthene, the 1,2,4,6,8,9-substituted isomer [63, 64].

PCDDs and PCDFs in Fly Ash. Recently, Olie et al. reported on the occurrence of PCDDs and PCDFs in fly ash and flue gases of municipal incinerators in the Netherlands [65]. No quantitative data were given in this report, but more recently Buser and Bosshardt made a quantification that the total amount of PCDDs and PCDFs in fly ash from a municipal incinerator in Switzerland was 0.2 μg/g and 0.1 μg/g, respectively, and the fly ash from an industrial heating facility, also in Switzerland, was found to have 0.6 μg/g and 0.3 μg/g, respectively [66]. In additional studies it has been shown that the number of individual isomers was quite large with up to 30 PCDD and over 60 PCDF isomers [67, 68]. The highly toxic PCDDs (2,3,7,8-tetra-, 1,2,3,7,8-penta-, and 1,2,3,6,7,8- and 1,2,3,7,8,9-hexa-CDD) were only minor constituents whereas the known toxic PCDFs (2,3,7,8-tetra-, 1,2,3,7,8- and 2,3,4,7,8-penta-CDF) were major constituents.

Formation of PCDDs and PCDFs

The photochemical dimerization of chlorophenols to PCDDs has been studied by Crosby et al. [16]. The only PCDD formed in this study was the octa-CDD. Other PCDDs can be formed by photochemical cyclization of chlorinated o-phenoxyphenols, so called predioxins [69]. These predioxins are very common impurities (1–8%) in commercial chlorophenols [6–9], but the cyclization is only a minor reaction pathway, the main reaction being the dechlorination of the predioxin [69].

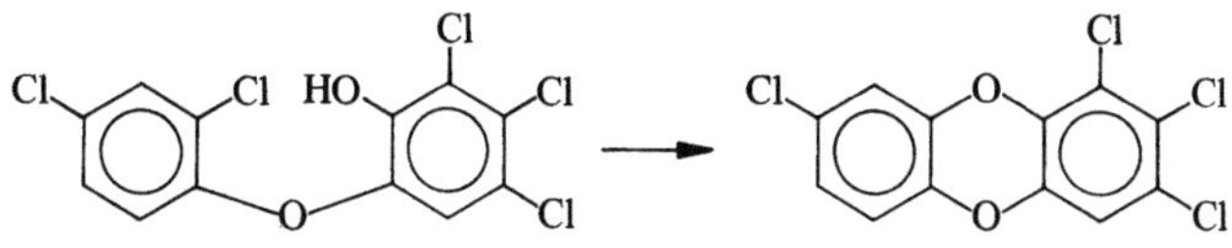

1, 2, 3, 8–tetra–CDD

Another photochemical process of potential environmental importance is dechlorination of the higher chlorinated PCDDs and PCDFs, octa-CDD and octa-CDF [70]. The products formed in solution photolysis of octa-CDD have now been identified. By comparison with authentic standards it was found that the main tetrachloro isomer was the 1,4,6,9-tetra-CDD; the major pentachloro compound is expected to be the 1,2,4,6,9-isomer. The main hexa- and heptachloro compounds were the 1,2,4,6,7,9- (or 1,2,4,6,8,9-) and the 1,2,3,4,6,7,9-isomer, respectively. The reaction scheme deduced from this data shows that the chlorine atoms are removed preferably from the lateral positions on the carbon rings. Consequently the most toxic PCDD isomers such as 2,3,7,8-tetra-CDD are not likely to be formed from the solution photolysis of the higher PCDDs [45]. In the case of the octa-CDF the photochemical loss of chlorine seems to be a non-specific reaction: all four possible hepta-CDFs were formed in about similar amounts [71].

The identification of PCDDs and PCDFs in fly ash and flue gases indicates that these hazardous compounds can be formed in pyrolytic processes or by burning. A recent report [72] referred to these compounds as possibly ubiquitous products of all combustion processes. Salts of chlorophenols have been found to undergo a pyrolytic dimerization yielding PCDDs. This reaction has been used for the preparation of a large number of PCDD standards [45]. The main PCDDs found in the fly ash are the same as those formed in a laboratory pyrolysis of a mixture of 2,4,5-tri-, 2,3,4,6-tetra- and pentachlorophenol, the most commonly used commercial chlorophenols, suggesting these products to be precursors to PCDDs in fly ash [67].

In a recent publications Rappe et al. have studied the burning of materials impregnated with various chlorophenates [18]. The burning of materials impregnated with commercial 2,3,4,6-tetrachlorophenates yielded a total of 150–1,000 μgPCDDs/g chlorophenate ranging from the tetra- to hepta-CDDs. The 2,3,7,8-tetra-CDD (peak 3, Fig. la) and 1,2,3,7,8-penta-CDD (peak 13) were both present as minor constituents. These two compounds are not the expected dimerization products of any chlorophenate present in these

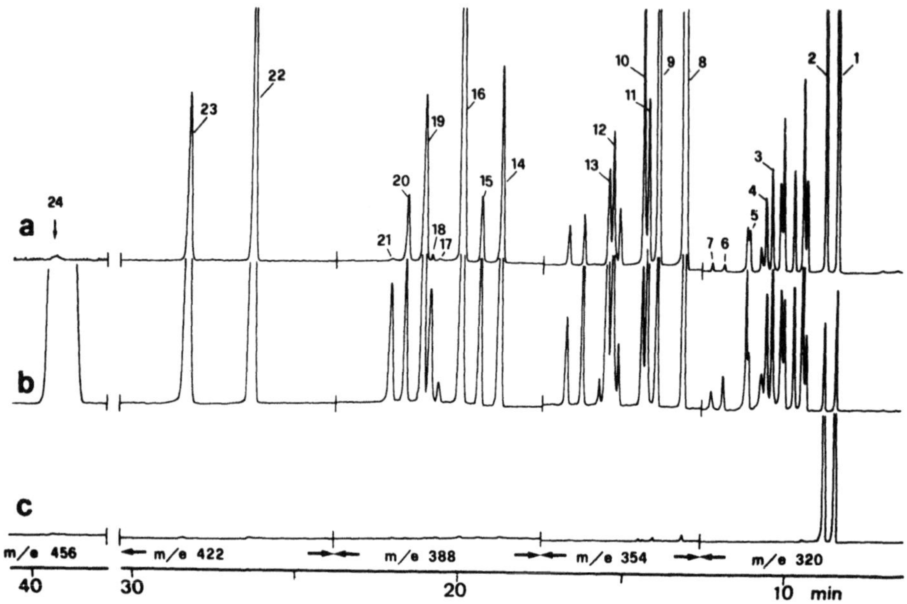

Fig. 1a–c. Mass fragmentograms showing elution of PCDDs from the burning of birch leaves of **a** commercial 2,3,4,6-tetrachlorophenate, **b** purified pentachlorophenate, **c** purified 2,4,6-trichlorophenate. (From Rappe et al. [18])

commercial mixtures and it was suggested that they are formed in cyclization reactions of impurities present in the commercial formulation [18].

The mass fragmentograms in Fig. 1b, c are from the burning of purified penta- and 2,4,6-trichlorophenate. In the case of the 2,4,6-trichlorophenate, the only tetra-CDDs were the 1,3,6,8- and 1,3,7,9-isomers (peaks 1 and 2), the expected dimerization products. In the case of the pure pentachlorophenate, the octa-CDD was the main product, but unexpectedly in addition, both hepta-CDDs, eight hexa-CDDs and several penta- and tetra-CDDs were detected and identified. Of special interest is the observation of the highly toxic 2,3,7,8-tetra-(peak 3) and 1,2,3,7,8-penta-CDD (peak 13) in these burning extracts, see Fig. 1. Although they are found to be minor constituents only, in individual burnings both have been found at levels exceeding 10 mg/kg chlorophenate. In this case, the formation of the lower chlorinated PCDDs takes place in a so far unknown nonspecific dechlorination process [18].

In other studies we have found that PCBs can be converted to PCDFs under pyrolytic conditions [68, 73]. The pyrolysis of commercial PCBs (Aroclor 1254 and 1260) yielded about 30 major and more than 30 minor PCDFs (see Fig. 2). One of the main constituents was 2,3,7,8-tetra-CDF, the most toxic of the PCDFs (peak 44). Taking into consideration the amount of PCB recovered after the pyrolysis, the yield of PCDFs was between 3–25% calculated on the amount of PCB decomposed. Consequently, uncontrolled burning of PCB can be an important environmental source of the hazardous PCDFs. Moreover, a comparison showed a striking similarity between the

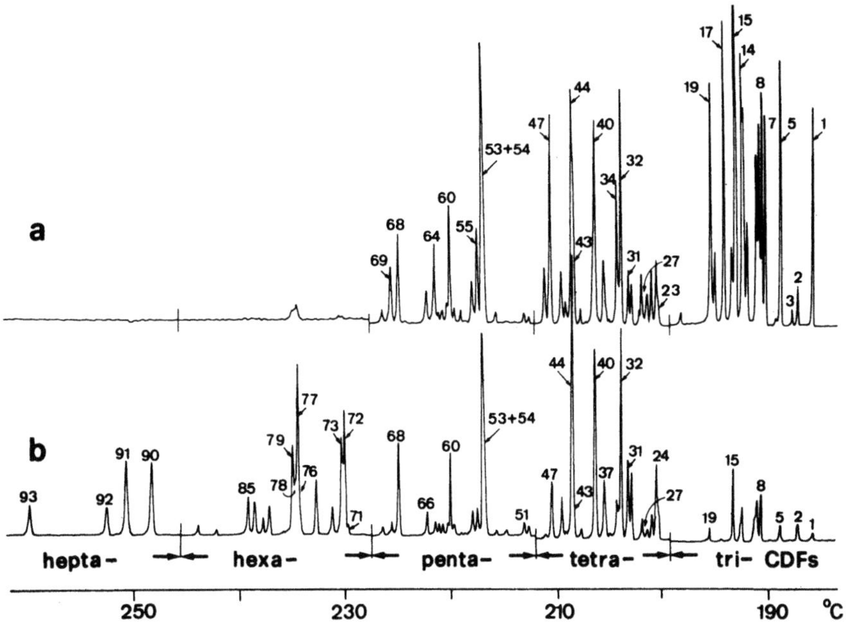

Fig. 2a, b. Mass fragmentograms showing elution of PCDFs in **a** pyrolyzed Aroclor 1254, **b** pyrolyzed Aroclor 1260. (From Buser et al. [68])

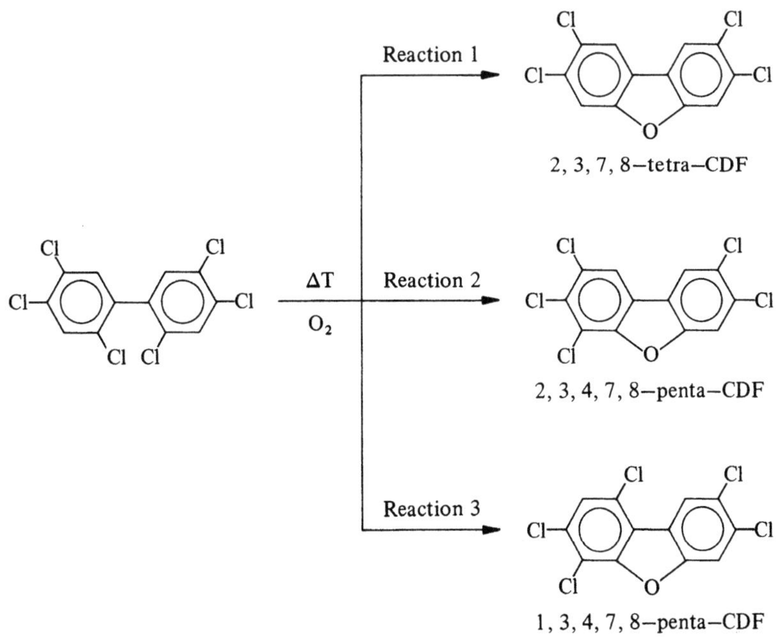

Fig. 3. Reaction routes leading to tetra- and penta-CDFs from the pyrolysis of 2,4,5,2′,4′,5′-hexachlorobiphenyl

pattern of the PCDFs in the fly ash and those formed in the pyrolysis of the commercial PCBs [68]. Therefore it is recommended that all destruction of PCB-containing wastes using incinerators must be carefully controlled, including monitoring of PCDFs in the exhaust.

Pyrolysis of individual synthetic PCB isomers showed that the formation of PCDFs can follow several reaction pathways [68]. In Fig. 3 the reaction routes leading to tetra- and penta-CDFs from 2,2′,4,4′,5,5′-hexachlorobiphenyl are illustrated, The reactions involve the loss of ortho-Cl_2 and ortho-HCl with and without a 2,3 chlorine shift. A fourth reaction route (loss of ortho-H_2) was later found to occur with some other PCB isomers [46]. Pyrolysis of the commercial flame retardant Fire Master BP 6, mainly 2,2′,4,4′,5,5′ -hexabromobiphenyl, yielded the very toxic 2,3,7,8-tetra-BDF at a yield in the percent range [73]. Consequently the use of this compound as a flame retardant should be discontinued. In the temperature range 300–400 °C however, the yield of the conversion of the halogenated biphenyls to the halogenated dibenzofurans seems to be only in the ppm range [74, 75].

In addition to PCBs, polychlorinated diphenyl ethers (PCDPEs) are found to be precursors to PCDFs. It has been shown that the thermal conversion of the ethers into PCDFs is of the same order as that from PCBs [38, 57]. In addition to PCDFs, some of the PCDPEs also yielded PCDDs [38], see Fig. 4.

It has recently been shown by Buser that the pyrolysis of chlorobenzenes yielded small amounts of both PCDFs and PCDDs. In both cases the known toxic isomers (see Biological effects) were present but not as main components [76]. The mechanisms for the synthesis of PCDDs and PCDFs from Cl-benzenes are not yet fully understood.

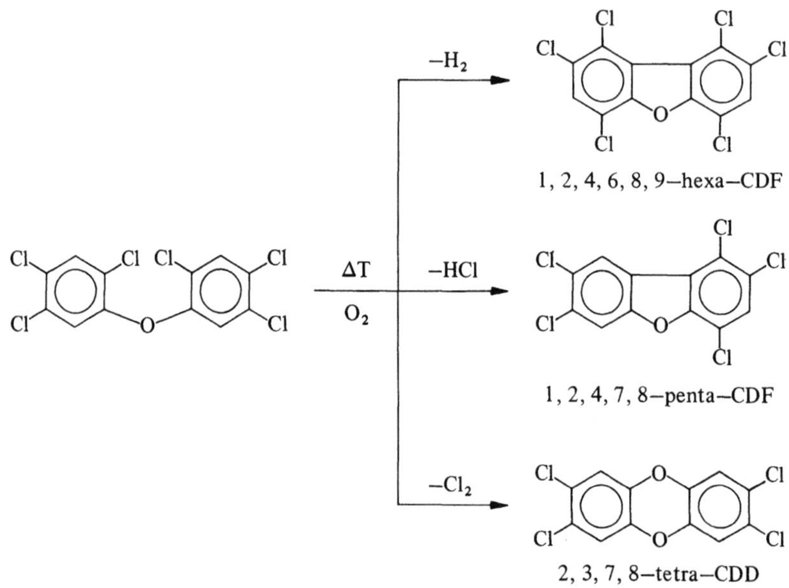

Fig. 4. Reaction routes leading to PCDFs and PCDDs from the pyrolysis of 2,4,5,2′,4′,5′-hexa-CDPE

The possible formation of 2,3,7,8-tetra-CDD and other PCDDs as the result of thermal reactions of the esters of the phenoxy acid 2,4,5-T has been the subject of much controversy. However, recent investigations have shown that this formation is only of very limited importance, if it takes place at all [77–79].

Recently 2,3,7,8-TCDD and higher chlorinated dioxins have been found in ashes from refuse incinerators, fossil-fueled power plants and fireplaces, charcoal grills, cigarettes and the emission of automobil engines both gasoline and diesel fueled [72, 80, 81]. These observations have been interpreted that dioxins can be generated in the combustion of any material. However, no data has been provided to show that TCDD and other dioxins and dibenzofurans are formed in thermal processes unless suitable precursors like chlorinated phenols, predioxins, diphenyl ethers, PCBs or chlorobenzenes are present, see Table 3.

Table 3. PCDFs formed in thermal processes

Starting material	Products	Yield (mg/kg)	Ref.
PCBs	PCDFs	10,000 - 250,000	[73]
Cl-phenols	PCDDs	1,000 - 100,000	[17, 18]
PCDPEs	PCDDs + PCDFs	10,000 + > 10,000	[38]
Cl-benzenes	PCDDs + PCDFs	2,000 + 10,000	[76]
Fly ash	PCDDs + PCDFs	0.1-1 + 0.1-1	[66]
Dow	PCDDs	< 0.001	[72]

The photochemical cyclization of PCDPEs to PCDFs has been studied by Norström et al. [34, 35] and by Choudhry et al. [36, 37]. A 20% yield of 2,8-di-CDF has been reported from the pyrolysis of 2,4,4′-tri-CDPE [34].

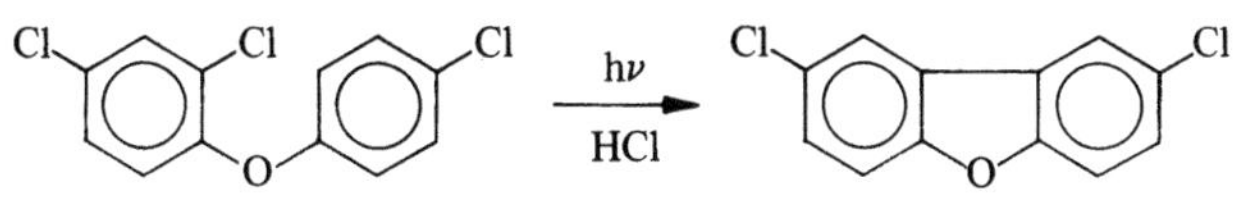

2, 4, 4′-tri-CDPE 2, 8-di-CDF

Analytical Methods

Due to the extreme toxicity of some of the PCDDs and PCDFs, very sensitive and highly specific analytical techniques are required. Detection levels in environmental and biological samples should be orders of magnitude below the usual detection limits obtained in pesticide analysis. Any analysis at such low levels is complicated by the presence of a multitude of other, possibly interfering compounds. The best available separation techniques followed by highly specific detection means have to be used for an accurate determination of these hazardous compounds. Different isomers of the PCDD or PCDF may vary significantly in their toxicological properties and therefore their separation and identification becomes important.

In recent years, many analytical methods were developed for the analysis of PCDDs, PCDFs and especially 2,3,7,8-tetra-CDD in environmental and industrial samples, the most specific methods making use of mass spectrometry [82]. Prerequisites for best analyses are efficient extraction and sample purification followed by good separation, ultra-sensitive detection and – very desirably – confirmation. A technique for analyzing individual PCDDs and PCDFs has recently been described and discussed in detail [57]. It involves one or two steps of column chromatographic clean-up followed by high-resolution gaschromatography using glass capillary columns and detection and quantification using mass spectrometry. Artifacts usually disturb the analytical work at extreme low concentration levels, but the risk can be minimized by a careful inspection of complete mass spectra. For a correct structure assignment of the PCDDs, a study of the low mass ions can be useful [45]. Positive and negative ion chemical ionization mass spectrometry (CI and NICI) has shown to be a useful complement to the normal electron impact (EI) technique for the quantification of trace amounts of PCDDs and PCDFs [25, 83, 84].

Transport in the Environment

Owing to analytical problems, no data are available on the transport of TCDD and other PCDDs or PCDFs in air and water. However, the accident at Seveso, Italy, clearly shows that such transport is of importance in local pollution. TCDD is practically insoluble in water (see Table 2), but because of its extreme toxicity even such low concentrations as 0.2 μg/1 (0.2 ppb) can be quite important. Most of the TCDD, if present in waterways, has been found in the sediments or attached to suspended particles. The half-life of TCDD in a lake sediment was found to be about 600 days [85].

The mobility of TCDD and of a dichlorodioxin in soils has been studied [86]. Both were found to be immobile in all soils and therefore would not be leached out by rainfall or irrigation, though lateral transport during surface erosion of the soil could occur.

The US Air Force conducted studies in an area in north-west Florida which had been heavily sprayed with the herbicide "Agent Orange" between 1962 and 1964. This herbicide mixture was contaminated with TCDD (see above). A 19.3 acre test grid received a total of 40 tons of 2,4,5-T between 1962 and 1964. When 6-inch core soil samples were taken in 1974, they showed TCDD concentrations ranging from 10 to 710 ppt. This study illustrates that significant levels of TCDD residues remained 10 years after the last herbicide application [87].

Chemical and Photochemical Reactions

PCDDs and PCDFs are considered to be stable compounds but due to the extreme toxicity of some of the isomers, their chemistry has not been fully evaluated. However, TCDD and other PCDDs undergoes substitution reactions [63, 88] and $SbCl_5$ has been found to react with PCDDs and PCDFs yielding higher chlorinated or perchlorinated isomers [89]. Thermally PCDDs and PCDFs are quite stable, and decomposition of 2,3,7,8-tetra-CDD occurs only at temperatures above 750 °C [90].

Under environmental conditions TCDD, like other PCDDs and PCDFs is not likely to be degraded at a significant rate by hydrolytic reactions. Earlier work indicated that pure TCDD was rather stable to photochemical degradation. Crosby et al. found the photochemical dechlorination of TCDD to be extremely slow on the soil surface [9], and Young et al. found the half-time of TCDD to be about half a year for soil [87]. Pure crystalline TCDD was stable to sunlight wavelengths when applied as thin films to glass or leaves or suspended in water [70, 91, 92].

However, several reports on rapid photochemical degradation of PCDDs and PCDFs under laboratory conditions make the situation more complicated. Rapid degradation of 2,7-di, 2,3,7,8-tetra-, and octa-CDD in solutions of methanol was shown with higher decomposition rates for the lower chlorinated species. Other experiments using 2,4,5-T ester formulations with known amounts of TCDD and exposed to natural sunlight on leaves, soil and glass plates showed that most of the TCDD was lost during a single day [91, 93]. In these two experiments a hydrogen donor, such as methanol or 2,4,5-T ester, highly enhanced the photochemical dechlorination [95]. It can be mentioned here that at Seveso the TCDD was released together with salts of 2,4,5-trichlorophenol, ethylene glycol and inorganic constituents [96], like water most of these are no potent hydrogen donors.

According to Bertoni et al., the addition of a solution of ethyl oleate in xylene enhances the breakdown of TCDD in soil by UV-light, more than 90% was degraded during 7-days exposure [97]. Similarly a cationic surfactant, 1-hexadecylpyridinium chloride was also reported to enhance photodecomposition [98].

Another experiments have shown that TCDD adsorbed on silica gel undergoes rapid photochemical degradation [99,100]. These experiments might be a good model for TCDD bound to dust particles during air transport.

Hutzinger et al. showed rapid dechlorination for di- and octa-CDF in solution and the formation of a series of lower chlorinated PCDFs was observed [101]. Similar results were obtained by Buser [71].

Metabolism and Biodegradation

Contrary to the chemical and physical effects, there is a pronounced difference in the biological effects between the different PCDD and PCDF isomers. The metabolic behaviour and biodegradation seem to be quite different for TCDD and for some of the other PCDDs and PCDFs.

Metabolism of TCDD. No metabolites of TCDD have been identified so far. Matsumura and Benezed [102] reported that most microorganism do not degrade TCDD. Of a total of 100 microbial strains with the ability to degrade persistent pesticides, only 5 strains showed some ability to degrade this compound. Ward and Matsumura [85] have studied the fate of TCDD using aquatic sediment and lake water under laboratory conditions. Evaporation is suggested to be the major mode of disappearance with metabolism playing only a minor role. The metabolic activities were enhanced under conditions which simulated microbial growth in the presence of sediment, and the unidentified metabolites were found to be released from the sediment to the ambient water.

It has recently been reported by Guenthner et al. [103] that TCDD can be metabolized by the mouse liver cytochrome P-450 system to reactive intermediates, which easily bind covalently to cellular proteins. It is suggested that this extreme reactivity inhibits the formation of normal metabolites like phenols, dihydrodiols, or conjugated products.

Following a single oral dose of ^{14}C-TCDD in rats, Rose et al. [104] were able to detect ^{14}C activity only in feces and not in urin. The half-life of ^{14}C activity in the body was about 31 days and the major part of the TCDD was stored in liver and fat.

After repeated oral doses the major route of excretion was again found to be feces, but the urin contained 3–18% of the total ^{14}C activity. The half-life of ^{14}C activity in these rats was about 24 days, and most of TCDD was found in liver and fat. The experiments indicated that materials other than TCDD constituted a significant fraction of the ^{14}C activity excreted in the feces, but no metabolite was identified [104].

Van Miller et al. [105] reported on the tissue distribution and excretion of ^{3}H TCDD in monkeys and rats. A marked difference was found in the tissue distribution in the two species. In monkeys, a large percentage of the dose was located in tissues that had a high lipid content, i.e. in skin, muscle, and fat; whereas in rats these tissues had much lower levels of TCDD.

Metabolism of Other PCDDs and PCDFs. Tulp and Hutzinger have studied the rat metabolism of a series of PCDDs. 1- and 2-Mono-, 2,3- and 2,7-di, 1,2,4-tri-, and 1,2,3,4-tetra-CDD are metabolized to mono- and dihydroxy derivatives, whilst in the case of the two monochloro isomers, also sulphur containing metabolites are excreted. It has also been shown that the primary hydroxylation exclusively takes place in the lateral positions (2-, 3-, 7- and/or 8-positions) in the molecule. In none of the experiments metabolites resulting from a fission fo the C–O–C bonds were detected. No metabolites were found from octa-CDD [105].

The results are rationalized in terms that the metabolism of the PCDDs occurs mainly via 2,3-epoxides. In the octa-CDD as in 2,3,7,8-tetra-CDD these positions are blocked, consequently the reaction is less likely to take place or takes place at a highly reduced rate [105].

Isomers retained:

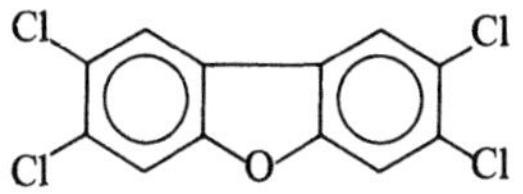

2, 3, 7, 8–

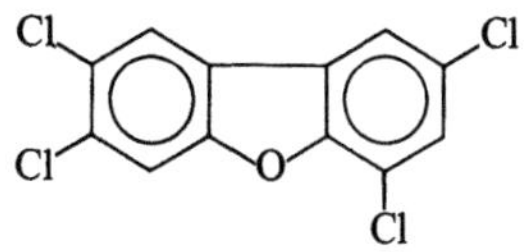

2, 3, 6, 8–

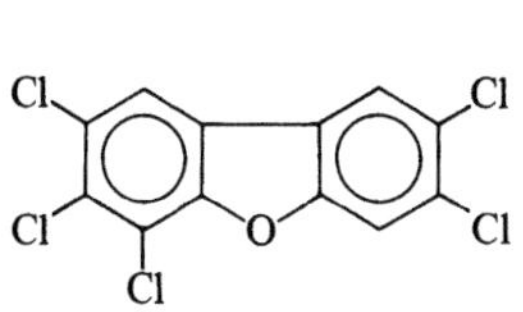

2, 3, 4, 7, 8–

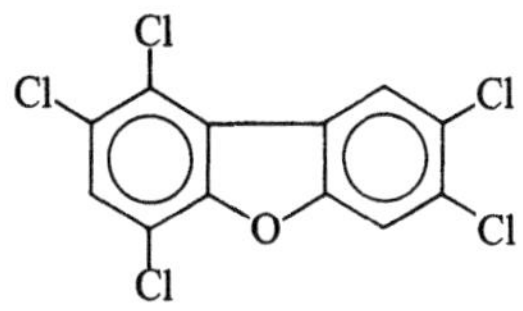

1, 2, 4, 7, 8–

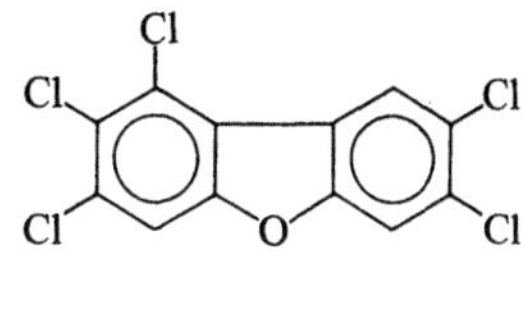

1, 2, 3, 7, 8–

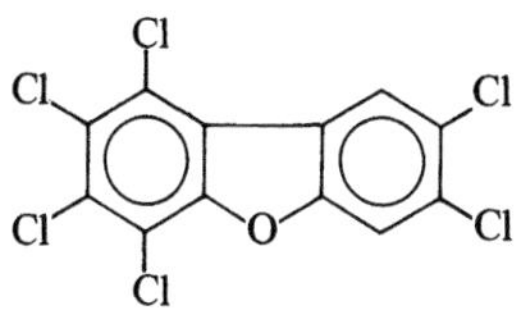

1, 2, 3, 4, 7, 8–

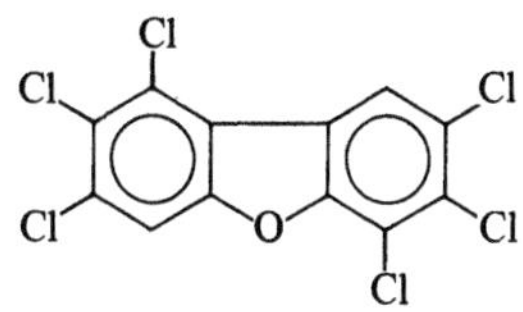

1, 2, 3, 6, 7, 8–

Isomers excreted:

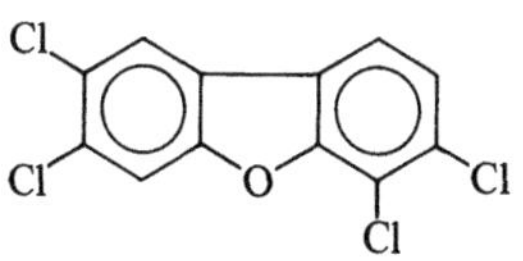

2, 3, 6, 7–

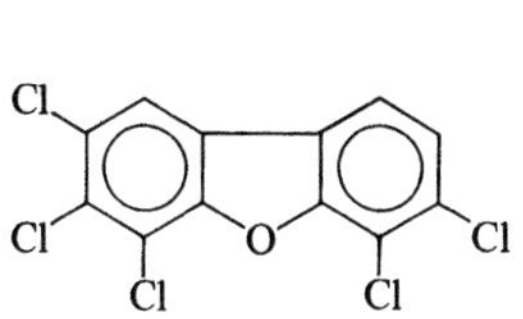

2, 3, 4, 6, 7–

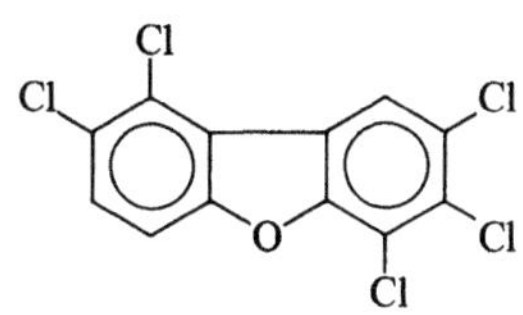

1, 2, 6, 7, 8–

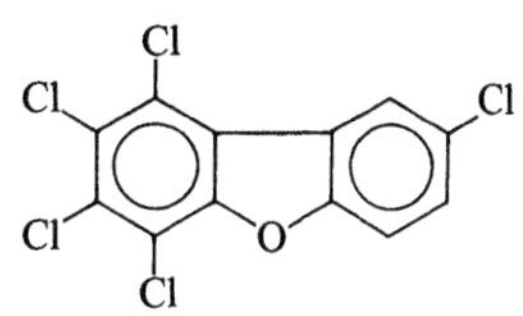

1, 2, 3, 4, 8–

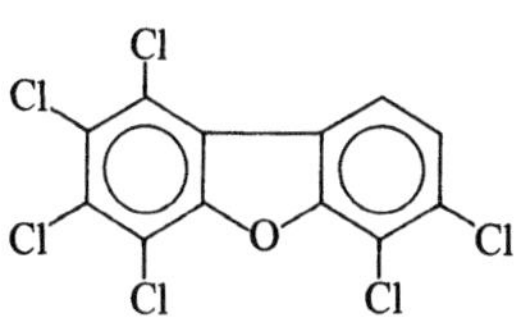

1, 2, 3, 4, 6, 7–

Fig. 5. PCDF isomers retained and excreted from the liver of Yusho patient. (From Rappe et al. [106])

A similar relationship between PCDF isomers retained and apparently excreted has been observed for patients with the Yusho disease, intoxication by a rice oil contaminated with PCBs and PCDFs. The contaminated rice oil and liver samples from two of the patients were analyzed by Rappe et al. [106] and all the major PCDFs were identified, see Fig. 5. A comparison revealed that none of the isomers retained had two *vicinal* hydrogenated C-atoms in any of the two C-rings of the benzofuran system. Most of these isomers had all lateral positions chlorinated. Contrary, all the PCDF isomers apparently excreted had two *vicinal* hydrogenated C-atoms in at least one of the two rings, and these unblocked positions are involved in the metabolism by forming epoxides, see Fig. 5.

Kuroki and Masuda [107] have estimated that 0.37% of 2,3,6,8-tetra, 0.006%–0.03% of 2,3,7,8-tetra and 0.9% of the 2,3,4,7,8-penta-CDF ingested were retained in the liver of one of the Yusho patients when he died 44 months after the use of the rice oil had been discontinued. Zitko et al. have shown that in fish, 2,8-di-CDF was metabolized to a hydroxylated derivative [108].

Accumulation and Persistence

When ^{14}C 2,7-di- and 2,3,7,8-tetra-CDD was added to soil, Isensee and Jones [92] found that both oats and soya beans accumulated small quantities of the dioxins. A maximum of 0.15% of the dioxins present in the soil was translocated to the aerial portion, but neither the grain nor the soya beans harvested showed any dioxins. Analyses quoted by Firestone [53] showed that TCDD analyses of vegetation from Seveso, Italy, gave values up to 50 mg/kg possibly due to direct contamination.

The bioaccumulation of ^{14}C-TCDD in aquatic organisms was investigated by Isensee and Jones [109], and the accumulation ratios were 2,000–7,000 times, which is about the same as those reported for many chlorinated hydrocarbon insecticides. Total amounts accumulated were directly related to water concentrations, and equilibrium concentrations were reached in tissues in 7–15 days.

Fish and shellfish taken from areas in South Vietnam that were heavily exposed to Herbicide Orange during military defoliation operations have been reported by Baughman and Meselson [110] to contain 18–810 ng TCDD/kg.

Young et al. [87] reported that in two creeks, associated with a military test area in Florida, USA, which also had been heavily sprayed with Herbicide Orange, 10 yr later the silt contained up to 35 ng TCDD/kg where eroded soil entered the water. Fish from the streams showed the presence of TCDD, the highest value reported was in the gut, 85 ng/kg. Young et al. [87] also reported on the analyses of terrestal animals (rodents reptiles) from the same area also collected 10 yr after the spraying, which were found to contain 130–1,300 ng TCDD/kg. The analytical method(s) used in these investigation is not specified.

After the accident near Seveso, Italy, over 1,100 animals were killed by direct exposure, and up to 225 μg TCDD/kg of tissue was found in the liver of dead rabbits from the most contaminated zone [53].

Using a direct probe and NICI technique, Dougherty et al. [111] have identified 2,3,7,8-tetra-CDD and found higher chlorinated PCDDs (Cl_5–Cl_8) in fish from dams in Tittabawassee River, Michigan, USA. These dams were located at Dow Chemical Inc. Contrary, in fish from Ohio River and Connecticut River, a series of PCDFs (Cl_4–Cl_7) was found, but no PCDDs. No identification of the discret isomers or quantification was possible using this analytical technique.

The US EPA initiated a TCDD monitoring programme of beef fat samples taken from cattle that had grazed on rangeland known to have been treated with 2,4,5-T. Three laboratories were involved in this programme. Of 52 samples, 19 were reported by one or more laboratory to have TCDD, in the range 5–66 ng/kg, and the overall average was 7 ng/kg [53, 82].

Tissues and milk from cattle possibly intoxicated by licking pentachlorophenol-treated timber has been analyzed by Hass et al. [84] for higher chlorinated PCDDs. Hexa-, hepta-, and octa-CDD were found in the µg-ng/kg range, the highest values found for the octa-CDD.

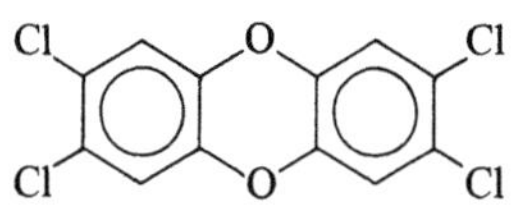

2, 3, 7, 8–tetra–CDD

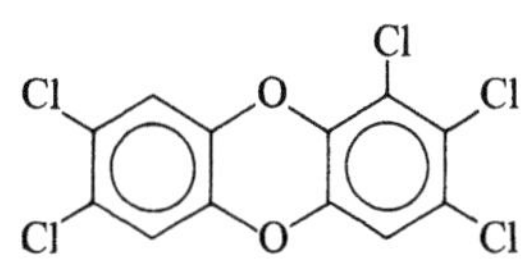

1, 2, 3, 7, 8–penta–CDD

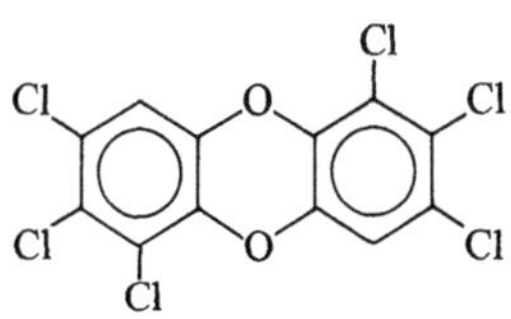

1, 2, 3, 6 7, 8–hexa–CDD

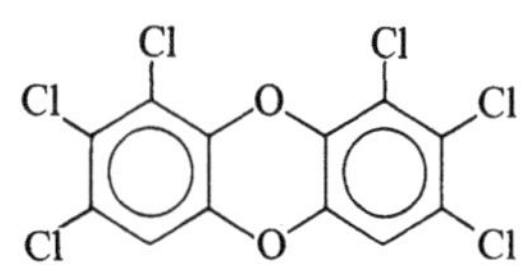

1, 2, 3, 7, 8, 9–hexa–CDD

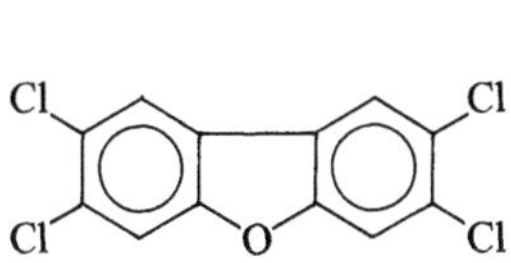

2, 3, 7, 8–tetra–CDF

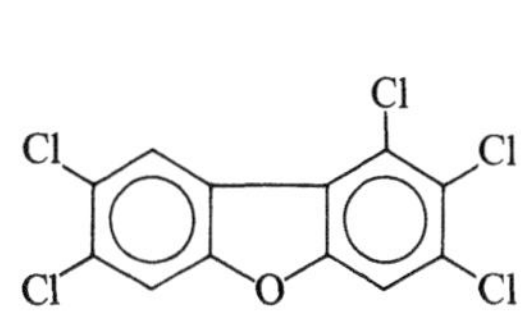

1, 2, 3, 7, 8–penta–CDF

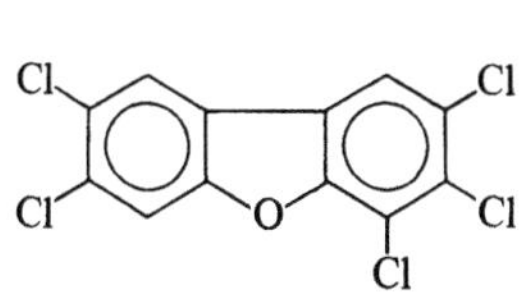

2, 3, 4, 7, 8–penta–CDF

Fig. 6. The most toxic PCDD and PCDF isomers

Biological Effects

Several books and reviews covering the toxicology and biological effects of PCDDs (mainly TCDD) and PCDFs have recently been published [42, 112 to 116].

The toxicity and biological properties of individual congeners is strikingly depending on number and position of chlorine substituents. The isomers with the highest acute toxicity appear to be the 2,3,7,8-tetra, 1,2,3,7,8-penta-, 1,2,3,6,7,8-, and 1,2,3,7,8,9-hexa-CDD and the 2,3,7,8-tetra, 1,2,3,7,8-, and 2,3,4,7,8-penta-CDF, see Fig. 6.

All these isomers have LD_{50}-values in the range 1–100 µg/kg for the most sensitive animal species [112, 113, 117, 118]. The bromine analogous like 2,3,7,8-tetra-BDD and 2,3,7,8-tetra-BDF seem to be equally toxic as the corresponding chlorine compounds [118, 119]. Recent work has shown that the positional isomers of PCDDs and PCDFs vary highly in their acute toxicity and biological activity. A factor of 1,000–10,000 can be found for closely related isomers such as 2,3,7,8- and 1,2,3,8-tetra-CDD [120, 121]. TCDD poisoning is characterized by loss of body weight with delayed lethality.

A large variety of sublethal effects have been identified after acute or chronic exposure to PCDDs and PCDFs, mainly TCDD [115, 117, 119]. TCDD has caused serious toxic effects in workers due to industrial exposure in 2,4,5-trichlorophenol plants resulting in irreversible liver damage, severe chlorance, hepatitis, porphyria and damage to the nervous system [116]. TCDD is a potent inducer of enzyme systems, particularily in the liver, it has mutagenic, teratogenic, carcinogenic and/or co-carcinogenic effects [115, 117, 119]. An increased incidence of liver cancer in people in Vietnam has been attributed to the TCDD present in Herbicide Orange sprayed in large quantities in this country [122, 123]. Exposure to PCDDs and PCDFs has been discussed in relation to malignant mesenchymal tumors and lymphomas among workers exposed to phenoxy acids and chlorophenols in Sweden [124–126].

References

1. Ranga Rao, K. (Ed.): Pentachlorophenol; Chemistry, Pharmacology and Environmental Toxicology. Plenum, New York 1978
2. Cirelli, D.P.: ibid., p. 13
3. Nilsson, C.-A. et al.: ibid., p. 313
4. Hoos, R.A.W.: ibid., p. 3
5. Ahlborg, U.G.: Metabolism of Chlorophenols; Studies on Dechlorination in Mammals. Swed. Environm. Prot. Board PM 895. Stockholm 1977
6. Rappe, C., Nilsson, C.-A.: J. Chromatogr. *67*, 247 (1972)
7. Jensen, S., Renberg, L.: Ambio *1*, 62 (1972)
8. Deinzer, M. et al.: Biomed. Mass. Spectrom. *5*, 566 (1978)
9. Firestone, D. et al.: J. Assoc. Offic. Anal. Chem. *55*, 85 (1972)
10. Buser, H.R., Bosshardt, H.-P.: ibid. *59*, 562 (1976)

11. Rappe, C., Garå, A., Buser, H.R.: Chemosphere *7*, 981 (1978)
12. Lindström, K., Nordin, J.: J. Chromatogr. *128*, 13 (1976)
13. Lamparski, L.L., Nestrick, T.J.: ibid. *156*, 143 (1978)
14. Edgerton, T.R., Moseman, R.F.: J. Agric. Food Chem. *27*, 197 (1979)
15. Wong, A.S., Crosby, D.G.: in Ranga Rao K., (Ed.).: Pentachlorophenol; Chemistry, Pharmacology and Environmental Toxicology. Plenum, New York 1978 p. 19
16. Crosby, D.G., Wong, A.S.: Chemosphere, *5*, 327 (1976)
17. Buser, H.R.: J. Chromatogr. *114*, 95 (1975)
18. Rappe, et al.: Chemosphere *7*, 269 (1978)
19. Reiner, E.A., Chu, J.P., Kirsch, E.J.: in Ranga Rao, K. (Ed.): Pentachlorophenol; Chemistry, Pharmacology and Environmental Toxicology. Plenum, New York 1978, p. 67
20. Kobayashi, K.: ibid. p. 89.
21. Glickman, A.H. et al.: Toxicol. Appl. Pharm. *41*, 469 (1977)
22. Larsen, R.V. et al., J. Pharm. Sci. *61*, 2004 (1972)
23. Pierce, R.H. Jr. et al.: Bull. Environ. Contam. Toxicol. *18*, 251 (1977)
24. Landner, L. et al.: ibid. *18*, 663 (1977)
25. Dougherty, R.C., Piotrowska, K.: Proc. Natl. Acad. Sci. USA *73*, 1977 (1976)
26. Kutz, F.W., Murphy, R.S., Strassman, S.C.: in Ranga Rao, K. (Ed.): Pentachlorophenol; Chemistry, Pharmacology and Endvironmental Toxicology. Plenum, New York 1978, p. 363
27. Pearson, J.E. et al.: Bull. Environ. Contain. Toxicol. *16*, 556 (1976)
28. Shafik, T.M.: ibid., *10*, 57 (1973)
29. Tagatz, M.E., Ivey, J. M., Tobia, M.: in Ranga Rao, K. (Ed.): Pentachlorophenol; Chemistry, Pharmacology and Environmental Toxicology. Plenum, New York 1978, p. 157
30. Adelman, I. R., Smith, L.L., Jr. Siesennop, G.D.: J. Fish Res. Board Can. *33*, 203 (1976)
31. Fahring, R., Nilsson, C.-A., Rappe, C.: in Ranga Rao, K. (Ed.): Pentachlorophenol; Chemistry, Pharmacology and Environmental Toxicology. Plenum, New York 1978, p. 325
32. Sundström, G., Hutzinger, O.: Chemosphere *5*, 187 (1976)
33. Norström, A., Andersson, K., Rappe, C.: ibid. *5*, 255 (1976)
34. Norström, A., Andersson, K., Rappe, C.: ibid. *5*, 21 (1976)
35. Norström, A., Andersson, K., Rappe, C.: ibid. *6*, 241 (1977)
36. Choudhry, G.G. et al.: J. Agric. Food Chem. *25*, 1371 (1977)
37. Choudhry, G.G. et al.: Chemosphere *6*, 327 (1977)
38. Rappe, C., Lindahl, R., Buser, H.R.: Chemosphere, in press.
39. Tulp, M. Th. M. et al.: Xenobiotica *9*, 65 (1979)
40. Neely, W.B., Branson, D.R., Blau, G.E.: Environm. Sci. Technol. *8*, 1113 (1974)
41. Zitko, V., Carlson, W.G.: Chemosphere *6*, 293 (1977)
42. Higuchi, K. (Ed.): PCB Poisoning and Pollution. Kodansha, Tokyo 1976
43. Carter, C.D. et al.: Science *188*, 738 (1975)
44. Pocciari, F.: Ecol. Bull. (Stockholm) *27*, 67 (1978)
45. Buser, H.R., Rappe, C.: Chemosphere *7*, 199 (1978)
46. Buser, H.R., Rappe, C.: ibid. *8*, 157 (1979)
47. Crummett, W.B., Stehl, R.H.: Environm. Health Perspect. *5*, 15 (1973)
48. Pohland, A.E., Yang, G.C.: J. Agric. Food Chem. *20*, 1093 (1972)
49. Sandermann, W., Stockmann, H., Carsten, R.: Chem. Ber. *90*, 690 (1957)
50. Langer, H.G., Brady, T.P., Briggs, P.R.: Environ. Health Perspect. *5*, 3 (1973)
51. Garå, A. et al.: Chemosphere, in press.
52. Courtney, K.D., Moore, J.A.: Toxicol. Appl. Pharm. *20*, 396 (1971)
53. Firestone, D.: Ecol. Bull. (Stockholm) *27*, 39 (1978)
54. Huckins, J.N., Stalling, D.L., Smith, W.A.: J. Assoc. Offic. Anal. Chem. *61*, 32 (1978)
55. Rappe, C., Buser, H.R., Bosshardt, H.P.: Chemosphere *7*, 431 (1978)
56. Goldstein, J. A. et al.: Biochem. Pharmacol. *26*, 1549 (1977)
57. Buser, H.R.: Thesis, Univers. Umeå, Sweden 1978
58. Vos, J.G. et al.: Food Cosmet. Toxicol. *8*, 625 (1970)
59. Roach, J.A.G. Pomerantz, I.H.: Bull. Environ. Contam. Toxicol. *12*, 338 (1974)
60. Bowes, G.W. et al.: Nature *256*, 305 (1975)
61. Rappe, C., Buser, H.R.: Manuscript in preparation

62. Buser, H.R., Rappe, C., Garå, A.: Chemosphere *7,* 439 (1978)
63. Baughman, R.W.: Thesis, Harvard Univers., Cambridge, Mass. 1974
64. Göthe, R., Wachtmeister, C.A.: Acta Chem. Scand. *26,* 2523 (1972)
65. Olie, K., Vermeulen, P.L., Hutzinger, O.: Chemosphere *6,* 455 (1977)
66. Buser, H.R., Bosshardt, H.-P.: Mitt. Gebrauchs. Lebensmittelunters. u. Hyg. *61,* 191 (1978)
67. Buser, H.R., Bosshardt, H.-P., Rappe, C.: Chemosphere *7,* 165 (1978)
68. Buser, H.R. et al.: ibid. *7,* 419 (1978)
69. Nilsson, C.-A. et al.: Chromatogr. *96,* 137 (1974)
70. Plimmer, J.R. et al.: Adv. Chem. Ser. *120,* 44 (1973)
71. Buser, H.R.: J. Chromatogr. *129,* 303 (1976)
72. Dow Chemical Company: The Trace Chemistries of Fire, Report, Nov. 1978
73. Buser, H.R., Bosshardt, H.-P., Rappe, C.: Chemosphere *7,* 109 (1978)
74. Morita, M., Nakagawa, J., Rappe, C.: Bull. Environ. Contam. Toxicol. *19,* 665 (1978)
75. O'Keefe, P.W.: Environ. Health Perspect. *23,* 347 (1978)
76. Buser, H.R.: Chemosphere *8,* 415 (1979)
77. Rappe, C.: Ecol. Bull. (Stockholm) *27,* 19 (1978)
78. Stehl, R.H., Lamparski, L.L.: Science *197,* 1008 (1977)
79. Ahling, B. et al.: Chemosphere *6,* 461 (1977)
80. Rawls, R.L.: Chem. Engng. News Feb. 12, 1979, p. 23
81. Smith, R.J.: Science *202,* 1166 (1978)
82. McKinney, J.D.: Ecol. Bull. (Stockholm) *27,* 53 (1978)
83. Hunt, D.F., Harvey, T.M., Russell, J.W.: J.C.S. Chem. Comm. 151 (1975)
84. Hass, J.R. et al.: Analyt. Chem. *50,* 1474 (1978)
85. Ward, C.T., Matsumura, F.: Arch. Environ. Contam. Toxicol. 7, 349 (1978)
86. Helling, C.S. et al.: J. Environ. Quality *2,* 171 (1973)
87. Young, A.L. et al.: Fate of 2,3,7,8,-Tetrachlorodibenzo-p-dioxin (TCDD) in the Environment: Summary and Decontamination Recommendations. USA FA-TR-76-18 Boulder, Colo, USA 1976
88. Grey, A.P. et al.: J. Org. Chem. *41,* 2435 (1976)
89. Buser, H.R., Rappe, C.: unpublished results 1978
90. Stehl, R.H. et al.: Adv. Chem. Ser. *120,* 119 (1973)
91. Crosby, D.G. et al.: Science *173,* 748 (1971)
92. Isensee, A.R., Jones, G.E.: J. Agric. Food Chem. *19,* 1210 (1971)
93. Crosby, D.G., Wong, A.S.: Science, *195,* 1337 (1971)
94. Crosby, D.G.: Amer. Chem. Soc. Symposium Ser. *73,* 1 (1978)
95. Liberti, A. et al.: *10,* 97 (1978)
96. Rappe, C.: in Cattabeni, F., Cavallaro, A., Galli, G., (Eds.): Dioxin, Toxicological and Chemical Aspects. SP Medical and Scientific Books, Jamaica 1978, Chap. 17
97. Bertoni, G. et al.: Analyt. Chem. *50,* 732 (1978)
98. Botré, C., Memoli, A., Alhairque, F.: Environ. Sci. Techn. *12,* 335 (1978)
99. Plimmer, J.R.: Bull. Environ. Contam. Toxicol. *20,* 87 (1978)
100. Gebefügi, I., Baumann, R., Korte, F.: Naturwissenschaften *64,* 486 (1977)
101. Hutzinger, O. et al.: Environ. Health Perspect. *5,* 267 (1973)
102. Matsumura, F., Benezet, H.J.: ibid. *5,* 253 (1973)
103. Guenthner, T.M., Fysh, J.M., Nebert, D.W.: Pharmacology, in press
104. Rose, J.Q. et al.: Toxicol. Appl. Pharmacol. *36,* 209 (1976)
105. Tulp. M.Th.M., Hutzinger, O.: Chemosphere *7,* 761 (1978)
106. Rappe, C. et al.: ibid. *4,* 259 (1979)
107. Kuroki, H., Masuda, Y.: ibid. *7,* 771 (1978)
108. Zitko, V. et al.: Environ. Health Perspect. *5,* 187 (1973)
109. Isensee, A.R., Jones, G.E.: Environm. Sci. Technol. *9,* 668 (1975)
110. Baughman, R., Meselson, M.: Environm. Health Perspect. *5,* 27 (1973)
111. Dougherty, R.C. et al.: ibid. in press.
112. IARC Monogr. Evaluation Carcinogenic Risk to Man *15,* 41 (1977)
113. IARC Internal Techn. Rep. 78/001, IARC, Lyon 1978
114. Ramel, C. (Ed.): Chlorinated Phenoxy Acids and Their Dioxins. Ecol. Bull. (Stockholm) *27,* (1978)

115. Cattabeni, F., Cavallaro, A., Galli, G., (Eds.): Dioxin, Toxicological and Chemical Aspects. SP Medical and Scientific Books, Jamaica 1978
116. Ann. New York Acad. Sci, *320,* 1 (1979)
117. McConnell, E.E. et al.: Toxicol. Appl. Pharmacol. *44,* 335 (1978)
118. Moore, J.A. et al.: Ann. New York Acad. Sci, *320,* 1 (1979)
119. Poland, A., Glover, E., Kende, A.S.: J. Biol. Chem. *251,* 4936 (1976)
120. Poland, A., Glover. E.: Mol. Pharmacol. *9,* 736 (1973)
121. Bradlaw, J.: FDA Washington, personal communication
122. Tung, T.T.: Rev. Médecine *14,* 653 (1977)
123. Wade, N.: Science *204,* 817 (1978)
124. Hardell, L., Sandström, A.: Läkartidningen *75,* 335 (1978)
125. Hardell, L., Sandström, A.: Brit. J. Cancer *39,* 711 (1979)
126. Hardell, L.: Lancet *55* (1979)

Organic Dyes and Pigments

E. A. Clarke, R. Anliker

Ecological and Toxicological Association
of the Dyestuffs Manufacturing Industry (ETAD)
CH-4005 Basel 5, Switzerland

Introduction

Color, which contributes so much to the beauty of Nature, is essential to the attractiveness and acceptability of most products used by modern society [1]. As long ago as the 25th century BC man colored his surroundings and clothes using a limited range of natural colorants of both animal and vegetable origin. Alizarin (*18*)[1] extracted as the glycoside rubierythric acid from madder, was used by the ancient Egyptians and Persians, the use of indigo (*16*) obtained from *Indigofera* dates back to 3000 BC, and Tyrian Purple (6,6′-dibromoindigo), prepared from the sea snail *Murex brandaris*, has been used since the Roman era. However, the preparation in 1856 of the first synthetic dyestuff, mauveine (*12*), by Perkin gave birth to the development of many other important sectors of the modern chemical industry. Compared with natural dyestuffs, synthetic colorants are better able to meet the increasingly rigorous technical demands of the present day in terms of stability, fastness, etc. Color can add not only aesthetic appeal, but frequently provides an almost irreplaceable safety feature (traffic lights and signs, drug identification, control systems) [2].

Dyes and pigments are substances which when applied to a substrate lead to selective reflection or transmission of incident daylight[2]. Substances which create the sensation of blackness or whiteness are also regarded as dyes or

1 The chemical structures of a selection of representative colorants are given in Fig. 6, preceding the references at the end of this chapter

2 Throughout this article the terms "dyestuffs" (or "dyes") and "pigments" have been used according to this definition. The term "colorants" is applied collectively when no distinction is required. Descriptive terms such as "pigment preparation", "commercial dyestuff", "technical grade colorants" are used to describe various qualities of commercial product

pigments. Characteristic of pigments is their extremely low solubility in water, and in the application substrate. They generally also exhibit low solubility in organic solvents. For this reason they remain essentially in the solid state during processing and when applied to the substrate. Colorants, not covered by this definition of pigments, are dyestuffs.

Synthetic organic dyestuffs and pigments exhibit an extremely wide variety of physical, chemical and biological properties, making any comprehensive review of the ecotoxicological properties of the several thousand commercially available products difficult. In the following sections an attempt has been made to provide a perspective of the environmental problems posed by synthetic organic colorants, and to outline the current efforts by the manufacturing industry to ensure that its products present no unreasonable risk to the ecosystem, including man.

Most commercial dyestuffs and pigments are in fact mixtures in which the content of the specific colored component normally lies in the range 10–98%. The other components are necessary to confer the desired physical properties and may of course influence the product's ecotoxicological behaviour.

Fluorescent whitening agents, which are frequently classified as dyestuffs, are not treated in detail in this chapter. An assessment of the ecological behaviour and the pollution aspects of these products as well as their chemistry, application and their toxicological properties have been comprehensively reviewed [3].

Chemistry and Uses

A detailed description of the chemical processes involved in manufacture of the wide variety of synthetic organic colorants is beyond the scope of this handbook and the authoritative texts edited by Venkatamaran [4] should be consulted.

An essential preliminary to any discussion of the chemistry of synthetic organic colorants is the classification of the numerous products which are commercially available. The Colour Index [5] lists an estimated 38,000 commercial colorants involving 7,000–8,000 different chemical structures. This unique system identifies each product by a generic name (e.g. Acid Yellow 23) which describes the application type (i.e. Acid) and color (Yellow 23) together with dyeing characteristics. In addition where the chemical structure (*1*) has been made known to the Colour Index, a five-digit constitution number is allocated which uniquely identifies the structures (in this case C.I. 19140). The specific constitution numbers are drawn from ranges which have been allocated to the various chemical types and a cross reference is given for each generic name to the known commercial brands (and vice-versa). Many of the allocated C.I. constitution numbers refer to products which are no longer used commercially.

A factor of particular relevance for the ecological behaviour of a dyestuff is its ionic character and Table 1 summarises the major applications of the major dyestuff and structural classes.

Table 1. Summary of ionic character, chemical type and applications

	Dyestuffs and Pigments												
	Water soluble							Water insoluble					
	Cationic	Anionic						Soluble in organic solvents		Low solubility in organic solvents			
	Basic	Acid	Mordant	Direct	Reactive	Solubilised Vat	Sulphur	Disperse	Solvent	Mordant	Vat	Sulphur	Pigment
Azo	x	x	x	x	x			x	x	x			x
Stilbene				x									
Di-, and Triphenylmethane	x	x							x	x			x
Xanthene	x	x							x	x			x
Acridine	x												x
Quinoline	x	x		x					x				x
Methine	x			x				x					x
Azine	x	x							x				
Oxazine	x			x						x			x
Thiazine	x								x	x	x		
Sulphur							x					x	
Anthraquinone	x	x	x		x	x		x	x	x	x		x
Indigoid		x				x					x		x
Phthalocyanine		x		x	x				x				x
Textile dyeing substrates	PAc PE PAm P C	P PAm PAc	P	C P PAm	C P PAm	C P	C	PE PP, PVC CE CE PAm	-	P C	C P	C	Textile Printing
Other substrates	Paper	Paper Leather		Paper Leather	Leather				Leather				
Other usages	Basic: Inks, manifold record systems Acid: Food, drugs, cosmetics, varnishes, inks, plastics, resins Disperse: Plastics Solvent: Wood stains, varnishes, lacquers, inks, polishes, plastics, colored smokes Pigment: Paints, inks, lacquers, plastics, cosmetics, waxes, writing utensils, carbon paper												

Key:
- C = Cellulose (cotton, linen)
- CE = Cellulose ester (acetate, triacetate)
- P = Protein (wool, silk)
- PAc = Polyacrylics
- PAm = Polyamide
- PE = Polyester
- PP = Polypropylene
- PVC = Polyvinylchloride

Production Data

Production or sales statistics are only available for a few countries, notably USA [6] and Japan, but total world dye production is now estimated to be 600,000–700,000 tonnes (based on active substance). The breakdowns of this total amount by use, production area, and application type are estimated in Tables 2–4 respectively.

Table 2. Estimated world production by use category (1978)

Use	Production (t)	%
Textile dyestuffs	360,000	56
Paper/leather dyestuffs	90,000	15
Organic pigments	150,000	23
Fluorescent whitening agents and others	40,000	6
	640,000	

Table 3. Analysis of estimated world production by industrial region (1978)

Production area	Production (t)	%
W. Europe	250,000	39
COMECON	147,000	23
USA	109,000	17
China	57,000	9
Japan	51,000	8
India	13,000	2
S. America and Mexico	13,000	2

Table 4. Analysis of world production by application type (1978)

Application type	Production (t)	%
Acid	60,000	9
Basic	45,000	7
Chrome	20,000	3
Direct	100,000	16
Disperse	75,000	12
Fluorescent whitening agents	30,000	5
Organic pigments	150,000	23
Reactive	35,000	6
Vat	65,000	10
Others	60,000	9

Analytical Methods

A comprehensive treatment of the analytical chemistry of dyestuffs has been edited by Venkatamaran [7] and this section seeks only to outline the importance of modern analytical techniques for the monitoring of the environment for dyestuffs or associated impurities which may result in adverse toxicological or ecological effects. Good analytical methodology is essential to most ecological and toxicological investigations and it is ironical that the remarkable advances in sensitivity of analytical techniques should have sharpened so many controversies because these advances far exceed the ability to interpret the results in terms of risk. Although no rational person believes that a "no risk" society can be achieved, such concepts as "zero tolerance" [8–10], "one-hit theory of carcinogenesis" and the "Delaney Amendment" [11, 12] still have a pervasive influence on regulatory decision-making.

The major areas of application within the scope of this article are as follows:

- monitoring of product quality
- monitoring trade effluent quality, (both solid and liquid waste). See also p. 10–11
- monitoring work-place environment
- monitoring ecological and toxicological experiments
- monitoring environmental levels of dyestuffs.

Apart from the need to control effluents within specific limits for color, the required analytical techniques are essentially for measurement of trace harmful impurities, particularly:

(i) Metals.[3] Discharge consents for trace metals are increasingly stringent in most developed countries. Extensive analytical surveys have been conducted on the trace metal content of commercial dyestuffs [13, 14] and pigments [15], and these confirm that with the exception of the metal complex dyestuffs the metal content is low (generally $\leqslant$ 100 ppm). Amongst the more important modern techniques available for trace metal analysis are atomic absorption spectrophotometry [16–18], neutron activation analysis [19], polarography [20], emission spectrophotometry [20] and spark source mass spectrophotometry [21].

---

3 In 1975 ETAD recommended to its member companies that their dyestuffs should meet the following trace metal limits (ppm):

As	50	Hg	25
Ba	100	Mn	1,000
Cd	50	Ni	250
Co	500	Pb	250
Cr	500	Sb	50
Cu	250	Se	50
Fe	2,500	Sn	250
		Zn	5,000

These limits, which do not apply to metal complex dyestuffs, were derived from currently known legal requirements for metals in effluents, on the basis of a 2% dyeing and a total dilution of the effluents in a ratio of 2500:1 in relation to the dyestuff used

(ii) Lipophilic aromatic amines. Some aromatic amines are recognized human carcinogens and several have shown carcinogenic activity in animal feeding studies. Some of these products are subject to special legislative controls and sensitive analytical methods have been developed to determine the presence of trace amounts in:
a) Working environments. Typically the method involves absorption from sampled air followed by gas-liquid chromatography [22].
b) Dilute aqueous solution. As applied to aqueous trade effluent and urine. Typically the method involves an extraction/enrichment stage from alkaline solution, followed by analysis using gas-liquid chromatography, thin-layer chromatography or high performance liquid chromatography [23–28].

(iii) PCBs. The wide environmental distribution of polychlorobiphenyl compounds (PCBs) has generated considerable interest in these persistent products [29]. Although their major usage is as thermally stable dielectric fluids in transformers they are also adventitious impurities in certain commercially important diarylide and phthalocyanine pigments. Although these trace amounts present no significant environmental burden, the development of several new analytical methods was necessary in response to regulatory pressure in the USA to meet a 50 ppm PCB limit [30, 31].

Ecological Aspects

Environmental Assessment of Colorants

Accurate data on the quantity of colorants discharged to the environment from the manufacturing and processing operations are not available. Percentage losses vary from product to product and are dependent on the processes and equipment employed. Table 5 gives a breakdown of estimated losses.

Table 5. Estimated losses of synthetic organic colorants from manufacturing and processing operations (1978)

		% Losses in		
	Production (t)	Production	Processing	Total loss (t)
Textile dyestuffs	360,000	2	10	43,000
Paper/leather	90,000	2	6	7,000
Organic pigments	150,000	1	1-2	4,000
Others	40,000	2	10	5,000

In most industrialized countries only about 20% or less of process losses will reach open waters due to effective adsorption in the primary and the biological treatment stages (see Effluent Treatment Processes). Possible environmental damage does not depend solely on the quantity released, but also on the ecotoxicological properties of the individual products involved, and their environmental transport characteristics.

An adequate reassurance that the environmental release of colorants, either individually or collectively, does not lead to any significant disturbance of the ecosystem, particularly man, is required.

Because of the multitude of colorants available, most satisfying specific technical requirements, it is simply not practical to undertake exhaustive evaluation of each individual product. A more effective approach is to concentrate available resources on

(i) reduction of release to the environment
(ii) environmental monitoring to detect any localized high concentrations and possible associated adverse effects
(iii) characterisation of certain ecological properties of selected members of important chemical classes, from which the likely environmental behaviour may be predicted
(iv) carrying out limited screening tests on individual products as a basis for identifying possible problem products for more intensive evaluation.

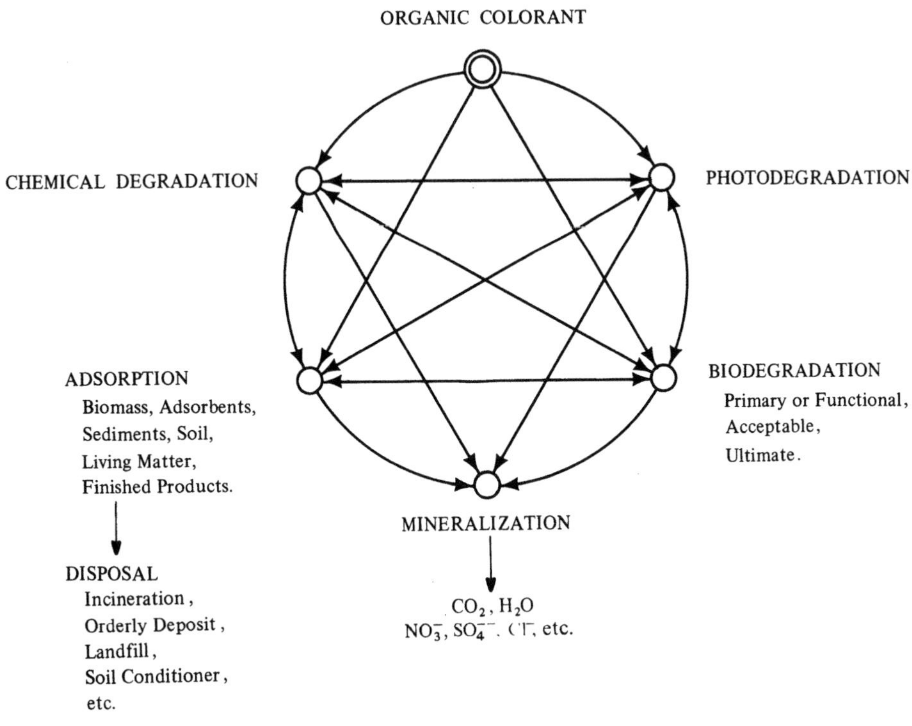

Fig. 1. Processes of elimination and degradation of waste organic colorants

Elimination and Degradation Cycle

Figure 1 displays the various processes, and their combinations, of elimination and degradation of organic colorants in the environment [32].

The term *primary* or *functional biodegradation* means biodegradation of a substance to an extent sufficient to remove some characteristic property of the molecule; in biological terms, this would be referred to as a biological transformation [33]. In the case of surfactant biodegradation this is usually taken as loss of surface activity, and in the case of a dyestuff it would typically be loss of color.

Under *environmentally acceptable biodegradation* one understands the biodegradation of a substance to such an extent that environmentally undesirable properties are lost. Although this clearly requires a judgement it may be possible to apply a simple bioassay such as acute toxicity to fish to demonstrate that biodegradation has resulted in degradation products which are of lower acute toxicity.

Ultimate biodegradation [33] is the breakdown of an organic compound to carbon dioxide, water, the oxides or mineral salts of other elements present or to small organic compounds which will be utilised for the synthesis of new cell material. This may involve complete mineralization, but does not necessarily do so. *Bioelimination* [33] is a term applied to the sum of various processes (e.g. sludge adsorption, chemical oxidation, volatilization) contributing to the removal of a compound from the aqueous phase during sewage treatment and may, or may not, include biodegradation.

Effluent Treatment Processes

General Aspects

Untreated effluent from dyestuff production and dyeing plants are usually highly colored and thus particularly objectionable if discharged to open waters. Many dyestuffs are easily visible in water at concentrations well below 1 ppm. As a result considerable attention has been focussed on polluting discharges with high color content. Extensive testing indicates that dyestuffs are generally adsorbed to the extent of 40–80% by the biomass and are thus partially eliminated in sewage treatment plants. They are not, however, biodegraded in this stage to any significant extent [32, 34, 35]. Furthermore, experience shows that either chemical or biological treatment alone is not sufficiently effective for decolorization and a combination of physical, chemical and biological processes is usually necessary to achieve adequate color removal of a mixture of dyestuffs. ADMI [36] and more recently McKay [37] have reviewed the various treatment processes applied so far for color removal from textile effluents. In an EPA study [14, 38] the treatment of wastewaters from selected typical dyebaths by a variety of processes has been examined. Twenty systems were selected to provide a wide cross-section of dye classes, fibres and application techniques. Table 6 summarises the removal capability of the most common treatment processes used to remove color from such wastewaters.

Table 6. Summary of effectiveness of effluent treatment processes for various dyestuff classes

Dyestuffs in dyeing wastewaters	Color removal by treatment processes					
	Coagulation alum	Activated carbon	Biological	Combination physico-chemical and biological	Ozone	Sludge adsorptions of dyestuffs
Azoic	O	+	O	+	+	-
Reactive	O	+	O	+	+ (s)	O(+)
Acid	O	+	O	+	+	O
Basic	O	+ (s)	+	+	+	+
Disperse	+	O	O	+	O	O
Vat	+	O	O	+	+	-
Sulphur	+	O	O	+	+	-
Direct	-	-	-	+	-	+

Color removal: O unsatisfactory, + good, s specially suitable, - not investigated

Although in the following text only the elimination of the dyestuff entities themselves from the generally complex composition of effluent from dyestuffs manufacturing or processing plants is considered, the broader aspects of the problem are considered elsewhere [39–52]. Dyestuffs generally exhibit low acute toxicity to warm-blooded animals, fish, and sewage works bacteria and have so far not caused any serious environmental problems. Even in the case of Basic dyestuffs, which show a somewhat higher toxicity compared with other dyestuff classes in acute toxicity studies in mammals, fish, algae and activated sludge bacteria [34, 35, 53], their high degree of exhaustion in the dyebath and their strong adsorption characteristics facilitate the achievement of low concentrations in the effluent.

The composition of dyestuff-containing effluents from manufacturing plants and dyeworks, normally characterised by the parameters BOD_5, COD and TOC[4] vary widely depending on the level of production, product mix, dilution procedures etc. In the case of dyestuffs manufacturing plant the effluent composition and pH is dependent on the nature of the particular manufacturing process, whereas in processing plants the effluent typically contains large quantities of surfactants, resins, textile auxiliaries etc. Within the framework of an EPA study [38] 20 different textile dye process effluents were systematically investigated. The average BOD_5 was 280 mg/l (range

4 BOD_5: biological oxygen demand over 5 days
COD : chemical oxygen demand
TOC : total organic carbon

12–1470 mg/l) and the average TOC was 276 mg/l (range 55–1120 mg/l). In a raw effluent from a tannery [40] the average BOD_5 was 1900 mg/l, and average COD 5200 mg/l whilst the dyestuff concentration varied between 22–56 ppm (COD $\leq$ 112 mg/l); i.e. the COD contribution of the dyestuff was less than 2%. Effluent from a dyestuff and chemical manufacturing plant [41] had a BOD_5 of 900–1400 mg/l and a TOC of 600–1000 mg/l. The corresponding values for domestic effluent are 90–200 and 40–160 mg/l respectively. In Table 7 the theoretical TOC and COD values for some selected dyestuffs are displayed.

Table 7. Theoretical TOC and COD values for selected dyestuffs [a]

Structure [b]	Colour index name	Constitution number	Theoretical TOC % C	Theoretical COD mg O_2/mg dyestuff
2	Disperse Yellow 3	C.I. 11855	67	2.56
4	Direct Yellow 12	C.I. 24895	57	1.94
9	Disperse Yellow 54	C.I. 47020	75	1.67
11	Basic Yellow 11	C.I. 48055	68	2.44
13	Basic Blue 3	C.I. 51004	67	2.62
15	Sulphur Black 1	C.I. 53185	39	1.22
17	Acid Green 25	C.I. 61570	58	1.94
20	Direct Blue 86	C.I. 74180	52	2.02

[a] Calculation based on 100% active dyestuff
[b] See Fig. 6

Because dyestuffs are not readily biodegradable they make little contribution to BOD. The contribution to the measured COD and TOC of textile dyeing plant effluents may amount to a few percent, but is unlikely to exceed 10%; typically the dyestuff concentration (as active ingredient) lies in the range 10–50 mg/l.

Decolorized effluents contain less than 1 mg/l dyestuff and the TOC contribution of dyestuff following the primary and biological treatment stages is normally considerably less than 0,5 mg/l.

Colored wastewaters usually contain a large number of individual dyestuffs and very often the identities of the colorants responsible for the color of the water are not known exactly. This makes the determination of the color concentration using individual dyestuff spectra almost impossible. To overcome these difficulties a standard method was introduced by APHA (American Public Health Association). As a refinement of the existing platinum-cobalt APHA standards [54], Allen et al. [55] introduced the ADMI color value which provides a measure of the color of aqueous solutions which is independent of hue and can be related to APHA values. The ADMI method requires relatively inexpensive instrumentation allowing the use of a wide variety of color measuring instruments[5]. Investigation of 45 commercial dyes showed that solutions measured as 50 ADMI units corresponded to commercial dyestuff concentrations varying from 0,1 mg/l to 16,5 mg/l, depending upon the inherent tinctorial strength and the active dyestuff content of the commercial dyestuffs involved. Although these values serve well to define color of wastewaters and to monitor decolorization, the ecotoxicological assessment of dyestuffs in wastewaters requires quantification of the actual amounts of individual dyestuffs and more specific analytical methods. Tincher [56] has analyzed the distribution of individual dyestuffs in the Coosa River basin on which over 50% of the US carpet dyeing industry is centred.

Physical and Chemical Treatment Processes (Abiotic Processes)

In the commonly used abiotic processes the decolorization is achieved either by removal of the intact dyestuff or by its destruction. Precipitation and flocculation procedures using lime [57], alum [38, 58], ferric chloride and ferric sulphate [59, 60], and organic agents [61] have given good results.

Activated charcoals of various origins and qualities seem particularly suitable for Acid, Basic and Reactive dyestuffs [38, 62–69]. Different adsorbents such as activated carbon and organic agents may be added at the biological stage [38, 61, 70]. Silica gel, Fuller's earth, and bauxite have shown good results as adsorbents for Basic dyes [62]. Other adsorbent materials including peat [37, 71] and wood [72] have also been investigated for possible application in decolorizing wastewaters. Ion-exchange resins have been used for the elimination of anionic and cationic dyestuffs [73, 74], but, in the case of dyestuff mixtures recovery is normally not economic and in the disposal of the desorbed concentrate additional high costs may be incurred. Other processes including foam-fractionation, dynamic membrane hyperfiltration (reverse osmosis), are still in the experimental stage [37, 75], but again the disposal of the concentrate may pose major problems.

Of the chemical processes for color removal, ozonisation has achieved the greatest practical importance. However, for an extensive degradation of dyestuff it is necessary to use a large amount of ozone [76]. In order to minimize costs the dyestuffs are only partially oxidized using ozone and then further oxidized catalytically [77, 78]. The oxidation products, mostly polycarboxylic acids, are then either removed by flocculation or subjected to a biological treatment stage. Cheaper than ozonisation, but less satisfactory, is bleaching using chlorine, chlorine dioxide, or chloramine. By heavy chlorination a complete decolorization can be achieved in many cases; however, the

5 The procedure includes the following steps:
1. Measurement of sample on a suitable spectrophotometer or colorimeter
2. Calculation of C.I.E. Tristimulus Values X, Y, Z (may be inherent in instrument)
3. Conversion of X, Y, Z to V_x, V_y, V_z (from published tables)
4. Calculation of Adams-Nickerson Color Difference (DE)
5. Conversion of DE to ADMI value

possible formation of chlorinated compounds, which may be less acceptable in terms of toxicity or biodegradability, should not be overlooked. The γ-radiation induced oxidation of dyestuffs in wastewaters is of potential interest in some cases [66, 79, 80]. For example, colored wastewaters of an anthraquinone dyestuffs manufacturing plant are totally decolorized by this means [81]. For some classes of dyestuffs, particularly azo-dyestuffs, decolorization with reductive agents such as hydrosulphite is a workable proposition [82]. In this case the reduction can be reversible and generally involves no real degradation of the dyestuff molecule.

A recently developed wet pressure oxidation process [83] has successfully dealt with non-biodegradable by-products from the production of dyestuff intermediates, but the applicability of this technique to dyestuffs has not yet been investigated.

Electrochemical oxidation and the electrolytic reductive precipitation have not yet obtained any practical importance [37, 84].

Biological Treatment Processes

Of the four most common biological treatment processes: stabilization ponds, aerated lagoons, trickling filters and activated sludge [37], the last named is the most widely used today. With the possible exception of Basic dyestuffs, these processes have proved in most cases to be insufficiently effective in removing dyestuffs from wastewaters. As already indicated, dyestuffs are practically not biodegraded in this stage, but may be adsorbed by the sludge to the extent of about 40–80%, or even more, depending on the individual dyestuff and treatment conditions. Complete removal or decolorization can only be achieved by combination with other treatment processes [85–90]. In many cases it is more economical to remove the residual color by a "polishing" treatment such as adsorption or coagulation after the biological stage.

Sludge Adsorption and Digestion

In the biological treatment plant, dyestuffs are eliminated essentially only by an adsorption process on the sludge. Investigations indicate that the extent of adsorption is determined by the dyestuff structure, the pH and the composition of the wastewater. Lower pH conditions favour adsorption. In the practical concentration range of 10–50 mg dyestuff per litre, there is an almost linear relationship between the concentration in solution and the amount adsorbed. The adsorptive capacity of activated sludge for the dyes investigated [91] was, in neutral media, in the range of 0.01–4% of dyestuff on dry weight sludge.

Table 6, last column, gives some indication of the adsorption behaviour of the various dyestuff classes [92]. In general it is found that adsorption is favoured by hydroxyl-, nitro- and azo-groups, as well as increase in length of the dyestuff molecule. Sulpho-groups reduce adsorption in the case of Acid dyestuffs whereas the number of sulpho-groups does not appear to influence adsorption in the case of Reactive and Direct dyes.

Following the biological treatment, the sludge containing adsorbed dyestuffs may be digested under anaerobic conditions. In an investigation of 42 dyestuffs by ADMI [35] only 4 dyestuffs had an inhibitory effect on the sludge bacteria when fed daily at a concentration of 150 mg/l to anaerobic digestors. Of the 29 soluble dyestuffs studied only 4 (Acid Black 1, Basic Blue 3 (*13*), Acid Green 25 (*17*), and Acid Blue 45) showed no signs of decolorization. The rest were either completely decolorized (16 out of 25) or underwent significant spectral changes. The mechanism of the reported decolorization has still to be determined. Little knowledge has been accumulated on the chemical steps of anaerobic degradation of dyestuffs. In the case of azo dyestuffs, this loss of color is almost certainly due to reductive cleavage of the azo groups (see section on biodegradation).

Environmental Elimination Processes

Photochemical Degradation

The photochemical properties of some naturally occurring dyestuffs are fundamental to photosynthesis and vision. Similarly the properties of synthetic organic colorants have been exploited by modern technology in photography, organic videcon tubes for television, photochromic plastics, dye-sensitized photo-tendering of textiles, etc. A high degree of stability to photochemical degradation is normally required of commercial dyestuffs when applied to a textile substrate, and it is known that this stability is dependent on such factors as humidity, temperature, presence of oxygen, substrate and spectral distribution [93]. Of relevance to the environmental fate of dyestuffs, however, is the photochemical behaviour of the dyestuff in the low concentrations in aqueous solution which may occur in lakes and streams, and this has not been widely investigated.

A recent review [94] concluded that in aquatic systems photoreduction of azo dyestuffs to hydrazines and amines is possible, but is likely to be very slow except in oxygen poor water.

Porter [95] studied the stability of 36 commercial dyes to visible and ultra-violet light and reported only slow degradation under the experimental conditions chosen. Normally such studies determine the rate of disappearance of color, but do not identify the reaction products. In the case of Basic Green 4, some degradation products were identified and the degradation mechanism given in Fig. 2 was proposed.

Heitz and Wilson [96] showed that the photodegradation of several xanthene dyes in dilute solution proceeds as a first order reaction, that the rate increases with increasing halogenation of the dyestuff, and that the degradation results in the loss of the phototoxic properties of the dyestuff to both insects and bacteria.

Since even low-levels of pollution of waters by dyestuffs can be readily detected visually, due to their high tinctorial value, the increasingly stringent legal restrictions on such pollution are based, not improperly, more on

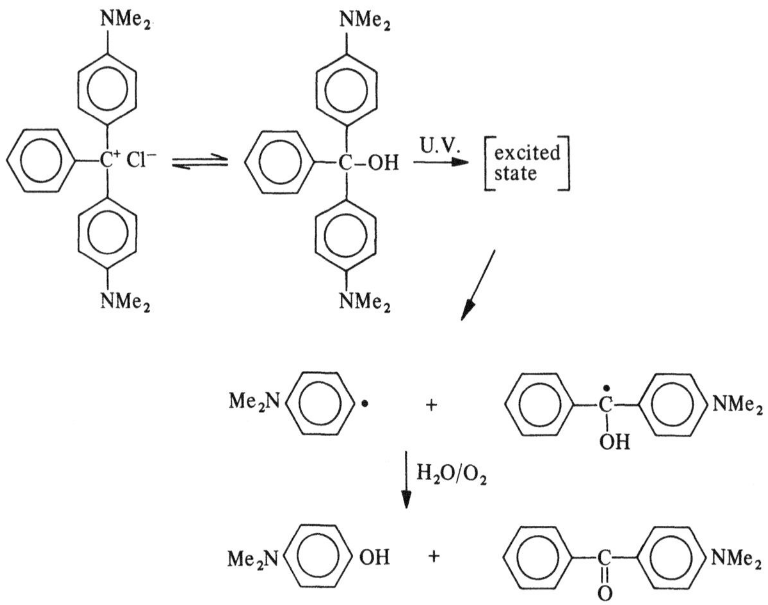

Fig. 2. Photodegradation of Basic Green 4

aesthetic than toxicological considerations. Practical considerations lead to the conclusion that photodegradation does not play a dominant rôle in the environmental fate of dyestuffs, although its contribution to the total mineralization of widely dispersed trace amounts may be underestimated.

Biodegradation – Metabolism

A knowledge of the biodegradation characteristics of a chemical is of primary importance in selecting what further testing is appropriate to evaluate its possible environmental effects. This is recognized for example in the Japanese Law "Control of Chemical Substances" under which it is only necessary to proceed sequentially to bioaccumulation and to fish toxicity testing if the product is of low biodegradability [97]. Dyestuffs, for satisfactory technical performance, must be resistant to change under aerobic conditions in contact with body fluids (perspiration, urine) and are, therefore, highly resistant to aerobic biodegradation. There are some advantages from this situation as the undegraded dyestuff molecules are more effectively removed in the biological treatment plant by adsorption processes, than would smaller, more water-soluble degradation products.

Under anaerobic conditions, such as in digesting sewage sludge, the indications are that dyestuffs degradation takes place at least slowly, as shown by decolorization. It is probable that subsequent total mineralization in the environment proceeds through a series of both aerobic and anaerobic degradation steps. However, studies of degradation pathways are so demanding of resources, that in the case of dyestuffs they have only been undertaken for a few model compounds, e.g. a series of substituted phenylazonaphthalene compounds [98].

The wide variety of biodegradability tests available and their usefulness was recently summarised by Gilbert [99]:

Table 8. Summary of biodegradation tests (After Gilbert, P.A. [99])

Test classification	Primary biodegradation	Ultimate biodegradation
I. Biodegradability potential tests		
a) Ready biodegradability	OECD screening test	Modified OECD screening test (DOC) Closed bottle test (O_2) Closed bottle test (O_2 saturation) (O_2) Sturm CO_2 evolution test (CO_2) AFNOR T90-302 (DOC) MITI test (O_2)
b) Inherent biodegradability	Bunch-Chambers test	Zahn-Wellens test (DOC)[a] SCAS with sewage feed (DOC)[a] Modified activated sludge test
II. Simulation tests		
a) Biological treatment (aerobic)	SDA semi-continuous activated sludge test[a] OECD confirmatory test[a] Porous pot test[a]	Coupled units test[a]
b) Biological treatment (anaerobic)		Anaerobic digestion test[a]
c) River	River die-away test General river elimination test	
d) Estuary		
e) Sea		
f) Soil		

[a] For these tests it may on occasion be difficult to distinguish between biodegradation and bio-elimination

All such tests need to be carefully interpreted as they grossly simplify the complex natural situation. It must of course be borne in mind, that commercial dyestuffs are frequently mixtures of different colorants and other non-colored constituents necessary for the satisfactory use of the dye. These non-colored compounds may be readily biodegradable, in which case the commercial dyestuff may show a substantial degree of biodegradability in a non-specific biodegradability test, even though the colorant itself may be unchanged.

Metabolic studies of dyestuffs have essentially been confined to those dyestuffs whose use involves a significant or deliberate human exposure, i.e. dyestuffs used as food additives, or in cosmetics or drugs. Because of the intensely colored nature of dyestuffs, only low concentrations are tolerable in waters, thus imposing a low ceiling on the concentration of subsequently formed biodegradation products, which in any case would generally be expected to undergo more rapid further degradation.

Azo Dyestuffs

By far the most intensively investigated class of dyestuffs are the azo dyes, and the literature up to 1969 has been comprehensively reviewed by Walker [100]. The discovery [101] that the ability of the azo dyestuff Prontosil to heal Streptococcal infections was due to the in vivo cleavage of the azo linkage to form sulphanilamide, which was the active antibacterial, is of historical importance in chemotherapy. This reductive cleavage is characteristic of the metabolism of azo dyestuffs and has been demonstrated using gut microflora [102], cell-free extracts of gut microflora [103], liver enzymes [104], and environmental bacteria [105], and indeed azo reductase activity has been reported in other tissues [106].

In the case of food dyestuffs, which are typically water-soluble, sulphonated, acid dyestuffs, little absorption from the gastrointestinal tract takes place and it appears that the gut microflora are more important than the liver reductase system in any metabolism which does occur. However, the primary metabolites, even the sulphonated primary amines, can be absorbed to a much greater extent from the gut and tend to appear in the urine or bile rather than the faeces [107].

The apparent generality of the azo reductive cleavage has prompted concern about the potential hazards associated with exposure to azo dyestuffs which could metabolize to recognized carcinogens. For example Rinde and Troll [25] demonstrated that some azo dyestuffs derived from benzidine are metabolized in Rhesus monkeys to benzidine, the human carcinogenicity of which is beyond doubt [108–110]. Benzidine and dianisidine have been detected [111] in the urine of workers exposed to dyestuffs manufactured from these intermediates. Such findings must be interpreted carefully as the detection of a carcinogenic metabolite in the urine does not provide conclusive proof of risk from the compound exposed [112]. Nevertheless it does pose the question of possible environmental risk from reductive cleavage of azo dyestuffs to toxic intermediates[6] either in dyeworks reductive stripping operations, or through microbial biodegradation of dyestuffs released to the environment in process effluent streams.

---

6 Most dyestuff manufacturing firms abandoned the manufacture of dyestuffs derived from benzidine in 1972

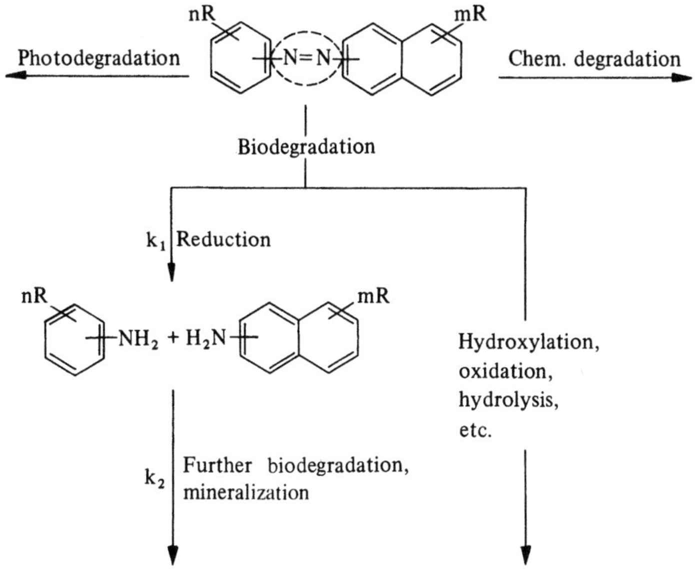

Fig. 3. Biodegradation of azo dyes. Each R is any of various substituents (typically SO_3H, COOH, OH, NO_2, NH_2, NH-, N=, N=N-, alkyl, halogen).

General consideration of the kinetics of this simplified scheme indicates that accumulation of the intermediate amines would only arise if the rate of dyestuff degradation (k_1) exceeds the subsequent amine degradation rate (k_2). Although there seems little doubt that $k_2 \geq k_1$ in many instances (e.g. dyestuffs which form aniline on degradation), the generality of this situation has not yet been substantiated. However, that amines can be expected to be degraded fairly readily by natural ecosystems is supported by the recent studies [113] of the aerobic degradation of benzidine.

Although reduction is undoubtedly the major metabolic step occurring in the gastrointestinal tract other modifications have been reported including hydrolysis of conjugates [114], acetylation [115], heterocyclic ring cleavage, e.g. for tartrazine [116]. In the liver, conjugation [117], acetylation [107], demethylation [118], and hydroxylation [119] are the main metabolic processes reported in addition to azoreductase activity.

Triphenylmethane Dyestuffs

Compared with the azo dyestuffs, the triphenylmethane dyestuffs have received little attention in terms of metabolic studies, although this group still contains several products which are used as food additives. These food additives are highly water soluble products containing sulphonic acid groups and they are poorly absorbed from the gastrointestinal tract [120]. The absorption, excretion and distribution of the dyestuff following oral ingestion and intravenous injection has been reported for Benzyl Violet 4B (C.I. 42640) [121] and Guinea Green B (C.I. 42085) [122].

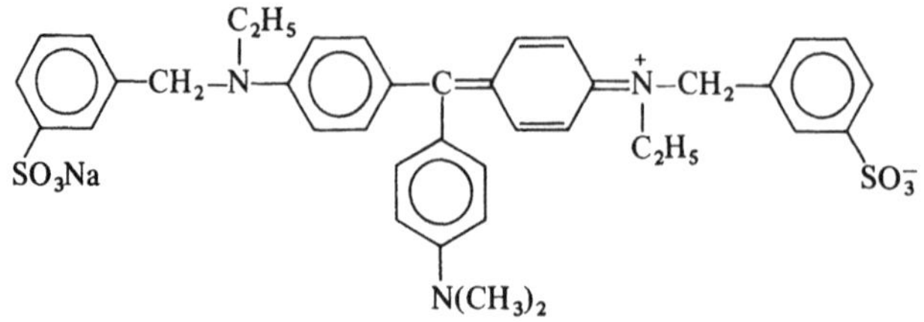

Fig. 4. Benzyl Violet 4B, C.I. Acid Violet 49, C.I. Food Violet 2, C.I. 42640

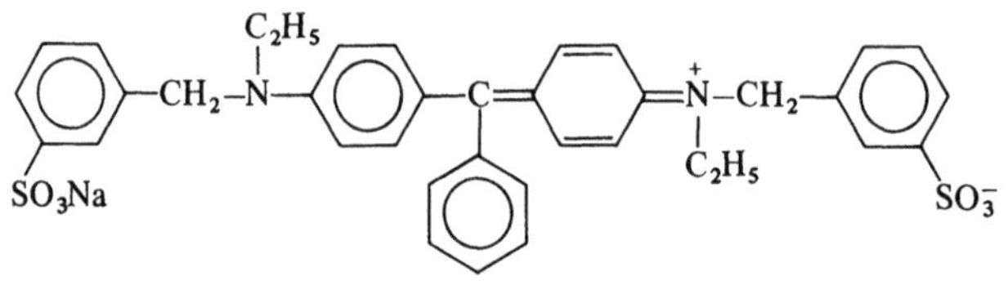

Fig. 5. Guinea Green B (FD & C. Green No. 1). C.I. Acid Green 3, C.I. Food Green 1, C.I. 42085

Xanthene Dyestuffs

The most important xanthene dyestuffs are fluorescein and its mono- and poly-halogenated derivatives, and it has been reported [123] that the mono-halogenated fluoresceins are degraded to fluorescein in rats whereas the poly-halogenated derivatives, e.g. eosin, are metabolically inert.

Webb and Hansen [124] demonstrated the stepwise de-ethylation of Rhodamine B to 3′,6′-diaminofluoran and subsequent studies [125] indicate that this process occurs in the liver cell microsomes.

Accumulation and Persistence

Dyestuffs in general must be classified as substances which biodegrade only slowly in the environment. This raises the question as to whether they persist to an extent which could present toxic hazards for the environment, and in particular whether there is any propensity to bioaccumulate with the possibility of affecting man by transport through the food cycle.

The "Yusho" incident [29], followed by the discovery of the wide environmental distribution of polychlorobiphenyls (PCBs) in Japan, prompted the enactment of the Chemical Substances Control Law which came into force in 1974 [97]. This law requires that all new substances to be marketed in Japan, which are not demonstrated to be biodegradable, must be subjected to a bioaccumulation test in fish. Although this requirement for an expensive (ca. $ 25,000.–) fish accumulation test is unique to Japan, the possible environmental build-up of a substance was also a concern of the US Environmental Protection Agency in formulating their criteria for identifying hazardous waste [126] and some indication of the tendency of a new product to accumulate will undoubtedly also be a feature of most new product control legislation.

In regard to the assessment of accumulation potential there are indications that in many instances a non-biological screening test may provide a reliable means of identifying products of low bioaccumulation potential. This test involves the determination of the n-octanol/water partition coefficient (P) [127, 128]. Neely [127] has demonstrated a linear correlation between log (bioconcentration) of several chemicals in trout muscle and log (partition coefficient).

The experimental determination of P can be difficult for products which are highly polar, insoluble, or highly lipophilic but it is also possible to derive P mathematically [129, 130] with sufficient reliability to assess whether there is any likelihood of the compound bioaccumulating to an unacceptable extent (a factor of 100 or even higher is generally considered acceptable).

In compliance with the Japanese Chemical Substances Control Law the dyestuffs manufacturing industry has conducted fish accumulation studies on a large number of new products. An investigation of these results confirmed that no products with $P_{calc.}$ value less than 1,000 showed an accumulation factor of over 100. Although it would be premature to use the partition coefficient as a decisive criterion of bioaccumulation potential, it does seem reasonable to conclude that such fish accumulation tests are superfluous in the case of highly polar water-soluble dyestuffs.

Products which are almost completely insoluble in water present particular experimental difficulties both in the fish accumulation test and for the measurement of partition coefficient. These difficulties have necessitated, for the purpose of compliance with the Japanese legislation, the development of methods for solubilizing these substances for bioaccumulation testing under conditions that have no relevance to the real situation in nature: because of the application technology and their extremely low solubility these substances do not reach open waters to any significant extent.

Toxicological Aspects

Toxicity to Aquatic Organisms

Fish

The fish is a particularly important test animal not only because its well being is a useful indicator of the general condition of waters, but also because it is an important source of food for human populations.

In 1973 Little and Lamb [131] reported extensive studies of the toxicity of 46 dyes to the fathead minnow (*Pimephales promelas*). A survey of available fish toxicity data on over 3,000 commercial products by ETAD member firms indicated that about 98% have LC_{50} values greater than 1 mg/l, a concentration at which colored pollution of a river would normally be observable.

The remaining 2% consisted of 27 different chemical structures including 16 Basic dyestuffs of which 10 were of the triphenylmethane type. In only one case was the LC_{50} as low as 0,01 mg/l, which is comparable to DDT (0,006 mg/l) and synthetic pyrethrin (0,025 mg/l) [132].

Although internationally little harmonisation has been achieved in terms of standardization in fish species, it is generally accepted that different species are unlikely to exhibit major differences in sensitivity and indeed a comparative study of three dyestuffs on minnows, trout, and golden orfe showed similar sensitivity [133].

Algae

As algae are important components of aquatic ecosystems, and algal photosynthesis is a critical source of oxygen supply, a knowledge of the effect of dyestuffs on algal activity is essential to the evaluation of the possible impact of discharges to oxidation ponds or receiving waters. In 1974 Little and Chillingworth [134] reported the effect of 56 selected dyestuffs on the growth of green alga (*Selenastrum capricornutum*). The results were in general agreement with the relative toxicities shown in the fish studies [131, 135], with the exception that many of the acid dyestuffs exhibit high toxicity to fish but do not significantly inhibit algal growth. The toxicity of 12 aminoanthraquinone dyes to *Selenastrum capricornutum* and *Pimephales promelas* has also been reported [136].

Mammalian Toxicity

General Aspects

The factors to be considered in the assessment of the toxicological risk of a colorant include: the total exposure potential, the seriousness of the toxic effect, and the fact or the probability of its occurrence. The complexity of assessing evidence of toxicity and specially carcinogenicity dictates that the evaluation of the potential human hazard of a given compound must be individualized in terms of the chemical and metabolic aspects of that specific agent, its intended uses, the data available at the time that a decision must be made, and other factors pertinent to the case under consideration. One can distinguish the following principal groups of colorants with regard to test requirements: colorants for drugs and food, for cosmetics, and for technical use for such as textiles, plastics, paints, leather, and paper. Toxicological studies of colorants for technical use are designed primarily to evaluate possible effects on exposed populations in the manufacturing and processing plants, and should be adequate to indicate any unacceptable risk of adverse effects at the much lower levels experienced by the consumer[7], or the general public through environmental pollution.

---

7 Special consideration is required in the case of higher exposure outlets, e.g. finger-paints, do-it-yourself products, children's and artists' paints

Acute Toxicity

The product safety data sheets now made available to customers by the major dyestuffs manufacturers [137] include the following information:
1. Acute oral LD_{50} in rats
2. Skin irritiation on rabbits
3. Eye irritation on rabbits

obtained using standardized methods.

These data provide a basis for recommendation of appropriate handling precautions, and under present day minimum acceptable standards of handling and working environment the acute toxicity of dyestuffs is not a problem.

Table 9. Determination of single administration toxicity of dyestuffs (commercial brands) by ETAD members[a]

No. of dyes tested	No. of dyes with LD_{50} values (mg/kg)			
	<250	250–2,000	2,000–5,000	>5,000
4461 (100)[b]	44 (1)	314 (7)	434 (9.7)	3,669 (82.3)

[a] Acute oral toxicity LD_{50} in rats. Only data published in the ETAD safety data sheets up to August 1977

[b] The values in parenthesis are percentages

A survey [34] of the data available in mid 1977 indicated that dyestuffs are of generally low acute toxicity (Table 9) as only 1% of the commercial products showed LD_{50} values under 250 mg/kg (none was less than 100 mg/kg). Closer investigation revealed that this more toxic 1% involved only 15 different chemical structures (Table 10).

Sensitization

Contact dermatitis or skin sensitization effects have been experienced with several specific dyestuffs both in terms of manufacturing or processing experience [138–140] and as a result of consumer exposure to dyed fabrics [138, 141, 142]. There are no known reports of such effects arising as a result of environmental pollution by dyestuffs.

Chronic Toxicity / Carcinogenicity

The question of most concern to managements, workers and public is whether a chemical product can produce a carcinogenic effect in humans as a result of chronic exposure at low levels. Epidemiological evidence in general supports the conclusion that, provided sensible working procedures are used, there is no significantly higher cancer incidence among exposed workers [143, 144].

Table 10. Analysis of LD_{50} values lower than 2,000 mg/kg by chemical type[a]

Chemical type	Total No. of products	No. of products with LD_{50} (mg/kg)	
		≤250	251–2,000
Monoazo	14	1 (125)	13
Monoazo/quaternary	16	0	16
Disazo	20	4 (199, 200, 240, 200)	16
Trisazo	1	0	1
Phthalocyanine	2	0	2
Diphenylmethane	1	0	1
Triphenylmethane	13	1 (100)	12
Xanthene	6	2 (220, 250)	4
Oxazine	4	1 (210)	3
Methine	20	4 (133, 213, 222, 224)	16
Anthraquinone	3	0	3
Indigoid	1	0	1
Stilbene azo	2	1 (150)	1
Miscellaneous	11	1 (221)	10
Total	114	15	99

[a] Criteria for analysis:
1. Mixture products were omitted
2. Azoic diazo components (28) were not included
3. Where products with the same basic structure, as identified by Colour Index No., have been tested by different member firms, only the lowest LD_{50} value was used
4. Where LD_{50}values were quoted as a range, the lower value has been used for classification. By this procedure the total number (358) of tested commercial products was reduced to 114 products with different structures

Evidence of adverse effects in human populations exposed to dyestuffs has however been reported [145–149]; for example, among kimono painters who had the habit of pointing their paint-brushes between their lips [148], and workers whose working conditions were apparently inadequate [149]. The primary prophylactic measure is to reduce exposure to a minimum by adopting good working procedures and personal hygiene, whilst at the same time seeking to detect any product which is a strong carcinogen. Evidence of carcinogenicity can be obtained from three sources [150]:

(1) Epidemiological evidence from exposed human populations. Although providing persuasive evidence of actual risk to man at the exposure levels experienced, this approach has the disadvantage of being retrospective. Furthermore, negative data cannot normally adequately establish non-carcinogenicity of a product although they can help define the upper limits of risk.

(2) Experimental animal studies. These provide the best available experimental evidence of carcinogenic potential, but the high cost (ca. $ 200,000.–) of a properly conducted bioassay, limited testing resources, and the problems of interpreting positive results obtained at high dosage levels in terms of risk at low practical exposure levels has limited their application in the case of textile dyestuffs. In this context the proposal by the Ameri-

can Industrial Health Council to categorize carcinogens by potency is a sensible approach [151].

(3) Evidence from studies of chemical structure, reactivity, and mutagenic effects as detected by various short-term tests. Such additional indication of carcinogenic potential must still be regarded as simply suggestive [152, 153]. However, the rapid developments in mutagenicity testing are of great interest to the dyestuff industry, which like other sectors of the chemical industry, would benefit from the availability of a rapid low-cost method for reliably assessing carcinogenic potential.

Several extensive reviews of the carcinogenicity of dyestuffs [154] particularly of azo dyestuffs [155–160] and food dyestuffs [120, 161] have been written. Evaluation of the available literature is made difficult by the fact that the identity and quality of the test substance are frequently not defined unequivocally, and often the route of exposure is not relevant (e.g. sub-cutaneous injection, pellet implantation). Most studies fall short of the criteria proposed by the Interagency Regulatory Liaison Groups [150] and indeed of the 58 organic colorants listed by NIOSH [162] as having been reported as carcinogenic or neoplastic only about 20–25% can be regarded as having been adequately tested.

Bioassay studies have concentrated largely on food dyestuffs and based on the evaluation of the results several products are no longer permitted as food additives. The finding that three benzidine-based dyestuffs produced pre-neoplastic hepatic lesions in rats in a 13-week subchronic study [149, 163] has led to concern about a possible hazard from azo dyestuffs based on benzidine or its derivatives [164], but no complete bioassay on a benzidine-based dyestuff has yet been reported.

Available evidence supports the conclusion that organic pigments are of low toxicity presumably because their generally very low solubility means that they are scarcely available for biological action. Few 2-year feeding studies have been completed although two pigments based on 3,3′-dichlorobenzidine (DCB) – Pigment Yellow 12 (*3*) and Pigment Yellow 83 – were found to be non-carcinogenic [165, 166]. These results are of particular interest because reductive cleavage of the azo linkages would have led to the regeneration of DCB, an animal carcinogen [167] and therefore a suspected human carcinogen, although epidemiological evidence in plants where DCB was handled for many years did not substantiate that there was a toxic risk under the conditions used [168–170]. No evidence of metabolism of these pigments was found in this work, or in a recent study of Pigment Yellow 13 [24], contrary to the earlier report by Akiyama [171].

No carcinogenic effects were found in feeding studies with Pigment Yellow 16 (based on o-tolidine) [165], or the monoazo pigments Pigment Red 49 [172], and Pigment Red 53:1 [173].

Mutagenicity

Although organic colorants have been tested in a variety of short-term mutagenicity procedures [174], notably the *Salmonella* microsome assay developed by Ames, none of these methods can be regarded as having been established as reliable indicators of carcinogenic potential. Although most carcinogenic azo dyestuffs of the p-dimethylaminoazo-benzene series are mutagenic [175], negative mutagenic results have been obtained on several other monoazo dyestuffs which have been reported to be animal carcinogens [176].

Mutagenic activity has been reported [177–179] in many anthraquinone derivatives and dyestuffs but a series of 22 acid dyes (6 anthraquinone, 14 azo and 2 nitro) were non-mutagenic [180] in *E. Coli* TM4 and *S. typhimurium* TA100.

Permitted food colors [181, 182] and cosmetic colors [183] have been found to be non-mutagenic with the exception of Pigment Orange 5 which is used in some lipsticks.

A study of colorants important in the graphic arts and painting industry [184] included 16 organic pigments of which 2 (Pigment Orange 5 and Pigment Red 1) were reported to be weakly mutagenic. In both cases the test substance was dissolved in DMSO and the interpretation of results in terms of risk under practical conditions, from products which are extremely insoluble in water and fat, needs clarification.

Legislation

In recent years there has been a dramatic increase in regulatory activities aimed at achieving safer manufacture, use and disposal of chemicals, including colorants [185]. The complexity of international laws and regulations is now such that their review is extremely difficult [186]. Regulations in themselves are not opposed by responsible industry provided they effectively serve certain specific needs and achieve their objectives in a cost-effective manner.

The regulation of gaseous and liquid effluents from manufacturing and processing plants, as well as the disposal of industrial waste, has evolved gradually in the developed countries and generally operates on a regional basis (State, local authority, or regional waste board) within a framework of national legislation. This allows some flexibility in approach and enables the requirements to be geared to the local situation. For the dyestuffs industry, where pollution problems are essentially localised, this approach operates satisfactorily, although an international approach is undoubtedly required to deal with major pollution of international waterways (e.g. Rhine river, W. Europe).

Dyestuffs are subject to the laws which apply to chemicals in general. The first stepwise efforts to exercise environmental controls were through national laws which applied to specific limited areas, e.g. the US national water pollution control laws. Table 11 shows the year of introduction of the major environmental laws in the OECD countries [187].

The complexity of the interaction of various Acts contributing to water pollution control is best illustrated by the US situation [188], involving the Federal Water Pollution Control Act, Marine Protection, Research and Sanctuaries Act (1972), Safe Drinking Water Act (1974), Resource Conserva-

Table 11. Date of introduction of major national water pollution control laws in the OECD countries

OECD country	Year	OECD country	Year
Australia	-	Japan	1958, 1970
Austria	1959	Luxembourg	-
Belgium	1971	Netherlands	1970, 1975
Canada	1970	New Zealand	1967, 1974
Denmark	-	Norway	1970
Finland	1961, 1965	Portugal	-
France	1964	Spain	-
Germany	1957, 1976	Sweden	1969
Greece	1978	Switzerland	1971
Iceland	-	Turkey	1971
Ireland	1977	United Kingdom	1961, 1974
Italy	1976	United States	1972, 1977

tion and Recovery Act (1976), Hazardous Materials Transportation Act (1974), Ports and Waterways Safety Act (1972), Federal Insecticide, Fungicide and Rodenticide Act (1972), Toxic Substances Control Act (1976), Atomic Energy Act (1954). The legislative process on an international level is even more complex. For example within the EEC the process of developing a directive for control of water quality was protracted over several years [189].

Examples of legislation concerned with the working environment are the Health and Safety at Work Act (1974) in the UK and the Occupational Safety and Health Act (1970) in the USA.

Some recent legislation has sought to be much wider in scope and to tackle the totality of the effects of chemicals on man and the environment. Probably the most far-reaching environmental legislation introduced so far is the US Toxic Substances Control Act (1976). In order to meet the requirements of this Act a register of existing commercial chemicals has been compiled and the introduction of a new chemical to the market must be notified to the authorities and an extensive dossier submitted which includes appropriate ecological and toxicological data. A similar notification system is foreseen in the EEC Directive "6th Amendment of the Council Directive of June 27, 1967 on the approximation of laws, regulations and administrative provisions relating to the classification packaging and labelling of dangerous substances" [190]. In Japan the Chemical Substances Control Law (1973) requires an expensive fish accumulation test for new substances which are not readily biodegradable. Provided newly introduced dyestuffs and pigments show no significant accumulation tendency in this accumulation test they are permitted under the terms of the law [97].

The increased incidence of bladder cancer among dyestuff manufacturing workers was attributable to the carcinogenic properties of some amines used as intermediates, particularly benzidine, 2-naphthylamine and 4-aminobiphenyl. These chemicals have been either banned or subjected to strict regulatory controls of handling and use in most industrial countries, but unfortunately

Table 12. Prohibition or control of various dyes and chemicals as carcinogens

Product	Australia	Belgium	Finland	W. Germany	Italy	Japan	Sweden	Switzerland	UK	USSR	USA
4-Aminobiphenyl	+	+	+	+	+	+	+	+	+		+
4-Aminobiphenyl salts						+			+		
Auramine			+			+	+		+		
Benzidine	+	+	+	+	+	+	+	+	+	+	+
Benzidine salts	+	+	+	+	+	+	+		+		+
Dianisidine			+			+	+		+		
Dianisidine salts						+			+		
3,3'-Dichlorobenzidine	+	+	+	+		+	+	+	+	+	+
3,3'-Dichlorobenzidine salts						+			+		
4-Dimethylaminoazobenzene	+	+	+			+	+	+		+	+
Magenta						+			+		
1-Naphthylamine	+[a]	+	+			+	+		+	+	+
1-Naphthylamine salts						+			+		
2-Naphthylamine	+	+	+	+	+	+	+	+	+		+
2-Naphthylamine salts						+			+		
o-Tolidine		+	+			+	+		+		
o-Tolidine salts						+			+		

[a] Addition proposed in draft "Carcinogenic Substances Regulations"

there has been much diversity in these controls from country to country [191] (Table 12).

This particular example underlines the urgent necessity for much closer international harmonisation of legislation and testing methodology: if the current tendency to insist on national preferences for test methodology persists, the limited resources and test capacity will be unnecessarily squandered with minimal benefits in terms of product safety.

Acknowledgement

The authors gratefully acknowledge the invaluable cooperation of the various members of the ETAD committees. Thanks are due in particular to Dr. D. Brown for his critical comments on the manuscript and to Dr. Dorothee Braun-Steinle for assistance in preparation of this contribution.

AZO

(a) *Monoazo*

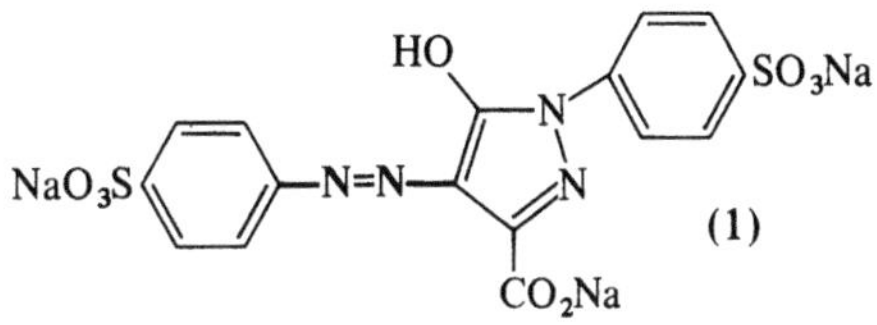

(1)

Acid Yellow 23
C.I. 19140
tartrazine

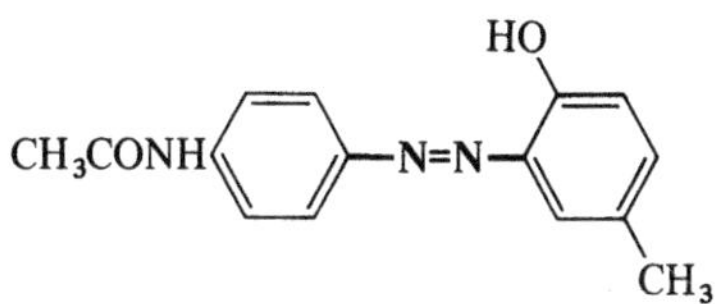

(2)

Disperse Yellow 3
C.I. 11855

(b) *Disazo*

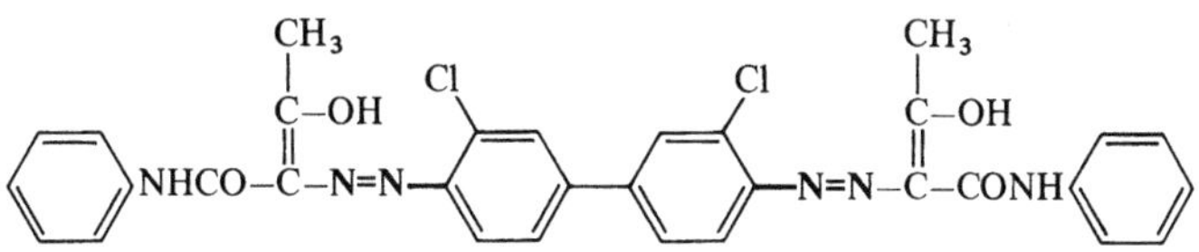

(3)

Pigment Yellow 12
C.I. 21090

Stilbene

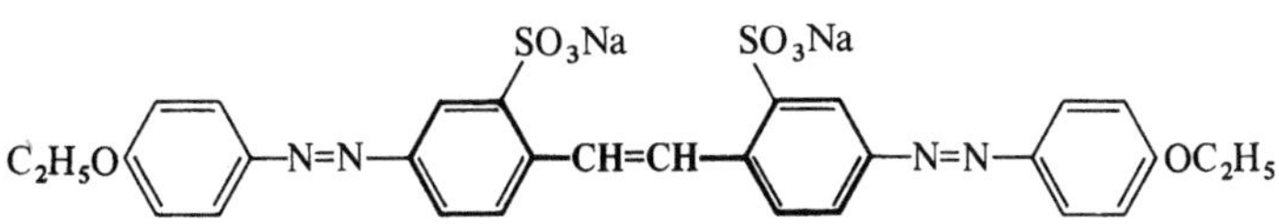

(4)

Direct Yellow 12
C.I. 24895

Diphenylmethane

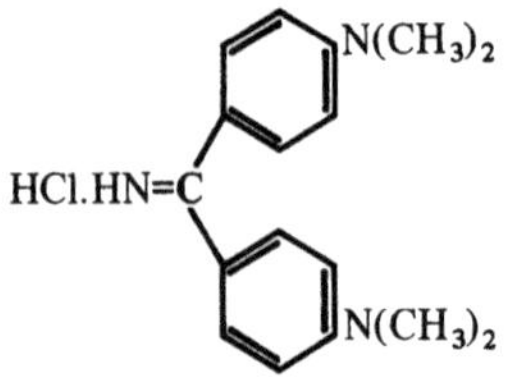

(5)

Basic Yellow 2
C.I. 41000
auramine

Triphenylmethane

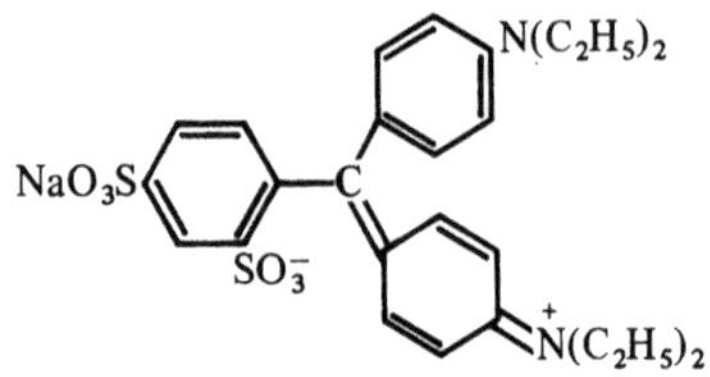

(6)

Acid Blue 1
C.I. 42045

Xanthene

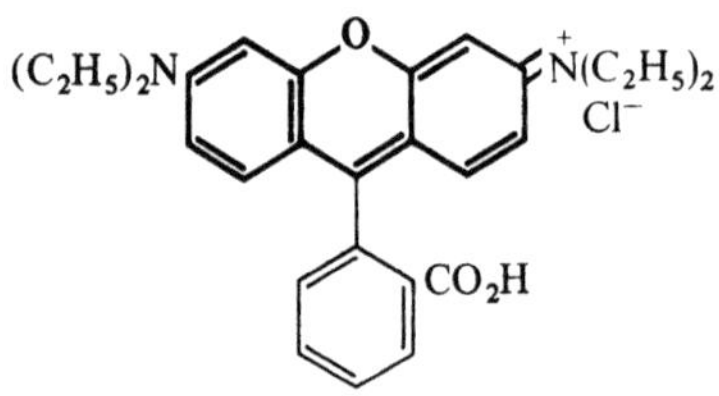

(7)

Basic Violet 10
C.I. 45170

Acridine

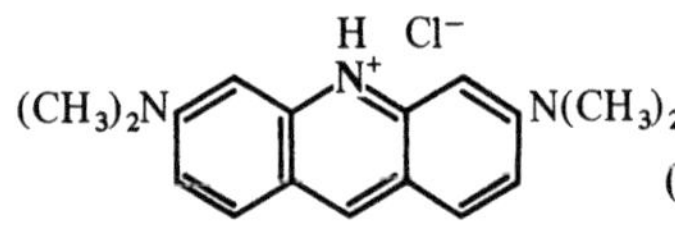

(8)

Basic Orange 14
C.I. 46005

Quinoline

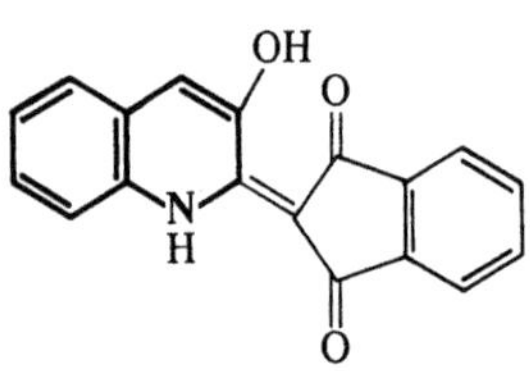

(9)

Disperse Yellow 54
C.I. 47020

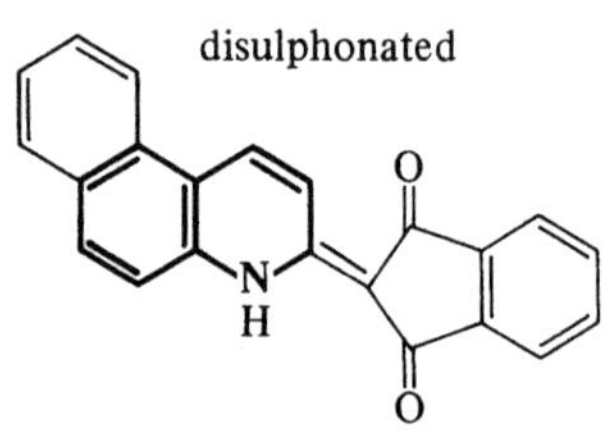

(10)

Acid Yellow 5
Direct Yellow 5
C.I. 47035

Methine

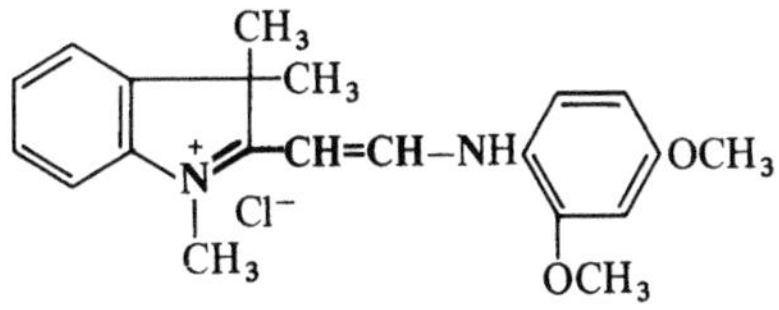

Basic Yellow 11
C.I. 48055 (11)

Azine

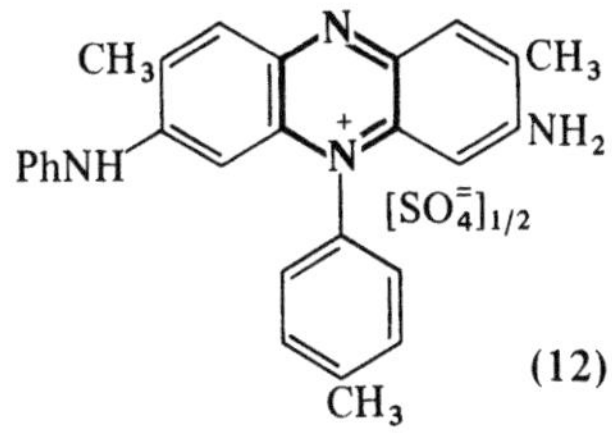

(12)
C.I. 50245
mauveine

Oxazine

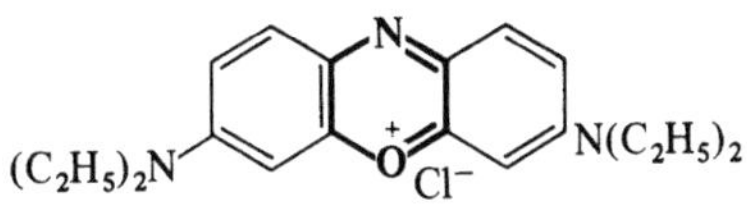

Basic Blue 3 (13)
C.I. 51004

Thiazine

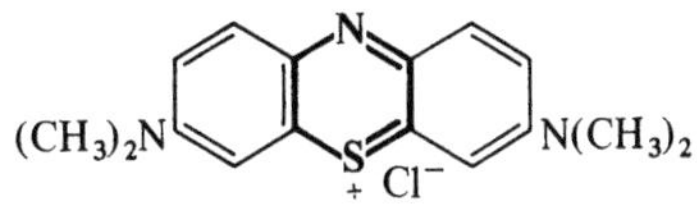

Basic Blue 9 (14)
Solvent Blue 8
C.I. 52015
methylene blue

Sulphur

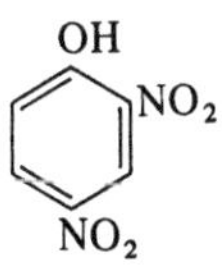

heated with sodium polysulphide (15)

Sulphur Black I
C.I. 53185

Indigoid

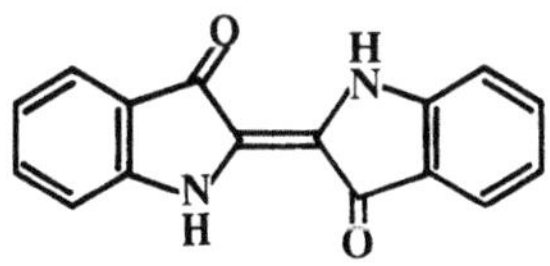

(16)
Vat Blue I
Pigment Blue 66
C.I. 73000
indigo

Anthraquinone

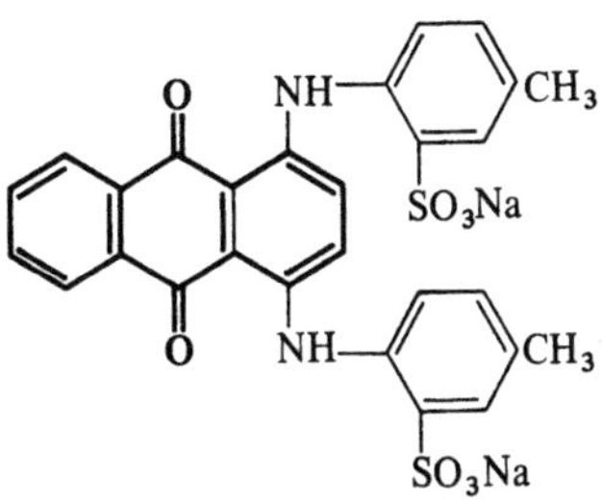

(17)
Acid Green 25
C.I. 61570

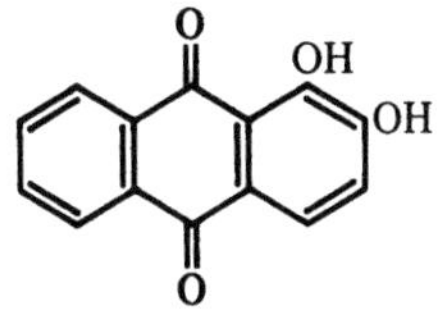

Mordant Red 11
Pigment Red 83 (18)
C.I. 58000
alizarin

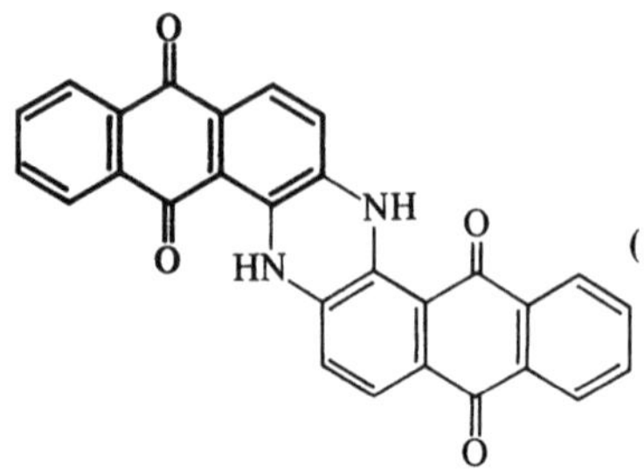

(19)

Pigment Blue 60
Vat Blue 4
C.I. 69800
indanthrone

Phthalocyanine

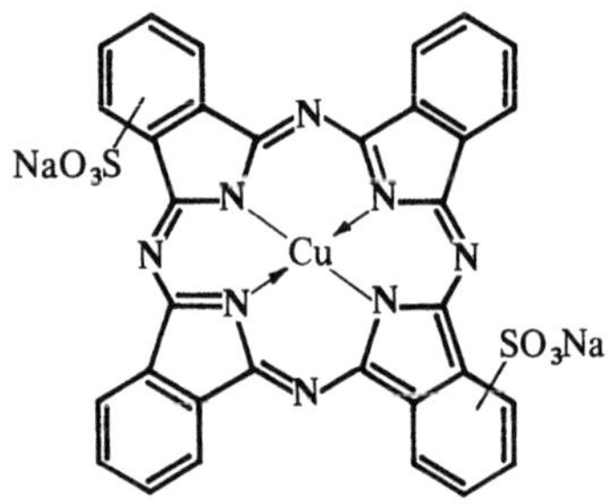

(20)

Direct Blue 86
C.I. 74180

Fig. 6. Selection of some representative chemical structures

References

1. Anliker, R.: Ecotox. Environ. Safety *1*, 211 (1977)
2. Wiedhaup, K.: Acta Pharm. Technol. Suppl. *8*, 67 (1979)
3. Anliker, R., Müller, G.: Fluorescent Whitening Agents. Environ. Qual. Safety Suppl. 4. G. Thieme, Stuttgart, Academic Press, New York 1975
4. Venkatamaran, K. (Ed.): The Chemistry of Synthetic Dyes. Academic Press, New York 1952 (Vol. I) to 1978 (Vol. VIII)
5. Colour Index, 3rd Edition. Soc. Dyers and Colourists. Bradford, England 1971 (Vol. 1–4), 1975 (Suppl. Vol. 5,6)
6. US Tariff Commission: Synthetic Organic Chemicals, U. S. Production and Sales. US Government Printing Office, Washington, issued annually
7. Venkatamaran, K. (Ed.): The Analytical Chemistry of Synthetic Dyes. J. Wiley & Sons, New York 1977
8. Weisburger, J.H., Weisburger, E.K.: Food Cosmet. Toxicol. *6*, 235 (1968)
9. Claus, G., Krisko, I., Bolander, K.: ibid. *12*, 737 (1974)

10. Falk, H.L.: Environ. Health Persp. *22*, 167 (1978)
11. Revision of US Federal Food Drug and Cosmetics Act, Section 409c, 3A, Oct. 1962
12. Jukes, T.H.: J. Am. Med. Assoc. *241*, 617 (1979)
13. Horning, R.H.: Text. Chem. Color. *4*, 275 (1972)
14. Horning, R.H.: ibid. *9*, 73 (1977)
15. Dry Color Manufacturers' Assoc.: Am. Ink Maker *51* (10), 31 (1973)
16. Annual Book of ASTM Standards, Part 31, Water. Amer. Soc. Testing and Materials, Philadelphia 1975
17. Methods for Chemical Analysis of Water and Wastes. US Environmental Protection Agency, Office of Technology Transfer, Washington, D.C. 1974
18. Burrell, D.C.: Atomic Spectrometric Analysis of Heavy Metal Pollutants in Water. Ann Arbor Science Publishers, Ann Arbor, Michigan 1975
19. Salbu, B., Steinnes, E., Pappas, A.C.: Analyt. Chem. *47*, 1011 (1975)
20. Standard Methods for the Examination of Water and Waste Water, 13th Ed. Amer. Public Health Assoc., New York 1971
21. Crocker, I.H., Merritt, W.F.: Water Res. *6*, 285 (1972)
22. NIOSH Manual of Analytical Methods, 2nd Edition. US Dept. of Health, Education, and Welfare. No. 77–157–A, April 1977
23. NCTR Techn. Rep. Experiment No. 196: Metabolism of Azo Dyes to Potentially Carcinogenic Aromatic Amines. Nat. Center for Toxicolog. Res. Jefferson, Ark., March 1979
24. Mondino, A. et al.: Medicina del Lavoro *69*, 693 (1978)
25. Rinde, E., Troll, W.: J. Nat. Cancer Inst. *55*, 181 (1975)
26. Rice, J.R., Kissinger, P.T.: J. Anal. Toxicol. *3*, 64 (1979)
27. Sciarini, L.J.: A.M.A. Arch. Ind. Health *11* (5), 420 (1955)
28. Jenkins, R., Kuo, Ch., Baird, R.: Determination of benzidine in wastewaters. Div. Environ. Chem., Am. Chem. Soc. *14*, 273 (1974)
29. Higuchi, K. (Ed.): PCB Poisoning and Pollution. Academic Press, New York 1976
30. EPA proposed rule banning manufacturing, processing, distribution in commerce and use of polychlorinated biphenyls (PCBs). US Fed. Register, *43*, 24803 (June 7, 1978)
31. EPA preamble to final rule on polychlorinated biphenyls (PCBs) manufacturing, processing, distribution in commerce and use prohibitions. US Fed. Register, *44*, 31514 (May 31, 1979)
32. Anliker, R.: Biodegradation of dyestuffs. Paper presented at a symposium on biodegradability, held in Watford, Herts., UK on April 25, 1979 (To be published in Chemosphere)
33. CEFIC Biodegradation Task Force: Biodegradation Status Report, Febr. 1978, p. 9
34. Anliker, R.: Ecotox. Environ. Safety *3*, 59 (1979)
35. ADMI: Dyes and the Environment, Vol. 1., Amer. Dye Manufacturers Inst. Inc., New York 1973
36. ADMI: A Literature Survey of Colored Waste, Amer. Dye Manufacturers Inst. Inc., New York 1971
37. McKay, G.: Am. Dyest, Rep. *68* (4), 29, 47 (1979)
38. Horning, R.H.: Textile Dyeing Wastewaters-characterization and treatment. US Environmental Protection Agency, EPA-600/2-78-098; Washington, D.C. 1978
39. Weeter, D.W., Hodgson, A.G.: Am. Dyest, Rep. *66* (8), 32, 61 (1977)
40. Radding, S.B. et al.: Assessment of potential toxic releases from leather industry dyeing operations, EPA-600/2-78-215; Washington, D.C. 1978
41. Roth, W.: Schweiz. Bauzeitung *95*, 620 (1977)
42. Gardiner, D.K., Borne, B.J.: J. Soc. Dyers Colour. *94*, 339 (1978)
43. Höke, B.: Textilveredlung *13*, 232 (1978)
44. Porter, J.J., Sargent, T.N.: Text. Chem. Color. *9*, 269 (1977)
45. Dürig, G.: Rev. Prog. Coloration *7*, 70 (1976)
46. Color removal and sludge disposal process for kraft mill effluents, US Environmental Protection Agency, EPA-660/2-74-008; Washington, D.C. 1974
47. Bailey, D.A.: J. Soc. Leather Techn. Chem. *62*, 75 (1978)
48. Shuttleworth, S.G.: ibid. *62*, 87 (1978)
49. Gaffney, P.E.: J. Water Pollut. Control Fed. *48*, 2590, 2731 (1976)
50. Yang, P.Y., Pescod, M.B.: Am. Dyest. Rep. *66* (12), 47 (1977)

51. Fischer, K.: Melliand Textilber. *59*, 487, 659 (1978)
52. Games, L.M., Hites, R.A.: Analyt. Chem. *49*, 1433 (1977)
53. ADMI: Dyes and the Environment, Vol. 2., Amer. Dye Manufacturers Inst. Inc., New York 1974
54. APHA: Standard Methods for the Examination of Water and Waste Water. 13th Ed. Amer. Public Health Assoc., New York 1971
55. Allen, W. et al.: Determination of color of water and waste water by means of ADMI color values. In Ref. 5, Chap. III
56. Tincher, W.C.: Survey of the Coosa Basin for Organic Contaminants from Carpet Processing. Environmental Protection Division, Dept. of Natural Resources, Contract No. E-27-630; Georgia, USA, Oct. 1978
57. Oswalt, Y.H., Land, Y.G.: Color removal from kraft pulp mill effluents by massive lime treatment, US Environmental Protection Agency, EPA-R2-73-086; Washington, D.C. 1973
58. Rinker, T.L.: Treatment of textile wastewater by activated sludge and alum coagulation; US Environmental Protection Agency, EPA-600/2/75-055; Washington, D.C. 1975
59. Olthof, M.G., Eckenfelder, W.W.: Water Res. *9*, 853 (1975)
60. Iida, H., Endo, M.: J. Appl. Chem. (Abs.) *18*, 333 (1968)
61. Richner, P., Kerres, B.: Melliand Textilber. *58*, 681 (1977)
62. McKay, G., Otterburn, M.S., Sweeney, A.G.: J. Soc. Dyers Colour. *93*, 357 (1978)
63. Rhys, O.G.: ibid. *93*, 293 (1978)
64. Digiano F., Frye, W.H., Natter, A.S.: Am. Dyest. Rep. *64* (8), 15 (1975)
65. Mitchell, M., Ernst, W.R., Lightsey, G.R.: Bull. Environ. Contam. Toxicol. *19*, 307 (1978)
66. Rodman, C.A.: Text. Chem. Color. *3*, 239 (1971)
67. DeJohn, P.B., Hutchins, R.A.: ibid. *8*, 69 (1976)
68. Shunney, E.L., Perrotti, A.E., Rodman, C.A.: Am. Dyest. Rep. *60* (6), 32 (1971)
69. Davies, R.A., Kaempf, H.J., Clemens, M.M.: Chem. Ind. (17), 827 (1973)
70. Porter, J.J.: Water Wastes Eng., Ind. 1972 (Jan.), A8-A10, A12
71. Ottemeyer, W.: Gesundh.-Ing. *53*, 185 (1930)
72. Poots, V.J.P., McKay, G., Healy, J.J.: Water Res. *10*, 1067 (1976)
73. Rock, S.L., Stevens, B.W.: Text. Chem. Color. *7*, 169 (1975)
74. Kennedy, D.C., Stevens, B., Kerner, J.W.: Am. Dyest. Rep. *63* (8), 11 (1974)
75. Brandon, C.A. et al.: Text. Chem. Color *5*, 134 (1973)
76. Snider, E.H., Porter, J.J.: Am. Dyest. Rep. *63* (8), 36 (1974)
77. Fiala, B., Villiger, K.: Textilveredlung *14*, 5 (1979)
78. Gerber, K.: Chem. Rundschau *30* (19), 1 (1977)
79. Croft, T.G., Eichholz, G.G.: Removal of color from dyework effluents by ionizing radiation and chemical oxidation, US Nat. Inform. Serv. PB Rep. 221486/4, 1973
80. Case, F.N., Ketchen, E.E., Alspaugh, T.A.: Text. Chem. Color. *5*, 62 (1973)
81. Gilbert, E., Güsten, H.: Chemiker Z. *101*, 22 (1977)
82. Gubser, H.: Colored effluents. CIBA-GEIGY Rev. (4), 31 (1972)
83. Bretscher, H., Eigenmann, G., Plattner, E.: Chimia *32*, 180 (1978)
84. Jones, D.L.: Am. Dyest. Rep. *61* (8), 28 (1972)
85. Shelley, M.L., Randall, C.W., King, P.H.: J. Water Pollut. Control Fed. *48*, 753 (1976)
86. Shriver, L.E., Daugue, R.R.: Am. Dyest. Rep. *67* (3), 34 (1978)
87. Alspaugh, T.A.: Text. Chem. Color. *5*, 255 (1973)
88. Masselli, J.W., Masselli, N.W., Burford, M.G.: Text. Ind. *135* (10), 84, 108 (1971)
89. Porter, J.J., Snider, E.H.: J. Water Pollut. Control Fed. *48*, 2198 (1976)
90. Weeter, D.W., Hodgson, A.G.: Am. Dyest. Rep. *66* (8), 32 (1977)
91. Dohányos, M., Maděra, V., Sedláček, M.: Prog. Wat. Tech. *10*, 559 (1978)
92. Hitz, H.R., Huber, W., Reed, R.H.: J. Soc. Dyers Colour. *94*, 71 (1978)
93. Meier, H.: Photochemistry of Dyes, in: The Chemistry of Synthetic Dyes, Vol. IV (Venkatamaran, K. (Ed.)), Academic Press, New York 1971
94. Radding, S.R. et al.: Review of the Environmental Fate of Selected Chemicals. Office of Toxic Substances, EPA 560/5-77-003; Washington, D.C. 1977
95. Porter, J.J.: A Study of the Photodegradation of Commercial Dyes. US Environmental Protection Agency, EPA-R2-73-058; Washington, D.C. 1973
96. Heitz, J.R., Wilson, W.W.: Photodegradation of halogenated xanthene dyes. ACS Sympo-

sium Ser. 73, Disposal and Decontamination of Pesticides (Kennedy M.V. (Ed.)), Washington, D.C. 1978
97. Kubota, Y.: Ecotox. Environ. Safety *3*, 256 (1979)
98. Meyer, U., Overney, G., von Wattenwyl, A.: Textilveredlung *14*, 15 (1979)
99. Gilbert, P.A.: Ecotox. Environ. Safety *3*, 111 (1979)
100. Walker, R.: Food Cosmet. Toxicol. *8*, 659 (1970)
101. Tréfouel, J. et al.: Compt. rend. soc. biol. *120*, 756 (1935)
102. Chung, K.T., Fulk, G.E., Egan, M.: Appl. Environ. Microbiol. *35*, 558 (1978)
103. Hartmann, C.P., Fulk, G.E., Andrews, A.W.: Mutat. Res. *58*, 125 (1978)
104. Manchon, Ph., Lowy, R.: Food Cosmet. Toxicol. *3*, 783 (1965)
105. Idaka, E. et al.: J. Soc. Dyers Colour. *94*, 91 (1978)
106. Juchau, M.R., Krasner, J., Yaffe, S.J.: Biochem. Pharmacol. *17*, 1969 (1968)
107. Daniel, J.W.: Toxicol. Appl. Pharmacol. *4*, 572 (1962)
108. Case, R.A.M. et al.: Brit. J. Ind. Med. *11*, 75 (1954)
109. Barsotti, M., Vigliani, E.C.: A.M.A. Arch. Ind. Hyg. Occup. Med. *5*, 234 (1952)
110. Scott, T.S.: Brit. J. Ind. Med. *9*, 127 (1952)
111. Genin, V.A.: Vopr. Onkol. *23* (9), 50 (1977)
112. Batten, P.L., Hathway, D.E.: Brit. J. Cancer *35*, 342 (1977)
113. Tabak, H.H., Barth, E.F.: J. Water Pollut. Control Fed. *50*, 552 (1978)
114. Williams, R.T., Milburn, P., Smith, R.L.: Ann. N. Y. Acad. Sci. *123*, 110 (1965)
115. Scheline, R.R., Longberg, B.: Acta Pharmacol. Toxicol. *23*, 1 (1965)
116. Roxon, J.J. et al.: Food Cosmet. Toxicol. *5*, 447 (1967)
117. Williams, R.T.: Detoxication Mechanisms. John Wiley Ltd, London 1959, p. 4
118. Müller, G.C., Miller, J.A., Glassner, M.: J. Biol. Chem. *202*, 579 (1953)
119. Larsen, J.C., Tarding, F.: Acta Pharmacol. Toxicol. *39*, 525 (1976)
120. Radomski, J.L.: Am. Rev. Pharmacol. *14*, 127 (1974)
121. Minegishi, K.I., Yamaha, T.: Toxicology *7*, 367 (1977)
122. Minegishi, K.I., Yamaha, T.: Chem. Pharm. Bull. *22*, 2042 (1974)
123. Webb, J.M., Fonda, M., Brouwer, E.A.: J. Pharmacol. Exp. Ther. *137*, 141 (1962)
124. Webb, J.M., Hansen, W.H.: Toxicol. Appl. Pharmacol. *3*, 86 (1961)
125. Webb, J.M. et al.: ibid. *3*, 696 (1961)
126. EPA Proposed Criteria for identifying hazardous waste; Standards for generators, and standards for management facilities, Sect. 250.14. US Fed. Register *43*, 58946 (Dec. 18, 1978)
127. Neely, W.B., Branson, D.R., Blau, G.E.: Environ. Sci. Technol. *8*, 1113 (1974)
128. Chiou, C.T. et al.: ibid. *11*, 475 (1977)
129. Fujita, T., Iwasa, J., Hansch, C.: J. Am. Chem. Soc. *86*, 5175 (1964)
130. Hansch, C. et al.: J. Med. Chem. *16*, 1207 (1973)
131. Little, L.W., Lamb, J.C.: in ref. 35, Chap. V
132. Tooby, T.E., Hursey, P.A., Alabaster, J.S.: Chem. Ind. (12), 523 (1975)
133. Hamburger, B., Häberling, H., Hitz, H.R.: Arch. Fisch Wiss. *28*, 45 (1977)
134. Little, L.W., Chillingworth, M.A.: in ref. 53, Chap. II
135. Little, L.W., Chillingworth, M.A.: in ref. 53, Chap. IV
136. Chillingworth, M.A.: in ref. 53, Chap. V
137. Anliker, R.: Swiss Chem. *1* (9), 34 (1979)
138. Cywie, P.L. et al.: Les eczemas allergiques professionnels dans l'industrie textile; Inst. Nat. Rech. de Sécurité, Rep. No. 244/R1. Paris, March 1977
139. Gardiner, J.S. et al.: Brit. J. Dermatol. *85*, 264 (1971)
140. Alanko, K. et al.: Clinical Allergy *8*, 25 (1978)
141. Cronin, E.: Trans. St. John's Hosp. Dermatol. Soc. *54*, 156 (1968)
142. Sim-Davies, D.: ibid. *58*, 251 (1972)
143. Ferber, K.H., Hill, W.J., Cobb, D.A.: J. Am. Ind. Hyg. Assoc. *37*, 61 (1976)
144. Proportional mortality study on members of the National Union of Dyers, Bleachers, and Textile Workers (NUDBTW), Bradford, England, unpublished
145. Anthony, H.M.: J. Soc. Occup. Med. *24*, 110 (1974)
146. Cole, P., Hoover, R., Friedell, G.H.: Cancer *29*, 1250 (1972)
147. Yoshida, O. et al.: Igaku no Ayumi *79*, 421 (1971)

148. Yoshida, O., Miyakawa, M.: Etiology of bladder cancer: Metabolic aspects, in: Analytic and Experimental Epidemiology of Cancer (Nakahara, W. et al. (Eds.)), University Park Press, Baltimore 1973
149. NIOSH/NCI Joint Current Intelligence Bulletin 24. Direct Black 38, Direct Blue 6, and Direct Brown 95, benzidine-derived dyes. US Dept. of Health, Education, and Welfare, April 17, 1978
150. IRLG Work Group Rep. on the Sci. Basis for identification of potential carcinogens and estimation of risks. US Fed. Register *44*, 39858 (July 6, 1979)
151. AIHC Recommended Alternatives to OSHA's Generic Carcinogen Proposal. Amer. Ind. Health Council. Jan. 9, 1978. Published in part in Chem. Eng. News *56* (5), 30 (1978)
152. Nat. Cancer Adv. Board. General criteria for assessing the evidence for carcinogenicity of chemical substances. Nat. Cancer Inst. *58*, 461 (1977)
153. Environmental Health Criteria 6. Principles and methods for evaluating the toxicity of chemicals. Part. I. WHO, Geneva 1978
154. IARC Monographs on the Evaluation of the Carcinogenic Risk of Chemicals to Man. Vol. 16. IARC, Lyon 1978
155. Ibid. Vol. 8. IARC, Lyon 1975
156. A literature survey oriented towards adverse environmental effects resultant from the use of azo compounds, brominated hydrocarbons, EDTA, formaldehyde resins and o-nitrochlorobenzene. US Environmental Protection Agency, EPA 560/2-76-005, Washington, D.C. 1976
157. Fishbein, L.: Potential Industrial Carcinogens and Mutagens. Elsevier, New York 1979
158. Burg, A.W., Charest, M.C.: An evaluation of the literature concerning the potential for carcinogenic properties of bisazobiphenyl compounds used as dyes. C-82353. A.D. Little Inc., Jan. 25, 1979
159. Auerbach Associates Inc.: Benzidine derived dyes and/or pigments, AAI-2434-100-TR-2, May 2, 1978. (Submitted to US Consumer Product Safety Commission under contract CPSC-C-77-0088)
160. Calculon: Monoazo dyes and pigments, CALC-2434-300-TR-2, Febr. 9, 1979. (Submitted to US Consumer Product Safety Commission under contract CPSC-C-77-0088)
161. Drake, J.J.P.: Toxicology *5*, 3 (1975)
162. NIOSH: Registry of Toxic Effects of Chemical Substances, Vol. I & II; US Dept. of Health, Education, and Welfare 1977
163. NCI: 13-week subchronic toxicity studies of Direct Blue 6, Direct Black 38 and Direct Brown 95 dyes, NCI-CG-TR-108, NCI Technical Rep. Ser. No. 108, 1978
164. Petition to OSHA by US Unions for emergency temporary standard for benzidine-derived dyes, May 16, 1978
165. Leuschner, F.: Toxicology Lett. *2*, 253 (1978)
166. NCI/NIH Report: Bioassay of Diarylanilide Yellow for possible carcinogenicity, DHEW publication no. (NIH) 77-830, 1977
167. IARC Monographs on the Evaluation of Carcinogenic Risk of Chemicals to Man. Vol. 4. IARC, Lyon 1974, p. 49
168. Gerarde, H.W., Gerarde, D.F.: J. Occup. Med. *16*, 322 (1974)
169. Gadian, T.: Chem. Ind. (19), 821 (1975)
170. MacIntyre, I.: J. Occup. Med. *17*, 23 (1975)
171. Akiyama, T.: Iikei Med. J. *17*, 1 (1970)
172. Davis, K.J., Fitzhugh, O.G.: Toxicol. Appl. Pharmacol. *5*, 728 (1963)
173. Davis, K.J., Fitzhugh, O.G.: ibid. *4*, 200 (1962)
174. Burg, A.W., Charest, M.C.: Mutagenicity results with implications for carcinogenicity, A.D. Little Inc., C-82875, March 9, 1979
175. Yahagi, T. et al.: Cancer Lett. *1*, 91 (1975)
176. Garner, R.C., Nutman, C.A.: Mutat. Res. *44*, 9 (1977)
177. Tamaro, M., Monti-Bragadin, C., Banfi, E.: Boll. Ist. Sieroter. Milanese *54*, 105 (1975)
178. Brown, J.P., Brown, R.J.: Mutat. Res. *40*, 203 (1976)
179. Brown, J.P., Dietrich, P.S.: ibid. *66*, 9 (1979)
180. Tamaro, M., Banfi, E.: Boll. Ist. Sieroter. Milanese *55*, 191 (1976)
181. Brown, J.P., Roehm, G.W., Brown, R.J.: Mutat. Res. *56*, 249 (1978)

182. Viola, M., Nosotti, A.: Boll. Chim. Farm. *117*, 402 (1978)
183. Muzzall, J.M., Cook, W.L.: Mutat. Res. *67*, 1 (1979)
184. Milvy, P., Kay, K.: J. Toxicol. Environ. Health *4*, 31 (1978)
185. Anliker, R.: Aquatic Ecological Chemistry (Japan) *1*, 211 (1979)
186. Alston, P.: Ecology Law Quart. 1978, 397
187. Anonymous: Environ. Sci. Technol. *13*, 786 (1979)
188. Barrett, B.R.: Environ. Sci. Technol. *12*, 154 (1978)
189. Tetlow, J.A.: Chem. Ind. (6), 183 (1979)
190. Smeets, J.: Ecotox. Environ. Safety *3*, 116 (1979)
191. Anliker, R.: J. Soc. Dyers Colour. *95*, 317 (1979)

Inorganic Pigments

W. Funke

II. Institut für Technische Chemie, Universität Stuttgart
D-7000 Stuttgart 80, Federal Republic of Germany

Introduction

In contrast to dyes inorganic colorants are generally used as pigments rather than in a molecular-dispersed state. Besides imparting color for decorative, indicatory and informational purposes, such pigments may serve various other or additional purposes like corrosion protection, filling or reinforcement.

Table 1 presents a survey of the more important fields of application of these colorants, but does not claim completeness.

From the large number of inorganic colorants some have been shown to cause hazardous or toxic effects when taken up by the human or animal organism. This may in principle, occur *via* the ingestive and respiratory systems, but sometimes also *via* skin contact. Important factors in estimating

Table 1. Inorganic colorants and powders in various applications

	Decorative	Protective	Indicatory	Compositional	Reinforcing	Filler
Plastics	+		+	+		
Rubber				+	+	+
Coatings	+	+	+	+	+	+
Building materials	+			+		+
Putties and sealants				+		+
Printing inks	+		+			
Artist paints	+					
Vitreous materials	+			+		
Ceramics	+			+		
Enamels	+			+		
Cosmetics	+			+		
Drugs			+			+

Table 2. Inorganic pigments, based on heavy metals

Name	Colloquial name	Formula	Density [g/cm³]	Solubility [g/100 ml]* cold H_2O	dil. HCl	dil. CH_3COOH	EtOH
Lead	*pigments*						
Red lead oxide	Red lead, Minium	Pb_3O_4 (+ PbO)	9,1	i	s	s	i
Calcium plumbate		Ca_2PbO_4	5,7	i	s	s	
Lead molybdate		$PbMoO_4$	6,9	i	s	s	i
Metallic lead	Lead powder	Pb	11,3	i	s	s	i
Chromate	*pigments*						
Lead chromates	Chrome yellow	$PbCrO_4$	6,1	0,0000058 ($gCrO_3/100cm^3$)	s	i	
	Chrome red	$PbCrO_4 \cdot PbO$	6,6	i	s	s	
	Chrome lemon	$PbCrO_4 \cdot PbSO_4$					
	Chrome orange	$PbCrO_4 \cdot Pb(OH)_2$					
	Chrome green	$PbCrO_4 + KFe[Fe(CN)_6]$	4,1				
Lead silico chromate		$PbCrO_4 + SiO_2$					
Zinc chromate	Zinc yellow	$ZnCrO_4$	3,5				
Zinc potassium chromate		$ZnCrO_4 \cdot K_2CrO_4$		0,06–0,15			
Zinc tetraoxychromate		$ZnCrO_4 \cdot 4Zn(OH)_2$	4,0	0,004			
Strontium chromate		$SrCrO_4$	3,9	0,04–0,1			
Barium chromate	Lemon yellow	$BaCrO_4$	4,5	0,0005	s		
Cadmium	*pigments*						
Cadmium sulfide		CdS	4,4–4,8	0,00013	s		i
Cadmium zinc sulfide		(Cd,Zn)S					
Cadmium sulfo selenide		Cd(S,Se)	5,8	i			i

Name	Colloquial name	Formula	Density [g/cm^3]	Solubility [g/100 ml] cold H_2O	dil. HCl	dil. CH_3COOH	EtOH
Silicogenous	*pigments*						
Asbestos		Hydro silicates of var. comp.	2,8–3,4				
Talc		$Mg_3 [Si_4O_{10}] (OH)_2$	2,8	0,008			
Silica		SiO_2	2,0–2,3			0,014–0,052	
Pigments and Fillers for special purposes							
Antimony trioxide		Sb_2O_3	5,2	vsls	vsls	s	0,03
Lead antimonate	Naples yellow	$Pb_3(SbO_4)_2$	6,6		i		
Copper aceto arsenate	Emerald green, Paris green	$Cu(CH_3COO)_2$ $3Cu(AsO_2)_2$	3,3				

* i = insoluble
s = soluble
vsls = very slightly soluble

hazardous effects is the total surface area of exposure where interactions take place. With the exception of some fillers or extenders the average particle diameter of inorganic colorants ranges between 0,15 and 2 μm. Though inorganic colorants are used for many purposes, considering the probability and intensity of possible exposures, it is reasonable to concentrate discussion on the pollution and hazards of their use in organic coatings and thin films. A competent and informative survey on legislation, standards and codes of practice and toxicology for the paint industry also includes inorganic colorants [1]. In the following discussion hazards encountered in the industrial production of inorganic colorants are mentioned only incidentally since these operations are usually subjected to special regulations and safety measures.

In Table 2 inorganic colorants are listed which have already been mentioned in the literature as being potentially hazardous, together with some data relevant to pollution aspects. Several surveys on ecological problems with inorganic colorants and metal oxides are known from the literature [2, 3, 4].

Considering the actual danger in handling objects containing hazardous colorants, its relevancy very much depends on their special use. Thus toxic inorganic pigments that have been used in old paintings hardly present any danger because it is highly improbable that non-specialists come into contact with them. On the other hand little attention is paid to possible hazards on paint removal and in the deposition and combustion of coated or painted subjects, which are quite safe when in normal use. It should be also realized that coating materials containing potentially hazardous pigments, which have been used long since by expert workers without significant accidents, are now considered dangerous after they get into the hands of amateurs in Do-it-Yourself projects.

The two important routes by which hazardous pigments may be absorbed by the body are inhalation or ingestion. The most dangerous is inhalation, because it allows toxic substances to be directly absorbed by the blood stream or deposited on the outer surface of the lung. In the latter case irritations may be caused with temporary or even permanent reduction in the area available for the gas exchange.

As inorganic pigments are only slightly soluble in water and even less in organic solvents, they are scarcely or not at all absorbed by intact skin or mucous membranes. Though acute poisoning is highly improbable (LD_{50} in most cases are > 5 g/kg) the danger of chronic effects must be considered.

The solubility of pure inorganic colorants may differ from that of corresponding technical pigments. The latter may contain varying amounts of soluble impurities from the production. In using such data it should be ascertained therefore, that soluble impurities are not more toxic than the soluble fraction of the pure substance. The water soluble fraction of a pigment may be determined by extraction with water at room temperature or at elevated temperatures [5]: The sample is dispersed in carbon dioxide-free water, thoroughly stirred, water is added to a fixed volume and the dispersion filtered after vigorous shaking. A part of the filtered solution is evaporated to dryness and after further drying at elevated temperature and cooling to room temperature in a desiccator the residue is weighed. This procedure is repeated until two successive weighings do not differ significantly.

In considering hazards by paint films one should allow for the fact that due to the encapsulation in the binder matrix, the amount of extractible or leachable hazardous colorants is normally lower than what is expected from solubility data of the free pigment. To obtain practically significant data, the amount of leachable pigment should be determined by some extraction method [6] applied to the paint film after its formation is completed.

Sources of Hazards in Using Inorganic Colorants

As far as organic coatings are concerned, exposure to hazardous colorants may occur under various circumstances.

Production Process

The production of inorganic colorants, paint and coating materials is subject to safety regulations. They are mainly concerned with preventing hazards from inhalation of pigment dusts by the workers during mixing, dispersing, storing and handling of these materials.

Application

Various methods of paint application differ considerably in ease and probability of exposure to hazardous pigments. Special protection measures are required in spraying to avoid inhalation of the paint spray. In other application methods like brushing, dipping or electro-coating, exposure is possible on cleaning the equipment after the application or on changing the coating material. This is also true for electrostatic powder coating.

Usually the overspray is caught by water curtains in the spraying booth or, in powder coating, by the air stream. Before disposal to the waste water system the solid waste material (paint sludge) must be removed from the water by coagulation and separation, both in spraying of conventional solvent-based paints and in the electro-deposition of water borne paints. In the case of electrostatic powder coatings the overspray powder is recovered for further use.

Performance

Even more than in paint application, exposure to paint films and coatings containing hazardous pigments differs widely, depending on the special function of the coated subject.

Accordingly attention has to be paid to all possible exposures in which children may be involved, e.g. indoor wall paintings, and paints for furniture, toys and pencils. Colorants for food and drugs are subjected to special laws and regulations, however in these materials inorganic pigments are of minor importance.

Removal

Removal of old paint layers before repainting or repair painting presents a special problem if hazardous pigments are involved. Mechanical methods like wire brushing or sand blasting may produce dust particles that may be inhaled by the worker or contaminate the soil. Detrimental effects may be expected if these waste materials are incorporated by plants or animals. Another hazard may arise in the chemical or physical removal of paint by paint removers or solvents. Care has to be taken on disposing the waste material produced in these processes.

Welding

A special kind of danger in using hazardous pigments is the welding and flame cutting of painted iron or steel constructions. Due to the high temperature some pigments may vaporize or decompose to volatile toxic products. Under unfavorable circumstances toxicological hazards may be encountered even with inherently non-toxic inorganic pigments like zinc oxide. Welding of zinc-rich primers may produce zinc-oxide fumes above the TLV-limit of 5 mg/m^3 [1, 7, 8, 9].

Waste Disposal

Discarded manufacturing batches, waste from cleaning tools and mixing, dispersing or application machines, used paint containers, coagulated paint from spraying booths as well as various kinds of painted waste materials and painted subjects may provide another hazard when deposited, combusted or disposed of in the waste water system.

Some more important possible exposures to potentially toxic colorants by which harmful effects may be expected after absorption or ingestion, are summarized as follows:

1. inhalation of pigment dust, paint spray or mechanical abrasion products on paint removal, and absorption through lung tissue;
2. Contamination of fingers on processing or manipulating paints; paint may become embedded under the finger nails and may be transmitted to the mouth or absorbed by the skin;
3. putting water color brushes in the mouth, as during artist's work;
4. smoking when handling hazardous pigments or paint containing them;
5. handling food, eating or drinking with contaminated hands resp. dishes;
6. allowing pigments or paints to contact injured or scratched skin where direct absorption into the blood stream is possible;
7. children chewing on toys, furniture or other internal equipment in dwelling houses, that have been painted with hazardous materials.

To prevent possible dangers by intoxication in most countries a series of directions and regulations exist which describe safety measurements on handling hazardous colorants and material containing them. They will be referred to in the following discussion of these colorants. The analytical detection and quantitative determination of inorganic colorants is largely identical with that common in respective heavy metals. It is therefore referred to the respective chapters on these metals.

Inorganic Colorants Based on Heavy Metals

Lead Pigments

The toxic nature of lead and its compounds has been known for a long time and has been thoroughly studied. Divalent lead may replace calcium and thus be retained in the body over long periods by accumulation in the bone. However, the mobile fraction in the blood is significant. The total lead content increases as long as the contact with soluble lead compounds continues [2]. Despite these facts, lead containing pigments are still widely used, especially as anticorrosive constituents in corrosion protective paints. The total production of lead containing pigments and oxides in West Germany during 1978 amounted to about 56,000 tons (calculated on PbO), half of which is red lead (Pb_3O_4) [10].

The toxic effect of lead containing pigments strongly depends on their solubility in water and dilute acids [11]. Basic lead carbonate $2PbCO_3 \cdot Pb(OH)_2$ (white lead, flake white or Chremnitz white) is soluble in dilute acids. Despite giving durable paint films e.g. for window frames and walls, it is the complete acid solubility which has made this pigment dangerous to use in any paints. In the past numerous cases have been reported of lead poisoning mainly in children who have chewed and swallowed flakes of dried lead paints detached from treated woodwork [1, 12].

One of the most important anticorrosive pigments is red lead, Pb_3O_4 (+PbO). As for all anticorrosive pigments some solubility is essential for its corrosion protective action in primers. Despite many efforts to substitute this pigment, it is still considered to be indispensible for a number of corrosion prevention measures which require protective coatings to be applied at the location of final use.

It is less well known that inhalation of lead or lead containing pigments is even more dangerous than ingestion. Below 5 μm diameter about 80% of the particles containing lead or its compounds are incorporated on inhalation, whereas only 10% are resorbed on ingestion and most of it is excreted via the bile [13]. The ready absorption of lead from inhaled dust justifies the stringent regulations covering almost every aspect of industrial handling of compounds containing lead.

Whereas hazards of dust inhalation are mainly limited to pigment and paint manufacturing and to mechanical paint film removal, care must also to be taken to prevent inhalation of lead containing particles on spraying of paints. However the general regulations for handling and spraying paints are also considered to be sufficient for the application of lead containing paints [14].

Besides red lead also dibasic lead phosphite ($2PbO \cdot PbHPO_3 \cdot \frac{1}{2}H_2O$), lead phosphate ($Pb_3(PO_4)_2.3H_2O$), calcium plumbate ($Ca_2PbO_4$) and lead powder are used to some extent as corrosion protective pigments. Lead cyanamide, which was manufactured for corrosion protective purposes at the beginning of 1950, is not used any more, but should be considered in connection with removal of paint films from old steel constructions. For these pigments the regulations and safety measures of working with red lead are also applicable. Some lead pigments used in artists painting, like lead antimonate (Naples Yellow), $Pb_3(SbO_4)$, are specially hazardous due to their high solubility.

Another source of danger is the volatilization of lead and lead oxides from paint films during welding. It has been proposed that silica should be added to anticorrosive paints containing red lead in order to transform these substances to the less volatile lead silicates [15].

For general information on numerous national and international legislation regulations and standards concerning the use of lead containing pigments the following literature is recommended [1, 4, 16, 17].

According to the Occupational Safety and Health Administration (OSHA) in the USA the permissible exposure level for lead in the air is 50 $\mu g/m^3$. The lead-in-paints regulations of the Consumers Product Safety Commission (CPSC) denotes paints and coatings containing more than 0,06% of lead as "lead containing paints". Such paints are prohibited in toys and other objects used by children.

As a consequence of the restrictions in using lead containing pigments, there is a tendency to use lead-free paints and coatings [18]. As has been recently stated [1] however, there are still some questions to be answered: the relative contributions of the different sources of lead as a poison to man; whether the hazards result from the total amount of lead or only the lead soluble in acid organic fluids; the necessity for the use of small amounts of lead compounds in paints as drying agents in air drying paints. It has been found [11] that under the same experimental conditions lead naphthenate, which is a common drying agent, migrates more easily from the film than lead chromate. Attempts have been made to substitute these drying agents by lead-free compounds [19]. In view of there being still no equivalent alternative to the use of red lead in anticorrosive paints that can be offered for various protective applications, its future use is justifiable, provided the safety regulations and restrictions are properly observed.

Chromate Pigments

The worldwide production is estimated to be about 150,000 tons annually [20] (During 1977 and 1978 almost 12,000 t of chromate pigments were produced in West Germany [10]).

The hazardous nature of chromate pigments have been known for many years [21, 22]. Soluble hexavalent chromium is toxic, potentially carcinogenic and a common contact dermatitic. As all chromate pigments used in paints are soluble to some extent in acidic fluids of the body, they are hazardous to the same extent. For chromates used as corrosion protective pigments in primers, like zinc chromate, zinc tetraoxy chromate or strontium chromate, some solubility is necessary for the protective action, as with all anticorrosive pigments. Correspondingly the vehicles used in anticorrosive primers must be slightly swellable in water and more diffusible for the dissolved fraction of the anticorrosive pigment than vehicles of other organic coatings. In the case of chromate pigments, such as the lead chromates, which are used as colorants only, some encapsulation effect may significantly decrease the extraction of soluble pigment on exposure to water or similar liquid media [23].

As chromate pigments are used in anticorrosive primers for car bodies, contact dermatitis has been observed with workers engaged in wet sand papering of these primers on car bodies [24]. There is strong evidence that lung cancer may be caused on inhalation of chromate pigment dust [25, 26].

Similar to other chromate pigments lead chromate has also been suspected of being potentially carcinogenic. Above average incidence of lung cancer has been observed in factories producing both zinc chromate and lead chromate. However, it is assumed that only zinc chromate has to be blamed for this hazard [27]. More recent epidemiological studies [28, 29] showed that no increased risk for lung cancer exists in the production and processing of lead chromate pigments if modern regulations for working hygiene are observed.

Despite the very low lead release from plastics pigmented with lead chromate, this pigment should not be used in paints for toys, in toys made from plastics and in packing or wrapping material coming in direct contact with food. For these purposes lead-free alternatives of colorants are used. In all other fields where colored plastics or paints are used, no economical alternative to lead chromate exists, and there is also no ecological reason for replacing it [30].

According to the OSHA-regulations in the USA 1 $\mu g/m^3$ air of chrome containing carcinogenic substances are tolerated. The current TLV-resp. MAK-values for airborne levels of chromate (calculated as CrO_3) are 0,1 mg/m^3 [31, 32].

Cadmium Pigments

Both cadmium and selenium in dissolved or soluble form are very toxic when directly taken up by the blood circulation system. Cadmium, especially as cadmium oxide, is a respiratory poison and industrial poisoning has been caused on exposure to fumes or dust.

In paints cadmium is used as cadmium sulfide or cadmium sulfoselenide, which present a range of high quality yellow to red pigments [33, 34]. There are also some composite cadmium pigments containing zinc sulphide or mercury sulfide. World production of cadmium pigments is estimated to 8,500 t annually [20], about 80% of which are used in plastics and 10% each in paints and ceramic products [35]. Cadmium pigments have superior technical properties and are widely applied in plastics, ceramics and some industrial coatings.

In contrast to some very toxic cadmium compounds like cadmium oxide, a series of more recent investigations have shown that the cadmium sulfide and sulfoselenide pigments are much less hazardous, especially when used in paints or plastics. Of course the usual hygienic regulations for handling industrial dusts must be observed.

The solubility of cadmium pigments in 0,1 N hydrochloric acid is less than 0,1% [1, 33, 36]. Purity requirements for cadmium pigments used in plastics and coatings coming in contact with foodstuffs may be checked by a solubility test according to DIN 53 770 [27]. Animal tests with cadmium chloride showed a low level of toxicity [37, 38] and in toxicological investigations it was found [39, 40] that as much as 30 ppm of Cd given as $CdCl_2$ in the food could be tolerated over a three months period without harm. A two-years feeding test with rats resulted in a no-effect-level of as low as 10 ppm. No carcinogenic effects on oral application of water soluble cadmium sulfate to animals were observed [41]. Also no toxic effects have been reported on using cadmium pigments in paints [42]. Obviously there are significant differences in toxicity depending on whether soluble cadmium compounds are taken up orally or via other pathways than the digestive system, e.g. by inhalation.

Colorants containing cadmium should not be used in food and cosmetics [43]. The use of cadmium pigments for packaging material – especially plastic packaging material – coming into contact with foodstuffs is, in general, permitted by the various European legislations. The legal regulations vary from country to country. The majority of purity requirements or maximum allowable migration values are, as a rule, met by cadmium-pigmented plastics. This is also true for plastic toys, whereas in a few countries restrictions have been imposed on the use of cadmium-pigmented paints intended for the coating of toys [33].

Emission of toxic cadmium compounds on combustion of waste or extraction in waste deposits has been estimated to be unimportant in comparison to other sources. [34].

The regulations on the use of cadmium pigments vary in different countries as do the MAK-TLV limits given below. As no corresponding values for pigments are known these values should serve for orientation only:

Fed. Rep. Germany	MAK-value CdO (smoke)	0,1 mg/m³
England	Inhalable fraction of total dust	0,05 mg/m³
	Total wt. of Cd-compounds soluble in 0,1 N HCl	0,2 mg/m³
Sweden	Total Cd	0,05 mg/m³
	Inhalable fraction of Cd	0,02 mg/m³
USA	TLV CdO (smoke)	0,05 mg/m³
Japan	MAK Cd	0,05 mg/m³

Many countries insist on the "non migration principle" according to which no colorant should migrate visibly to the food or to the test solution. As cadmium pigments do not migrate at all, they can be used as colorants for plastics except in swellable ones like polyamide in contact with acids [34].

Silica, Silicates and Asbestos

These powders are used as fillers and extenders rather than as colorants. Although not being toxic in a chemical sense, inhalation of such dust particles on handling, spraying of paints or on mechanical paint film removal should be avoided because of the risk of silicosis or asbestosis and associated lung cancer. The hazards depend on the particle size. Silica is most dangerous with particle sizes between 0,5 and 5 μm [2, 44]. Smaller particles remain suspended in the streaming air and may leave the lung. Larger particles are usually filtered by some other protective mechanism and pass to the intestines where they are harmless.

Various kinds of asbestos in fibrous form are specifically harmful to the lung tissues [45]. The most hazardous powder of this group is blue asbestos (crocidolite). Only asbestos fibers above 5 μm in length are dangerous. Shorter ones remain suspended in the respired air [2].

As talc is similar to asbestos it has also become suspicious as being carcinogenic, however evidence is still controversial [46].

Although there is no indication that industrially produced highly disperse amorphous silica causes silicosis even under extreme working conditions [47], the usual safety measures on handling industrial dusts should be obeyed when paint films containing them have to be removed mechanically.

In West Germany the maximum concentration of inert fine dusts (MAK) is 8 mg/m^3, which can also be considered as a limiting value in using the fillers mentioned above. In England asbestos regulations [48] have to be applied when the average atmospheric concentration of asbestos dust (other then crocidolite) is 2 $fibres/cm^3$ or 0,2 fibres of $crocidolite/cm^3$.

The American Conference of Governmental Industrial Hygienists (ACGIH) has related the TLV to the percentage of quartz to determine the TLV (VSHS standard) for silica bearing dust by the formula (30 mg/m^3) / (% quartz in dust + 2) [2]. Thc asbcstos content in powders may be determined by X-ray diffraction or by a microscopic dispersion staining technique [49].

Miscellaneous Inorganic Colorants

Some inorganic pigments for special purposes, containing antimony, arsenic or barium have been also discussed as hazardous compounds.

Antimony

As soluble antimony compounds are known to be toxic, antimony oxide has been suspected of being hazardous. However, this oxide, which is used in high performance flame retardant paints, is only slightly soluble in water and in hydrochloric acid. No harmful effects could be detected with workers in a plant manufacturing Sb_2O_3 even on prolonged severe exposure [50, 51]. Possibly the only hazard that may arise is by inhalation of fumes from the burning or welding of painted surfaces [2]. Lead antimonate, also known as Naples

Yellow, is soluble in acids. It has been mainly used in artists paints. Its toxicity, which may be equally well be ascribed to the presence of the lead, is well known [3].

Arsenic

Copper aceto-arsenate $Cu(CH_3COO)_2.3Cu(AsO_2)_2$, (Emerald or Paris green) is one of the earliest examples of dangerous pigments in paints [3]. However its use in artist paints is negligible today.

Barium

Soluble barium salts are highly toxic. The lowest toxic dose reported for humans is 80 mg $BaCl_2$/kg body weight [2]. Apart from barium chromate, which has been mentioned with the chromate pigments, barium metaborate is the only pigment to be considered in this connection [1]. This pigment is used on account of its fungicidal and anticorrosive properties. As far as the limit of acid soluble barium content is concerned its use would be excluded in most paint specifications.

References

1. O'Neill, L.A.: Health and safety environmental pollution and the paint industry – a survey covering legislation, standards, codes of practice and toxicology. England: Paint Research Association, Jan. 1977
2. Morrison, R.: Hazardous Paint Pigments. Australian OCCA, Proc. and News, Oct. 1975, 5
3. Mansell, H.: ICCM Bulletin, Pigment Toxicity. *3*, No. 2, 11, June 1977
4. Dunn, M. J.: Paint and Varnish Production, Aug. 1973, pg. 49
5. Deutsche Norm, DIN 53 197, Nov. 1971
6. Zorll, U.: Dtsch. Farbenztschr. *11*, 495 (1976)
7. Chmielewski, J. et al.: Bull.-Inst. Mar. Med. Gdansk *25*, 43 (1974)
8. Inchingolo, P. et al.: Industr. Vern. *30*, (8), 3 (1976)
9. Douglas, C.P., Plummer, R.M.: Protection *13*, (4), 3 (1976)
10. Farbe + Lack *85*, 597 (1979/7)
11. Brezinski, D.R.: Coatings Technol. *48/4*, 48 (1976)
12. Gage, J.C., Litchfield, M.H.: J. Oil Col. Chem. Ass. *52*, 236 (1969)
13. Konietzko, H., Elster, I., Reill, G.: Zbl. Arbeitsmed. *1978/6*, 163
14. Niemann, E.: I-Lack, *46*, 390 (1978/11)
15. Schatz, H.: Korrosionsschutz *1978/8*, 13, (Ed. Verein Dtsch. Bleioxid Hersteller, Köln)
16. Niemann, E., ibid, March 1979
17. Umwelt-Bundes-Amt: Ber. *76/3*, 116, West Germany
18. Schneider, W.F.: Am. Paint a. Coatings J., Convention Daily, 30. Oct. 1976, pg. 34
19. Mann, A.: Mod. Paint a. Coatings, Febr. *67/2*, 21 (1977)
20. Farbe + Lack *85*, 598 (1979/7)
21. Gross, E., Kölsch, F.: Arch. Gewerbepath *12*, 164 (1943)
22. Langård, S., Norseth, T.: Brit. J. Industr. Med. *32*, 62 (1975)
23. Am. Paint a. Coatings J., 25. Oct. pg 9 (1976)
24. Engel, H.O., Calnan, C.D.: Brit. J. Industr. Med. *20*, 192 (1963)
25. Nat. Paint & Coatings Ass., Safety & Health Bulletin, 1975, No. 27
26. Brit. Colour Makers Ass., Polymer Paint Col. J., *166*, 933 (1976)
27. Davies, J.M.: J. Oil Col. Chem. Assos. *62*, 157 (1979)

28. Davies, J.M.: The Lancet, Febr. 18 (1978)
29. Sperfeld, R.: Farbe + Lack *84,* 137 (1978)
30. Endriß, H.: Kunststoffe *69,* 403 (1979)
31. Health and Safety Executive: Threshold Limit Values for 1975, USA. Techn. Data Note 2/75
32. Deutsche Forschungsgemeinschaft, Maximale Arbeitsplatzkonzentrationen, Mitt. XIII u. Mitt. XIV, 1977/78, West Germany
33. Technical notes on cadmium, Cadmium Pigments. Cadmium Association, London, and Cadmium Council, New York 1978
34. Endriß, H.: Kunststoffe *69,* 39 (1979)
35. Polymer Paint a. Colour J. June 13, 1979, 595
36. Chem. Ind., XXXI, (6), 369 (1979)
37. Gabby, J.L.: Ind. Hyg. & Occup. Med. *1,* 677 (1950)
38. Krynskaya, I.L. et al.: Plast. Massy *1,* 65 (1975)
39. Streatfield, G.R.: Pigment Res. Technol. *6,* 18 (1976)
40. Loeser, E., Lorke, D.: Toxicology *7,* 215 and 225 (1977)
41. Occup. Hyg. *17,* 205 (1975)
42. Pigmente – Toxikologie, Ullmanns Encyklop. techn. Chem. *13,* 822 (1962)
43. Umwelt-Bundes-Amt, Luftqualitätskriterien für Cadmium, Berichte *77/4,* 149
44. Über das physiologische Verhalten von hochdispersen Oxiden des Siliciums, Aluminiums und Titans. Schriftenreihe DEGUSSA "Pigmente", 1977, No. 64, (9)
45. Buckup, H.: Zbl. Arbeitsmedizin *16,* 203 (1966)
46. Pelfrène, A., Shubik, P.: Nouvelle Presse Med. *4,* 301 (1975)
47. Hofmann, W.: Gummi, Asbest, Kunststoff, *1974,* 624
48. The Asbestos Regulations 1969. Statutory Instrument No. 690, H.M.S.O., Great Britain
49. Julian, Y., McCrone, W.C.: Microscope *18,* 1 (1970)
50. Oliver, T.: Brit. Med. J., *1,* 1094 (1933)
51. Fairhall, L.T., Hyslop, F.: US Treasure Dpmt. Publ. Health, Rep. *1947,* Suppl. No. 195

Radioactive Substances

G.C. Butler, C. Hyslop

Division of Biological Sciences
National Research Council of Canada
Ottawa, Canada K1A 0R6

Glossary

Radiation Protection Concepts

ALI	Annual Limit on Intake: the activity of a radionuclide which, taken alone, would irradiate a person, represented by Reference Man, to the limit set by the ICRP [7]
Class D, W, Y	(days), (weeks), (years): a classification scheme for inhaled material according to its rate of clearance from the pulmonary region of the lung [7]
DAC	Derived air concentration: equals the ALI for inhalation (of a radionuclide) divided by the volume of air inhaled by Reference Man in a working year (i.e. 2.4×10^3 m^3) (Bq m^{-3}) [7]
IL	Investigation Level: a value of dose equivalent or intake above which the results are considered sufficiently important to justify further investigation ([3], par. 151, p. 29)
Reference Man	A person with the anatomical and physiological characteristics defined in the report of the ICRP Task Group on Reference Man [30]

Nuclear Reactors

AGR	Advanced gas cooled reactor
BWR	Boiling water reactor
CANDU	Canadian deuterium uranium reactor
GCR	Gas cooled reactor
HTGR	High temperature gas cooled reactor
HWR	Heavy water reactor
LMFBR	Liquid metal fast breeder reactor
LWR	Light water reactor
NPD	Nuclear Power Demonstration reactor
NFS	Nuclear Fuel Services reactor
PWR	Pressurized water reactor

Organizations and Groups

BEIR	Advisory Committee on the Biological Effects of Ionizing Radiations, National Academy of Sciences-National Research Council (USA)
BNWL	Battelle Pacific Northwest Laboratory (USA)

EML	Environmental Measurements Laboratory (USA) (formerly HASL)
HASL	Health and Safety Laboratory (USAEC)
IAEA	International Atomic Energy Agency
ICRP	International Commission on Radiological Protection
ICRU	International Commission on Radiation Units and Measurements
LASL	Los Alamos Scientific Laboratory (USA)
MRC	Medical Research Council (UK)
NCRP	National Council on Radiation Protection and Measurements (USA)
ORNL	Oak Ridge National Laboratory (USA)
SCOPE	Scientific Committee on Problems of the Environment of the International Council of Scientific Unions
USAEC	United States Atomic Energy Commission
UNSCEAR	United Nations Scientific Committee on the Effects of Atomic Radiation
USNAS	United States National Academy of Sciences
WASH-1400	USAEC Reactor Safety Study (draft) 1974

Introduction

The purpose of this chapter is to show how to assess the detriment resulting from the release of radioactive materials to the environment. Because of the wide range of the subject and the limitation of space the chapter consists of little more than a listing of principles and concepts. A more adequate examination of these will require consulting the literature cited.

The minimum information required for the assessments is given for seven radionuclides of interest from the point of view of environmental contamination.

Basic Concepts

Radiation Doses and Units

Recently new units in the International System of Units (SI) have been introduced to quantify ionizing radiation [1, 2]. They are given below along with the older units they replace.

Exposure:	1 *roentgen* (R)	$= 2.58 \times 10^{4}$ coulombs per kilogram of air (C kg^{-1})
Activity:	1 *becquerel* (Bq) (new unit)	= 1 radioactive transformation per second (tr s^{-1})
		$= 2.7 \times 10^{-11}$ *curies* (old unit)
	1 *curie* (Ci)	$= 3.7 \times 10^{10}$ radioactive transformations per second (tr s^{-1})
Absorbed dose:	1 *gray* (Gy) (new unit)	= 1 joule per kilogram (J kg^{-1})
		= 100 *rads* (old unit)
	1 *rad*	$= 10^{-2}$ joules per kilogram (J kg^{-1})

Dose equivalent:	1 *sievert* (Sv) (new unit)	= 1 *gray* × quality factor (Q) (J kg^{-1})
		= 100 *rems* (old unit)
	1 *rem*	= 1 *rad* × quality factor (Q).

Effects of Radiation and Dose-Effect Functions

Radiation doses as low as those usually encountered in the environment result in "stochastic" detrimental effects ([3], Sect. 7, p. 2). These comprise malignant and hereditary diseases for which the probability of occurrence, rather than the severity, is proportional to the dose (e.g. cancers, lethal mutations). For these effects it is assumed that there is a "linear non-threshold" dose-effect relationship ([3], Sect. 27, p. 6; [4, 5]; [6], Sect. 36, p. 366; Sect. 143, p. 592). This means that all doses greater than zero received during a lifetime contribute, according to their magnitude, to causing biological effects.

Dose Equivalent (H)

All radiations do not have the same effectiveness, *gray* for *gray*, in producing stochastic effects, thus the concept of dose equivalent has been introduced and defined as follows [1]:

$$H = D\,Q\,N$$

where H = the dose equivalent at a point in tissue, expressed in *sieverts*
D = the absorbed dose, expressed in *grays*
Q = a quality factor dependent on density of ionization in tissue, produced by the radiation
N = the product of any other modifying factors such as rate of irradiation.

Mean values of Q adopted for the purposes of radiation protection are ([3], Sect. 20, p. 4):

Type of radiation	*Q*
X-rays, γ-rays, electrons	1
thermal neutrons	2.3
fission neutrons and protons	10
α-particles and other multiply-charged particles	20

Committed Dose Equivalent (H_{50})

The total dose equivalent accumulated by a given organ or tissue during an individual's working lifetime of 50 yr, from a single bodily intake of radioactive material, is called the committed dose equivalent. It is defined as follows ([3], Sect. 26, p. 6):

$$H_{50} = \int_{t_0}^{t_0 + 50\,y} \dot{H}(t)\,dt,$$

where H_{50} = the 50-year committed dose equivalent
$\dot{H}(t)$ = the dose equivalent rate at time t
t_0 = the time of intake.

The calculations of committed dose equivalent from a single intake are described in ICRP Publication 30 [7] and from multiple intakes in ICRP Publication 10A [8].

Dose-Equivalent Commitment (H_c)

The dose-equivalent commitment to an individual (H_c), resulting from a given decision or practice, is the infinite time integral of the *per caput* dose-equivalent rate ($\dot{H}(t)$) in a given organ or tissue for a specified population ([3], Sect. 25, p. 6; [6], Sect. 16, p. 28). It may be expressed as:

$$H_C = \int_0^{\infty} \dot{\overline{H}}(t)\, dt,$$

Table 1. Global dose equivalent commitments from various radiation sources. (From Table 3 of UNSCEAR ([6], p. 16))

Source of exposure	Annual absorbed dose (man-Gy)	Annual dose equivalent (mSv)	Global dose equivalent commitment (days)[a]
Natural Irradiation			
a) One-year exposure to natural sources	2×10^6	1	365
Natural Irradiation Enhanced by Technology			
b) One year of commercial air travel	3×10^3		0.4
c) Use of one year's production of phosphate fertilizers at the present production rate	10^2		0.04
d) One-year global production of electric energy by coal-fired power plants at the present global installed capacity [10^6 MW(e)]	50		0.02
e) Mining (α-irradiation of lungs)	50		0.02
Man-Made Sources of Radiation			
f) One-year exposure to radiation-emitting consumer products	10^4		3
g) One-year production of nuclear power at the present global installed capacity [8×10^4 MW(e)]	10^3		0.6
h) One year of nuclear explosions averaged over the period 1951–1976	10^5	0.07	30
i) One year's use of radiation in medical practice	4×10^5	0.35[b]	70

[a] The global dose commitment is expressed as the duration of exposure of the world population to natural radiation which would cause the same dose commitment. The occupational contribution is included

[b] In the most technologically developed countries [9]

Table 1 ([6], p. 16; [9]) shows global dose-equivalent commitments from various sources.

Risk Estimates

Estimates of the risk of biological effects of ionizing radiation have been published by UNSCEAR [6, 10, 11], ICRP [12, 13] and in the "BEIR Report" [14].

The most recent estimates of genetic risks are given by UNSCEAR ([6], pp. 425–564) and compared with USNAS values [14] in Table 2 ([6], p. 539). The risks of malignancies published by UNSCEAR [6] and ICRP [3] are shown in Table 3. The rates of incidence of these malignancies in Canada in 1975 are given in Table 4 [15].

Effective Dose Equivalent (H_E)

To estimate the total harm from an intake of radionuclides it is necessary to know the annual dose equivalents to specific high risk tissues and to multiply these by weighting factors proportional to the risks of stochastic effects ([3], Sect. 104, p. 21; [16]). The sum of all these weighted dose equivalents is called the effective dose equivalent (H_E) and is described algebraically as:

$$H_E = \sum_T H_T w_T$$

where H_T = the annual dose equivalent for tissue T
w_T = the weighting factor representing the ratio of the stochastic risk arising from tissue T to the total risk when the whole body is irradiated uniformly.

The values of w_T for the tissues at greatest risk, assigned by ICRP ([3], Sect. 105, p. 21) on the basis of the risks listed in Table 3, are listed in Table 5.

The effective committed dose equivalent (H_{50E}) is defined by the equation [7]:

$$H_{50E} = \sum_T H_{50T} \times w_T$$

where H_{50T} = the committed dose equivalent for tissue T.

Collective Dose Equivalent

The detriment to a population resulting from ionizing radiation may be proportional to the collective dose equivalent (S) defined by the equation ([3], Sect. 22, p. 5)

$$S = \sum_i H_i \times P_i$$

where H_i = the *per caput* dose equivalent to the whole body or an organ or tissue in sub-group i of the exposed population
P_i = the number of people in the sub-group i of the exposed population.

Table 2. Estimated effect of 10^{-2} Gy (1 rad) per generation of low-dose, low dose-rate, low-LET irradiation on a population of one million live-born individuals. Assumed doubling dose, 1 Gy (100 rad) (Table 50 of UNSCEAR [16], p. 539)

Disease classification[a]	Current incidence[b]	Effect of 10^{-2} Gy (1 rad) per generation: First generation[c]	Equilibrium
Autosomal dominant and X-linked diseases	10,000[d]	20	100
Recessive diseases	1,100	Relatively slight	Very slow increase
Chromosomal diseases	4,000[e]	38[f]	40
Congenital anomalies; Anomalies expressed later; Constitutional and degenerative diseases	90,000[g]	5[h]	45[h]
Total	105,200	63	185
Percentage of current incidence		0.06	0.17
Recalculated BEIR assessments			
Autosomal dominant and X-linked diseases	10,000	20	100
Recessive and chromosomal diseases	10,000	Relatively slight	Very slow increase
Congenital anomalies; Anomalies expressed later; Constitutional and degenerative diseases	40,000	2–20[i]	20–200[i]
Total	60,000	25–40[j]	125–300[j]
Percentage of current incidence	100	0.04–0.07	0.21–0.50

[a] Follows that given in the BEIR Report [14]
[b] Based on the results of the British Columbia Survey with certain modifications; see Table 9 in [6], p. 519
[c] The first generation incidence is assumed to be about one fifth of the equilibrium incidence for autosomal dominant and X-linked diseases; for those included under the heading "congenital anomalies etc." it is one tenth of the equilibrium incidence. For rationale see [14]
[d] See Table 9 in [6], p. 519
[e] Based on the pooled values cited in Nielsen and Sillesen (363 in [6], p. 553) includes mosaics but excludes balanced translocations
[f] The first generation incidence is assumed to include all the numerical anomalies and three fifths of the unbalanced translocations (the remaining two fifths being derived from a balanced translocation in one parent)
[g] Includes an unknown proportion of numerical (other than Down's syndrome) and structural chromosomal anomalies
[h] Based on the assumption of a 5% mutational component
[i] The range reflects the assumption of 5 and 50% mutational components; see [6] for explanation
[j] Rounded-off figures

The collective dose equivalent (S_k) resulting from a practice or source (k) is defined by the expression ([3], par. 23, p. 5)

$$S_k = \int_0^\infty H \times P(H)\, dH$$

Table 3. Estimated effects of one unit of low-dose, low dose-rate irradiation on a population of one million persons

		Risk	
Tissue	Effect	UNSCEAR (per 10^{-2} Gy)	ICRP (per Sv)
1. Gonads	Mutations	63 (f_1)	10,000 (f_1+f_2)
2. Body	All cancers	200	–
3. Breast	Fatal cancer	50 (population)	2,500 (workers)
4. Red bone marrow	Leukemia	20–50	2,000
5. Lung	Fatal cancer	25–50	2,000
6. Thyroid	Fatal cancer	10	500
7. Bone surfaces	Fatal cancer	2–5	500
8. Remainder (2–3, 4, 5, 6, 7)	Fatal cancer	35–93	5,000

For the qualifications concerning these numerical estimates UNSCEAR [6] and ICRP [3] should be consulted

Table 4. Rate of reporting of malignant neoplasms in Canada, 1975 [15]

	Incidence per million	
Tissue or effect	Cases	Deaths
Total body	1,900	1,500
Breast	350	140
Leukemia	60	60
Lung	300	300
Thyroid	20	0
Bone	8	8

Table 5. Values of w_T recommended by ICRP [3]

Tissue	w_T
Gonads	0.25
Breast	0.15
Red bone marrow	0.12
Lung	0.12
Thyroid	0.03
Bone surfaces	0.03
Remainder	0.30
Total	1.00

where H = the dose equivalent received
P(H) = the number of individuals receiving a dose equivalent in the range from H to H + dH.

Collective Dose Commitment (S_k^c)

To assess the dose equivalents received by a population and the resulting total detriment, from single exposures to long-lived radionuclides or repeated exposures to short- or long-lived ones, UNSCEAR ([6], Sect. 15, p. 29) has developed the concept of collective dose commitment. The collective dose commitment (S_k^c) due to a given event, decision, or finite practice k is defined as:

$$S_k^c = \int_0^\infty \dot{S}_k(t)\, dt$$

where $\dot{S}_k$ = the collective dose rate from source k.

In the case where releases of relatively short-lived radionuclides continue long enough for concentrations in environmental compartments to become constant, the collective dose equivalent resulting from one year of a practice is equal to the collective dose commitment of the amount released in one year ([17], pp. 102, 107; [18]).

Detriment and Dose Limits

One of the bases of the ICRP system of radiological protection is that any human activity should produce more benefit than detriment ([3], Sect. 69, p. 14). Detriment in a population may be defined as ([3], Sect. 16, p. 3): "... the mathematical 'expectation' of the harm incurred from an exposure to radiation, taking into account not only the probability of each type of deleterious effect, but also the severity of the effect."

The most recent recommendations of ICRP on dose limitation ([3], Sect. 104, p. 21) are based on the principle that, for stochastic effects, the risks resulting from the limit of dose should be equal for uniform and non-uniform irradiation of the body and its tissues. The annual limit recommended by ICRP for H_E and H_{50E} for workers is 50 mSv (5 *rem*).

Two other occupational dose limits, for non-stochastic effects ([3], Sect. 103, p. 21), are 0.3 Sv (30 *rem*) for the lens of the eye and 0.5 Sv (50 *rem*) for any other tissues; these are reported for completeness only since they would not likely be relevant to an environmental situation.

The ICRP ([3], Sect. 119, p. 23) recommends a dose-equivalent limit of 5 mSv (0.5 *rem*) per year for critical groups[1] or individual members of the public.

Transfer to Man

The pathways by which man is irradiated as a result of the presence of radioactive materials in the environment are complex and differ depending on

1 A critical group has been described by ICRP ([3], Sect. 85, p. 17) as a group within the population small enough to be relatively homogeneous, yet representative of those individuals in the population expected to receive the highest dose equivalents

whether the radioactivity is airborne or waterborne. The pathways have been described diagrammatically by ICRP Committee 4 [19]; their diagrams are reproduced as Figs. 1 and 2. UNSCEAR ([6], pp. 27–34) has described in a

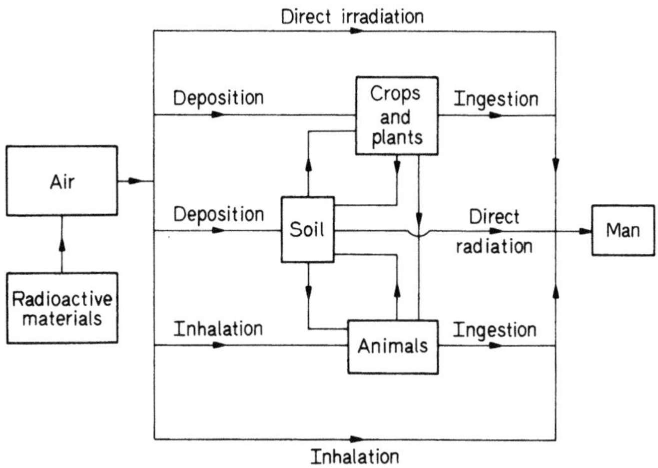

Fig. 1. Simplified pathways between radioactive materials released to atmosphere and man [19]

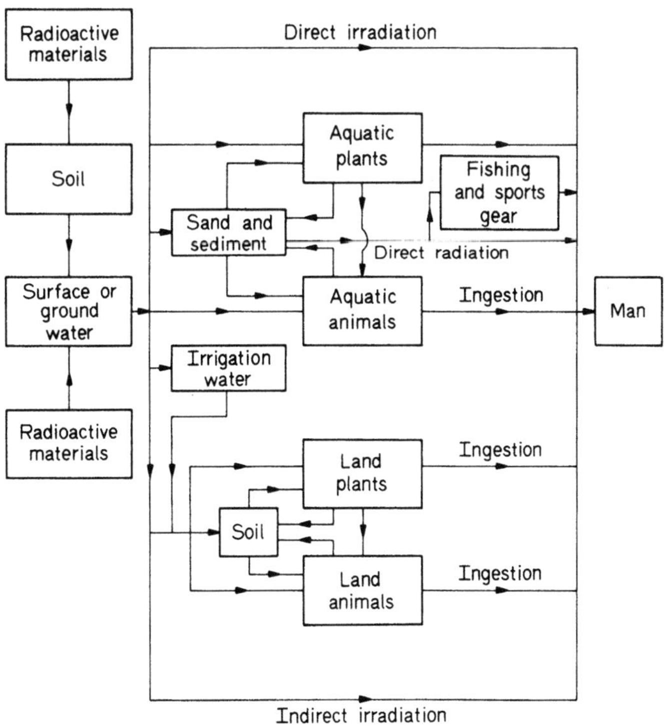

Fig. 2. Simplified pathways between radioactive materials released to ground or surface waters (including oceans) and man [19]

general way the concept of transport through these environmental compartments, and the resulting tissue and organ doses. The transfer factor from compartment i to compartment j has been defined by UNSCEAR as ([6], Sect. 29, p. 31):

$$P_{ij} = \frac{\int_{-\infty}^{\infty} C_j(t)\,dt}{\int_{-\infty}^{\infty} C_i(t)\,dt} \simeq \frac{\int_{0}^{\infty} C_j(t)\,dt}{\int_{0}^{\infty} C_i(t)\,dt}$$

where C_i and C_j are the quantities (e.g. activity concentrations) in the respective compartments at time t. Under conditions of constant release and under constant environmental conditions concentrations in the compartments may become constant, when

$$P_{ij} = \frac{C_j}{C_i}$$

where C_i and C_j are the constant concentrations in compartments i and j ([18], p. 15).

Exposures of Non-Human Biota

The ICRP ([3], Sect. 14, p. 3) "... believes that if man is adequately protected then other living things are also likely to be sufficiently protected." IAEA points out that, for humans, great importance is placed on the long-term effects on individual members of a population whereas for other organisms the long-term structure and fate of the populations are the main concern [20].

a) Doses received. For large groups the average dose and dose rates will nearly always be less than those due to natural sources, viz., 1 mSv (100 *mrem*)/y. At such dose levels only stochastic effects or late cumulative effects of low dose rates will be involved.

Some organisms may receive larger-than-average doses of direct radiation because of their location. Examples of non-human exposure that will be mentioned later in the chapter are:

i) terrestrial exposure
   - contamination of plants by fallout, e.g. lichens (^{137}Cs) ([11], Vol. 1, pp. 52–53)
   - plants and animals living in certain regions of India and Brazil ([6], pp. 48–49)
   - plants and animals living around reactors [21]

ii) aquatic exposure
   - organisms inhabiting bottom sediments accumulate relatively high levels of plutonium [22, 23]
   - organisms near the outfall of nuclear effluents [24, 25]

For indirect radiation, larger doses may be received because of some metabolic factor or a special niche in a food chain:

- caribou eating lichens contaminated with ^{137}Cs ([11], Vol. 1, pp. 52–53)
- domestic animals eating grass contaminated with radioiodine. At Wind-

scale ^{131}I contents of cows' and sheep's thyroids were measured following the accidental release in 1958 and the highest total radiation dose to the thyroid gland was around 10 Gy (1,000 rads) ([26], p. 136).
- if the radionuclide in question is readily absorbed and has a long half-life of retention it will accumulate in higher limnological levels such as piscivorous fish [27].
- animals feeding directly off bottom sediments of lakes and rivers, such as molluscs, usually contain high levels of radionuclides such as plutonium [22].

b) Radiosensitivity. Although lethal doses are not encountered in the environment the radiosensitivity of species may be compared in terms of $LD_{50/30}$ (lethal dose for 50% of organisms in 30 days). Mammals are generally more radiosensitive than other vertebrates, including birds, reptiles, amphibians and fish. $LD_{50/30}$ for dogs is about 3.4 Sv (335 *rem*) (X-rays) while for goldfish it is 6.7 Sv (670 *rem*) (X-rays) ([28], pp. 299–310). The dose of X- or γ-rays needed to kill an insect is at least 100 × greater than that needed to kill a mammal. Adult *Drosophila* are not killed by 64,000 R from ^{60}Co γ-rays, but are sterilized. Unicellular organisms may be less sensitive yet. $LD_{50/30}$ for *Amoeba* is 1,000 Sv (100,000 *rem*) (X-rays) ([28], pp. 299–310). In aquatic systems, teleost fish (especially developing eggs) are the organisms most sensitive to radiation [20, 29].

The genetic character of a species or strain is a major determinant of the carcinogenic response to radiation exposure ([6], Sect. 334, p. 622). Differences in susceptibility are especially manifest at low doses and tend to disappear with increasing doses and dose rates. In experimental animals such as mice, a dose of at least 50 rad is generally required to detect an increase over the natural tumor incidence ([6], Sect. 328, p. 621).

Mutation rates per locus per Gy for low-LET irradiation are in the range of 10^{-5} to 10^{-7} for organisms as diverse as mice, *Drosophila* and barley ([6], Table 44, p. 535). The dose of radiation needed to double the natural mutation rate when given in a single dose (doubling dose) is about 0.3 Gy (30 rads) in mice, 0.5–4 Gy (50–400 rads) in *Drosophila* and 0.3–0.6 Gy (30–60 rads) in plants ([28], p. 258).

Studies of the effects of irradiation in fetal rodents consistently show a reduction in sensitivity with advancement of fetal age ([6], Sect. 342, p. 709; [28], pp. 299–310). Insect larvae also become less radiosensitive with age. For *Drosophila* eggs 3 h old the LD_{50} is 200 R, for 4-h eggs it is 500 R and for pupae, 2800 R ([28], pp. 299–310).

Selected Radionuclides

Introduction

In this section seven radionuclides have received detailed discussion. The choice was made because of their practical importance, public interest or suitability for illustration. As far as possible, data are given which permit the calculation of the risks to human health resulting from a unit of practice. The reviews depend heavily on the most recent publication of UNSCEAR [6] where, on p. 116, the elements of the assessment are illustrated as

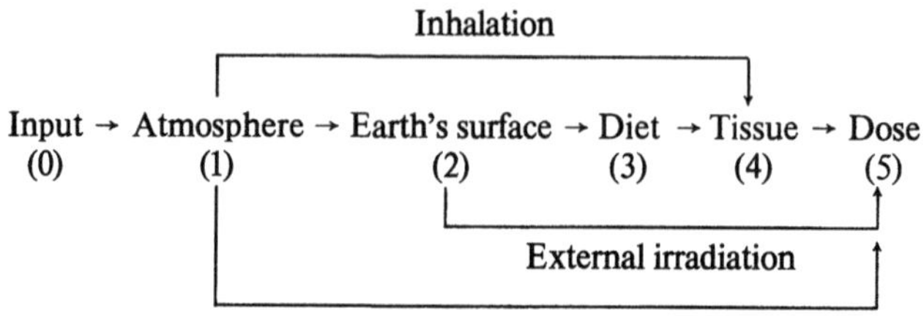

UNSCEAR has its own methods for calculating, from the rate of intake and the equilibrium body content, the resulting tissue concentrations and dose rates (in *grays*) to the tissues. In the present reviews, the UNSCEAR data are used to calculate intakes and thereafter when the resulting dose equivalents to tissues (in *sieverts*) are calculated, the data of ICRP Committee 2 are used. According to ICRP Publication 30 [7] the Annual Limits on Intake (by either inhalation or ingestion) give to all the tissues of the body an effective dose equivalent of 50 mSv or to a single tissue a dose equivalent of 500 mSv, whichever is the lesser intake. From these limiting intakes can be calculated the dose equivalent or the effective dose equivalent resulting from unit intake.

Tritium Oxide

Exposure Due to Natural Sources

a) Production and Release. Tritium occurs naturally, principally in the atmosphere where it is produced by cosmic ray protons and neutrons reacting with nitrogen, oxygen and argon. The reaction producing most of the tritium is

$$^{14}N + n \rightarrow {}^{12}C + {}^{3}H_1$$

where the energy of the neutrons is >4.4 MeV ([6], Sect. 82, p. 54).

The most recent estimates of production rate of ^3H and corresponding world inventory are 0.20 atoms per cm^2 of earth's surface per second and 1×10^{18} Bq (30 MCi), respectively ([6], Table 11, p. 55).

More than 99% of the tritium produced either by natural processes or human technology, when released to the environment, appears as tritiated water (HTO) and hereafter the tritium discussed will be assumed to be in that form, unless specified otherwise.

b) Pathways to Man. As mentioned above, most of the HTO produced in nature is found in surface waters of the earth. Concentrations of HTO in continental surface waters before nuclear explosions began were 0.2–0.9 Bq (6–24 pCi)/L ([6], Sect. 84, p. 55) and, assuming that the hydrogen of the body of Reference Man had the same proportion of tritium as had the surface waters, this would give a whole body dose of 1×10^{-8} Gy (1 μrad)/yr. (According to UNSCEAR ([6], Sect. 19, p. 118), 3.7×10^4 Bq (1 μCi)/L gives 9.5×10^{-4} Gy (95 mrad)/yr.)

Exposure Due to Man-Made Sources

Tritium arises from ternary fission in nuclear explosives or nuclear fuel and also by neutron activation reactions with isotopes of light elements such as lithium and boron.

Whenever water (which contains deuterium) is irradiated with a high flux of neutrons, tritium is produced according to the reaction

$$^2H_1 + n \rightarrow {}^3H_1 + \gamma.$$

Thus the chief sources of tritium production by man will be nuclear bomb explosions and nuclear reactors.

1. Exposure Due to Nuclear Bombs

a) Production and Release. UNSCEAR ([6], p. 117) has summarized the estimates of total tritium production in nuclear bomb explosions and the resulting world inventory. Since the total production is released the quantities given will serve for estimates of the release. The best estimates for the total release up to 1970 lie between 1.3×10^{20} and 1.7×10^{20} Bq (3,500 and 4,500 MCi), with 20% of this in the southern hemisphere and the remaining 80% in the northern hemisphere.

b) Deposition. The HTO released from above-ground nuclear explosions is injected into the stratosphere where the average residence time is about one year. It then passes to the troposphere and atmosphere and enters the earth's hydrological cycle.

According to UNSCEAR ([6], Sect. 17 and Fig. I, p. 117), between 1963 and 1969, 6.6×10^{19} Bq (1,780 MCi) of HTO were deposited in the northern hemisphere and 1.5×10^{19} Bq (400 MCi) in the southern hemisphere. The latitudinal distribution of HTO in the top 500 m of the Pacific Ocean between 1965 and 1972 was about 4.5×10^{19} atoms per km^2 at 30° south latitude and 32×10^{19} atoms per km^2 at 30°–40° north.

From the data acquired by the IAEA world network for monitoring HTO in precipitation ([6], p. 118) it was calculated that, at marine stations, the concentration of HTO in rain doubled as the latitude increased by 13° and that the concentration was 3.6 times as high over land as over water. The mean concentrations in surface waters of the USA varied from a low of about 0.6 Bq (15 pCi)/L in 1951–1953 to a high of about 185 Bq (5,000 pCi)/L in the 1960's. Concentrations in the Ottawa River (approx. 46° north latitude) were about twice as great.

c) Pathways to Man. The dose commitments from nuclear explosions have been calculated by UNSCEAR ([6], Sect. 18–27, pp. 118–119) to be 2×10^{-5} Gy (2 mrad) for the northern hemisphere and 2×10^{-6} Gy (0.2 mrad) for the southern hemisphere. For the population of the USA the dose commitment was calculated to be 1.5×10^{-5} Gy (1.5 mrad) and, for inhabitants of the Ottawa Valley, 2.8×10^{-5} Gy (2.8 mrad).

The collective dose commitment for explosions in the northern hemisphere is estimated to be 8×10^{4} man-Gy (8×10^{6} man-rad), corresponding to 8.1×10^{-16} man-Gy per Bq (3×10^{-3} man-rad per Ci) of HTO released.

2. Exposure Due to Nuclear Reactor Operations

a) Production. The rates of production given by UNSCEAR for various types of reactors are ([6], Table 10, p. 178):

BWR, PWR, GCR	7.4×10^{11} Bq (20 Ci) per MW(e)y
HWR	2.2×10^{13} Bq (600 Ci) per MW(e)y.

The reference LWR fuel described in [6], (Table 25, p. 202), irradiated to 33,000 MWd per tonne, and cooled 150 days, contained 7.9×10^{15} Bq (213 kCi) of tritium per tonne which corresponds to a production rate of 2.6×10^{14} Bq (7.1 kCi) per MW(e)y. In HWR of the CANDU type, operated by Ontario Hydro, the production rate is 8.9×10^{13} Bq (2,400 Ci) per MW(e)y [31].

b) Release and Deposition. The releases of HTO from various types of reactors reported by UNSCEAR ([6], Table 10, p. 178; Table 11, p. 179) are summarized in Table 6.

It can be seen that by far the greatest releases come from HWR and this merits some comment. The amounts released depend on the amounts leaking or escaping from the system during normal operations and minor accidents and on the amount of neutron irradiation received by the heavy water moderator and coolant. These factors will vary from reactor to reactor due to differences in design and operating experience.

Ontario Hydro has the greatest body of experience in operating heavy water power reactors and in 1978 [31] they reported the following release rates:

Installation	Release rate [Bq per MW(e)y]
NPD	2.7×10^{13} (730 Ci)
Douglas Point	7×10^{12} (190 Ci)
Pickering	8.9×10^{11} (24 Ci)

They have estimated that 20% of current releases are to liquid effluents and 80% to airborne effluents.

The Pickering generating station of Ontario Hydro is located on Lake Ontario; for this site it is assumed [31] that all the releases to liquid effluents, and that half the airborne releases, enter the lake.

Table 6. Releases of tritium from reactor operations

Type of reactor	Normalized release [Bq per MW(e)y] In liquid effluent	In airborne effluent (1974)
BWR	7.4×10^9 (0.2 Ci)	1.9×10^9 (0.05 Ci)
PWR	3.7×10^{10} (1 Ci)	7.4×10^9 (0.2 Ci)
GCR	1.1×10^{10} (0.3 Ci)	1.5×10^9 (0.04 Ci)
HWR (Pickering)	7.4×10^{11} (20 Ci)	5.6×10^{11} (15 Ci)

c) Pathways to Man. UNSCEAR ([6], Sect. 103–104, pp. 193–194) estimates that for airborne HTO released from an operating reactor the collective dose commitment is 5.4×10^{-17} man-Gy per Bq (2×10^{-4} man-rad per Ci). Thus, for the releases quoted above, the different types of reactors would give the following collective dose commitments:

Reactor type	Collective dose commitment [man-Gy/MW(e)y]
PWR	4×10^{-7} (4×10^{-5} man-rad)
GCR	4×10^{-7} (4×10^{-5} man-rad)
BWR	1×10^{-7} (1×10^{-5} man-rad)
HWR	4×10^{-5} (4×10^{-3} man-rad)

For HTO released in liquid effluents to a body of water providing drinking water, the calculated collective dose commitment is 1.9×10^{-15} man-Gy per Bq (0.007 man-rad per Ci) released, resulting in the following ([6], Sect. 105, p. 194):

Reactor type	Collective dose commitment [man-Gy/MW(e)y]
PWR	7×10^{-5} (0.007 man-rad)
BWR	7×10^{-6} (0.0007 man-rad)
HWR	3×10^{-4} (0.03 man-rad)

On the basis of operating experience, including environmental monitoring, Ontario Hydro has estimated the individual doses resulting from both airborne and waterborne releases at a typical station with CANDU-type reactors. The results of their calculations [31] are reproduced in Table 7, below.

Since the major, and unavoidable, part of the population dose resulting from a release of HTO arises through inhalation and skin absorption of airborne activity, this could form the basis of calculations. The most important information required is the dilution factor,

$$K_a = \frac{\text{air concentration at the target (Bq/m}^3\text{)}}{\text{release rate (Bq/s)}},$$

for the site in question. Ontario Hydro [31] has measured values of K_a at 1 km from four of their reactor stations and found it to average $2.8 \pm 1.3 \times 10^{-7}$ s/m³. It can be assumed that, beyond 1 km, the air concentration diminishes as the distance (in km) raised to the power -1.5, or, $C_d = C_l d^{-1.5}$, where C_d is the air concentration at d km, C_l is the air concentration at 1 km, d is the distance downwind from the source in km.

Table 7. Estimated annual dose equivalents from HTO released to air and water from a CANDU station

	Annual individual dose equivalent (μSv) as a function of distance from the station			
Pathway	1 km	3 km	5 km	10 km
1. Inhalation and skin absorption	5	1	0.6	0.2
2. Ingestion of milk[a]	1	1	1	1
3. Ingestion of home grown fruits and vegetables	1	0.2	0.1	0.09
4. Drinking water	0.6	0.6	0.6	0.6
Total	8	3	2	2

[a] Milk is from farms at a distance of 10–15 km from station

With a knowledge of the air concentration, the population density and the ICRP dose coefficient [7], viz. 2,000 h exposure to a concentration of 8×10^5 Bq/m³ (DAC) gives an effective dose equivalent of 50 mSv, the collective dose commitment from an accidental release may be calculated.

3. Exposure Due to Fuel Reprocessing

a) Production and Release. As reported by UNSCEAR ([6], Table 24, p. 201) operating experience in fuel reprocessing at two plants has given the following results:

	Tritium release rate [Bq/MW(e)y]	
Installation	Airborne	Liquid
Windscale (UK)	-	4.8×10^{11} (13 Ci)
NFS (USA)	3.7×10^{10} (1 Ci)	2.2×10^{11} (6 Ci)

Grathwohl from Karlsruhe, quoted in [6] (Sect. 146, p. 203), estimated that, for a PWR, the production rate of tritium would be 7×10^{11} Bq (19 Ci)/ MW(e)y and of this about 3.7×10^{10} Bq (1 Ci) would be released during reactor operation and 5.9×10^{11} (16 Ci)/MW(e)y released during fuel reprocessing. Similar amounts were estimated for HWR fuel, one-half of this amount for AGR fuel and one-tenth for HTGR fuel.

b) Pathways to Man. Using the same coefficients for collective dose commitment as for reactor operation (man-Gy per Bq released), the following collective dose commitments were calculated by UNSCEAR ([6], Sect. 159, p. 204):

Installation	Type of release	Collective dose commitment [man-Gy/MW(e)y]
NFS (USA)	airborne	2×10^{-6} (2×10^{-4} man-rad)
Windscale (UK)	to salt water	2×10^{-8} (2×10^{-6} man-rad)
NFS (USA)	to fresh water	4×10^{-4} (4×10^{-2} man-rad)

4. Exposure Due to Occupation

Special mention needs to be made of the occcupational exposure to two groups of workers: (a) staff of HWR, and (b) tritium luminizers.

a) Staff of HWR. The most informative statistics on this subject are provided by Ontario Hydro ([6], Table 11, p. 238) from more than 10 yr experience with operative CANDU type reactors. The mean collective occupational dose was 9×10^{-3} man-Gy (0.9 man-rad)/MW(e)y, similar to that for other types of power reactors in the USA, where most of the exposure is to external radiation. The exceptional feature of the CANDU statistics is that 26% of the collective dose was due to internal contamination with HTO.

b) Luminizers. Radium has been largely replaced by promethium and tritium for luminizing the dials of watches, although in the USA radium is still much in use for clocks, according to UNSCEAR ([6], Sect. 277 and Table 46, p. 96). These new luminous paints emit only soft β-particles and thus give smaller doses to the wearer than radium which emits γ-rays and also generates radon which leaks out.

Some luminous paints contain tritiated organic compounds which may leak out slowly giving internal doses to the wearer of a watch. UNSCEAR ([6], Sect. 286, p. 97) reported a study of the HTO content of the urine of eight persons wearing tritium-luminized watches; the content averaged 1.2×10^2 Bq (3.2 nCi)/L above background which corresponds to a whole-body dose of 3×10^{-6} Gy (0.3 mrad)/yr.

Workers with tritium luminous paint may be monitored by measuring the HTO content of urine. The results from four countries in 1975 ([6], Sect. 122–123, p. 255; Tables 81–84, pp. 289–290) are as follows:

Country	No. of workers	Average annual dose equivalent
UK	136	7×10^{-3} Sv (0.7 rem)
Switzerland	235	1×10^{-2} Sv (1 rem)
France	80	3.5×10^{-3} Sv (0.35 rem)
Germany	56	1.35×10^{-2} Sv (1.35 rem)

5. Annual Limits on Intake

ICRP Committee 2 [7] has calculated that the following intakes will give an effective dose equivalent, H_E, of 50 mSv:

Ingestion	3×10^9 Bq (50 mCi)
Inhalation (Class D)	3×10^9 Bq (50 mCi)

Krypton-85

Exposure Due to Man-Made Sources

1. Exposure Due to Nuclear Bombs

a) Production and Release. The $^{85}Kr/^{90}Sr$ ratio of fission yield = 0.07; thus about 1.1×10^{17} Bq (3 MCi) have been produced in nuclear explosions ([6], Sect. 41, p. 121). Another estimate [32] has given 2×10^{17} Bq (5 MCi).

b) Pathways to Man. Radioactive krypton, being a noble gas, is not deposited and does not enter into metabolic processes in the food chain nor in man. Thus, population doses are calculated by multiplying the air concentration by a coefficient (one for each tissue of interest) to give the corresponding dose rate. To calculate population doses, one needs to know only the concentration in the air surrounding the population.

UNSCEAR ([6], Sect. 158, p. 204; Sect. 191, p. 209) and NCRP [32] give the following dose rate coefficients for an air concentration of 3.7×10^{10} Bq (1 Ci)/m^3:

Organ or tissue	Dose rate (Gy/yr)
Testes	60 (0.6×10^4 rad)
Ovaries	160 (1.6×10^4 rad)
Red bone marrow	180 (1.8×10^4 rad)
Skin	18,000 (1.8×10^6 rad)
Lung	310 (3.1×10^4 rad)
Total body	150 (1.5×10^4 rad)

WASH-1400 ([33], Table VI-17, p. 60) gives a value of 1.1×10^4 Sv (1.1×10^6 rem)/yr (3.6×10^{-4} Sv (0.036 rem)/s) from an air concentration of 3.7×10^{10} Bq/m^3 but the tissue receiving the dose is not specified.

Assuming that the 1.1×10^{17} Bq (3 MCi) released are uniformly mixed in the earth's troposphere (5×10^{21} g of air) the resulting concentration at NTP $= 3 \times 10^{-2}$ Bq (0.8 pCi)/m^3. Assuming that the ^{85}Kr concentration declines with the radioactive half-life of 10.7 years, the average life is 15 yr and the exposure is 0.4 Bq-yr (12 pCi-yr)/m^3. When this is multiplied by the dose rate coefficients given above the following individual dose commitments and collective dose commitments result ([6], Sect. 42, p. 121):

Organ or tissue	Dose commitments	
	Individual (nGy)	Collective[a] (man-Gy)
Gonads	1.4 (0.14 μrad)	7.7 (770 man-rad)
Red bone marrow	2.2 (0.22 μrad)	12.6 (1260 man-rad)
Skin	220 (22 μrad)	1.26×10^3 (1.26×10^5 man-rad)
Lungs	3.7 (0.37 μrad)	21.7 (2170 man-rad)

[a] Based on a present world population of 4×10^9 that increases by 2% per year

2. *Exposure Due to Nuclear Reactor Operations*

a) Production. ^{85}Kr is only one of several radioactive noble gases produced in reactor operation; a number of isotopes of krypton and xenon are produced in fission and ^{41}Ar is a neutron activation product of the argon in air ([6], Sect. 50, p. 172).

About 25 cm^3 of Kr and Xe are produced in reactor fuel per MWd thermal. This creates pressure inside the fuel canister and any cladding failure results in an escape of the gas ([6], Sect. 51, p. 172).

The thermal fission yield of ^{85}Kr is 0.29% for ^{235}U and 0.14% for ^{239}Pu, corresponding to 1.9×10^{13} and 9.3×10^{12} Bq (500 and 250 Ci)/MW(e)y for the two fuels respectively. More detailed calculations give estimates falling between these two rates ([6], Sect. 144, p. 203; [32]).

The amount of ^{85}Kr in the LWR "reference" fuel of UNSCEAR is 1.4×10^{13} Bq (375 Ci)/MW(e)y ([6], Table 25, p. 202).

b) Release. The amount of radioactive noble gases escaping will depend on the number of fuel cladding failures, the design of the cooling and ventilating systems and operating procedures. Thus there are tremendous individual variations contributing to the overall normalized releases given by NCRP and UNSCEAR ([6], pp. 172–178; [32]) as follows:

Type of reactor	Normalized release [Bq ^{85}Kr per MW(e)y]	% of total noble gases
PWR	6.3×10^9 (0.17 Ci)	1
BWR	1.1×10^{12} (30 Ci)	2
GCR	Insignificant	
AGR and HTGR	Negligible	

c) Pathways to Man. Zuker et al. [34] used trajectory analysis based on historic wind data to calculate the 100-day-average ground concentrations of ^{85}Kr released at a constant rate from the Ontario Hydro Pickering Station. These ground-level concentrations were calculated for every point on a 70 km grid from Toronto to the east coast of North America. Population density figures, obtained from electoral districts and county censuses, were applied to the same grid. Multiplying the concentration at a square on the grid by the population and one of the dose conversion coefficients given above gave the annual collective dose for each square on the grid. When expressed as a function of distance from the source it was found that, for a constant release of 3.7×10^{10} Bq/s, the total annual collective dose was received within a radius of 1000 km and that more than 95% of this was within 600 km.

UNSCEAR ([6], pp. 191–193) reported the collective doses due to radioactive noble gases released from various reactor sites. The collective dose depends on the population density around the reactor site so there will be great variation from site to site. The following table summarizes the dose commitments from the normalized releases reported by UNSCEAR:

Reactor type	Collective dose [man-Gy (gonad) per Bq released]	Collective dose [total man-Gy (whole body) per MW(e)y]	% due to ^{85}Kr
BWR	1.2×10^{-16} (4.5×10^{-4} man-rad/Ci)	9×10^{-4} (0.09 man-rad)	0.4
PWR	4.1×10^{-17} (1.5×10^{-4} man-rad/Ci)	2.5×10^{-5}–5×10^{-5} (0.0025–0.005 man-rad)	0.1

From these figures it is apparent that ^{85}Kr is not a significant contributor to local or regional collective doses but, because it has the longest half-life of the radioactive noble gases, it may make the greatest contribution to the global dose commitment.

3. *Exposure Due to Fuel Reprocessing*

a) Production and Release. More than 90% of the ^{85}Kr generated by fission in fuel with intact cladding is released at the fuel reprocessing plant. The rates of release were 1.5×10^{13} Bq (400 Ci)/MW(e)y for Windscale (UK) and 1.3×10^{13} Bq (340 Ci)/MW(e)y for NFS (USA) ([6], Table 24, p. 201).

The NCRP predicted [32] that by the year 2000, when the world nuclear electric power generation would be 4,500 GW, the annual production of ^{85}Kr would be about 3.7×10^{19} Bq (1,000 MCi) and the amount accumulated in the world about 2.2×10^{20} Bq (6,000 MCi).

b) Pathways to Man. UNSCEAR ([6], par. 158, p. 204) estimated that reprocessing spent fuel after cooling 150 d would yield the following collective tissue doses:

Organ or tissue	Collective dose [man-Gy/MW(e)y]
Gonads	7×10^{-6} (7×10^{-4} man-rad)
Red bone marrow	1×10^{-5} (1×10^{-3} man-rad)
Lungs	2×10^{-5} (2×10^{-3} man-rad)
Skin	1×10^{-3} (1×10^{-1} man-rad)

The following assumptions were made:
i) All the ^{85}Kr in the fuel was released.
ii) Dispersion factor at 1 km = 5×10^{-7} s m^{-3} ([32] gives 1×10^{-7} s m^{-3}).
iii) $C_d = C_{1\,km}\, d^{-1.5}$, d in km.
iv) Population density = 100 km^{-2}.
v) Dose conversion coefficients as above.

4. Maximum Permissible Concentration (MPC)

Since the dose to skin is about two orders of magnitude higher than that to any other tissue the permitted concentration for continuous exposure would probably be that giving an annual dose of 0.5 Sv to the skin; this is 9.3×10^5 Bq (25 μCi)/m^3. If the dose limit is based on an annual dose of 0.3 Sv to the lens of the eye, the MPC would be 5.5×10^5 Bq/m^3 (15 μCi/m^3).

Strontium-90

Exposure Due to Man-Made Sources

1. Exposure Due to Nuclear Bombs

a) Production. ^{90}Sr is a fission product, the yield varying with the fissile material and with the method of fission, from about 1–9% [35].

The production in nuclear bombs is estimated to be 3.7×10^{15} Bq (0.1 MCi) per megaton of explosive energy [36] but this may vary greatly in individual tests.

b) Release. The explosion of nuclear bombs in the atmosphere results in some local fallout of fission products which has not been documented in the open literature. The remaining fission products are carried aloft to the troposphere and stratosphere, circulate around the globe and slowly deposit on the earth. The half-life of ^{90}Sr in the stratosphere is about one year ([6], Sect. 13, p. 117).

The variation of stratospheric inventory of ^{90}Sr from 1962 to 1975 for the world as well as northern and southern hemispheres has been published by UNSCEAR ([6], Fig. V, p. 121). The total in the stratosphere has declined from 2.3×10^{17} Bq (6.3 MCi) in 1962 to 3.7×10^{15} Bq (0.1 MCi) in 1974. In the northern hemisphere it has declined from 2×10^{17} Bq (5.4 MCi) in 1962 to 1×10^{15} Bq (0.03 MCi) in 1975.

c) Deposition. The deposition of bomb-produced stratospheric ^{90}Sr varies with latitude ([6], Table 3, p. 122), the maximum occurring at 40°–50° north.

The annual worldwide deposition of ^{90}Sr has been tabulated by UNSCEAR ([6], Table 2, p. 122). Part of the data are reproduced here as Table 8.

The deposition velocity of an airborne material may be calculated by dividing the rate of deposition (Bq/cm²/s) by the air concentration above the surface (Bq/cm³) or by dividing the integrated deposit (Bq/cm²) by the time-integrated air concentration (Bq-s/cm³). This gives the deposition velocity in

Table 8. Annual deposition of strontium-90. (From [6], Table 2, p. 122; [38])

	Annual deposition in Bq × 10^{16} (MCi)		
	Northern hemisphere	Southern hemisphere	Global
Pre-1958	6.7 (1.80)	2.4 (0.65)	9.1 (2.45)
1958	2.3 (0.63)	0.9 (0.25)	3.3 (0.88)
1959	3.9 (1.05)	0.7 (0.18)	4.6 (1.23)
1960	1.0 (0.26)	0.6 (0.17)	1.6 (0.43)
1961	1.3 (0.35)	0.6 (0.17)	1.9 (0.52)
1962	5.3 (1.44)	1.0 (0.26)	6.3 (1.70)
1963	9.7 (2.62)	1.1 (0.31)	10.8 (2.93)
1964	6.1 (1.66)	1.6 (0.42)	7.7 (2.08)
1965	2.8 (0.77)	1.3 (0.36)	4.2 (1.13)
1966	1.2 (0.33)	0.8 (0.21)	2.0 (0.54)
1967	0.6 (0.17)	0.4 (0.11)	1.0 (0.28)
1968	0.7 (0.20)	0.4 (0.10)	1.1 (0.30)
1969	0.6 (0.15)	0.5 (0.14)	1.1 (0.29)
1970	0.8 (0.21)	0.5 (0.13)	1.3 (0.34)
1971	0.7 (0.19)	0.6 (0.15)	1.3 (0.34)
1972	0.3 (0.09)	0.4 (0.10)	0.7 (0.19)
1973	0.1 (0.03)	0.1 (0.03)	0.2 (0.06)
1974	0.4 (0.12)	0.1 (0.04)	0.6 (0.16)
1975	0.2 (0.06)	0.1 (0.03)	0.3 (0.09)
1976	0.1 (0.03)	0.007 (0.02)	0.2 (0.05)
Integrated deposition	45 (12.16)	14.2 (3.83)	59.2 (15.99)
Stratospheric inventory	0.9 (0.23)	0.04 (0.01)	0.9 (0.24)
Total injection to January 1977	45.8 (12.39)	14.2 (3.84)	60.1 (16.23)

the usual dimensions of cm/s. When data like those contained in [37] for New York City are used for such calculations a value of 4 cm/s results. Similar data for other sites may yield somewhat lower values (1–3 cm/s).

d) Pathways to Man. The radioactive material released to the stratosphere is transported in the air, from which it may enter the body directly by inhalation or indirectly by deposition on the earth's surface and entry into the body in drinking water and food. This latter indirect route gives most of the dose to tissues.

UNSCEAR ([6], Sect. 68–69, p. 131) has published estimates of the transfer of fallout ^{90}Sr to diet in various parts of the world and for a number of foods. The numbers for P_{23} for total diet range from 0.1 to 0.4 Bq (3–10 pCi)y/g Ca per 3.7×10^7 Bq/km^2 (mean value of P_{23} is about 5).

Assuming that the daily dietary intake of Ca = 1 g for people of all ages ([30], p. 365), the daily ingestion of ^{90}Sr can be calculated from the rate of deposition of ^{90}Sr. For population intake by inhalation assume 100 people/km^2.

2. *Exposure Due to Nuclear Reactor Operations*

a) Production. Some examples of the estimated amounts of ^{90}Sr present in irradiated reactor fuel are:

- fuel irradiated to 20,000 MWd/tonne contains 1.9×10^{15} Bq (50,000 Ci) ^{90}Sr per tonne ([39], Table 3, p. 15);
- in the "reference" irradiated fuel of WASH-1400, 0.13% of the total activity is due to ^{90}Sr ([33], Table VI-1, p. 6);
- in a LWR, fuel irradiated to 33,000 MWd/tonne contained 2.8×10^{15} Bq (77,000 Ci) ^{90} Sr per tonne, corresponding to 9.6×10^{13} Bq (2,600 Ci) per MW(e)y [40].

b) Release. Releases of ^{90}Sr to the environment from nuclear power production are of current interest; they fall into two categories.

i) Normal operations. The releases from normal operations vary with the type of reactor and its containment but they have been documented by UNSCEAR for BWR ([6], pp. 188–190) as:

- to water, 3.7×10^6 Bq (100 μCi)/MW(e)y;
- to air, 1.9×10^5 Bq (5 μCi)/MW(e)y.

ii) Accidents. For accidental fuel melt-down the MRC ([39], Table 3, p. 15) has estimated that 1% of the ^{90}Sr in the fuel would escape to the atmosphere, although this could be reduced to a fraction of a percent by fuel-cladding ([39, p. 16). There could, in an event, be large variations in this number depending on many relevant influences. In WASH-1400 ([33], Table VI-2, p. 9) a wide variety of fractional releases (from a few percent to negligible) of strontium isotopes in irradiated fuel, along with their probabilities, have been estimated for different PWR and BWR.

c) Pathways to man. For releases to air the intake may be calculated as for bomb fallout. For releases to water, calculations [38] show that releases to salt water result in negligible population dose commitments and that for fresh water, most of the population dose comes from drinking the water rather than eating the fish that live in it. From the knowledge that Reference Man drinks 2–3 L of water per day ([30], p. 360), the daily intake may be calculated.

The principal routes of entry of ^{90}Sr into the body for the population most affected by an accidental release of fission products will be air transport resulting in inhalation and ingestion of contaminated milk through the air → forage → cow → milk food chain.

Concentrations in air must be measured or calculated according to [34] or [41] or from a knowledge of K_a (see ^{3}H, Sect. 2c). From the air concentration the amount inhaled can be calculated from the fact that Reference Man inhales about 2.2×10^4 L of air per day ([30], p. 346). The deposition rate may be calculated as described above. The MRC ([39], Table 12, p. 40) has calculated that, following the deposition of 3.7×10^4 Bq (1 μCi)/m^2 on pasture the integrated concentration of ^{90}Sr in milk is 1.7×10^9 tr (0.53 μCi-d)/L in the first year and a total of 4.8×10^9 tr (1.5 μCi-d)/L. From the daily intake of milk (0.7 L for an infant and 0.5 L for an adult) the ingestion of ^{90}Sr resulting from deposition on pasture can be calculated.

3. *Exposure Due to Fuel Reprocessing*

a) Production and Release. For fuel reprocessing UNSCEAR ([6], p. 201) has published the following typical release rates:

	UK	USA
To air	1.5×10^6 Bq (4×10^{-5} Ci)/MW(e)y	2×10^6 Bq (6×10^{-5} Ci)/MW(e)y
To water	2×10^{11} Bq (6 Ci)/MW(e)y	3.7×10^8 Bq (1×10^{-2} Ci)/MW(e)y

b) Pathways to Man. See Sect. 2c above.

4. *Annual Limits on Intake*

As described in [7], Committee 2 of ICRP has calculated the effective dose equivalent (H_E) resulting from the ingestion or inhalation of 1 Bq of ^{90}Sr and from this the intakes (ALI) that result in an effective dose equivalent of 50 mSv to all tissues of the body or that give a dose of 0.5 Sv to the most irradiated tissue. For ^{90}Sr these are:

Ingestion		1×10^6 Bq (27 μCi)
Inhalation	(Class D)	8×10^5 Bq (21.6 μCi)
	(Class Y)	1×10^5 Bq (2.7 μCi)

From these can be calculated the effective dose equivalents resulting from the intakes estimated above.

Iodine-131

Exposure Due to Man-Made Sources

1. Exposure Due to Nuclear Bombs

a) Production. Several radioactive isotopes of iodine are produced in appreciable yields in nuclear fission or as daughters arising from the radioactive transformation of other fission products (e.g. Te). The iodine isotope produced in fission which is of greatest concern in environmental contamination is ^{131}I. The fission yield of ^{131}I is 3% ([42], p. 255), half that of ^{137}Cs.

b) Release and Deposition. UNSCEAR ([6], Table 16, p. 139) has published integrated milk concentrations of ^{131}I for several locations from 1966 to 1976.

Reference [43] (Table 1.1, p. 59) reports values of deposition velocity from 0.1–5 cm/s and adopts a nominal value of 1 cm/s.

c) Pathways to Man. If it is assumed that dairy cows obtain all their fodder by grazing grass contaminated with radioiodine, the food chain air → grass → cow → milk → child outweighs the inhalation dose by a factor of 3 for an adult and 60 for an infant. Thus the dose commitments to human populations are often assessed by monitoring levels of ^{131}I in commercial fresh milk and calculating doses to the thyroid gland of Reference Man. The dose to a "reference child" would be ten times higher because the infant thyroid has a mass of about 2 g whereas the adult gland has a mass of about 20 g ([6], Table 18, p. 195).

UNSCEAR has published ([6], Table 16, p. 139) calculated dose commitments to infant thyroids for several places in the northern and southern hemispheres for the decade following 1966.

2. Exposure Due to Nuclear Reactor Operations

a) Production. Because of the relatively short half-life of ^{131}I the content in reactor fuel does not continue to increase with time of irradiation but soon reaches a constant equilibrium level which has been reported by UNSCEAR as 1×10^{15} Bq (30 kCi)/MW ([6], Sect. 82, p. 181). In the "reference fuel" of WASH-1400 ([33], Table VI-1, p. 6) 2.2% of the fission product activity was due to ^{131}I. The postulated fuel of MRC, 1975 ([39], Table 3, p. 15), irradiated to 20,000 MWd/tonne contained 2×10^{16} Bq (6×10^5 Ci)/tonne of ^{131}I. As reported in ORNL-4451 the LWR fuel irradiated to 33,000 MWd/tonne, and cooled 150 days, contained 8×10^{10} Bq (2.2 Ci)/t [40].

b) Release. UNSCEAR ([6], Sect. 84, p. 184; Table 13, p. 185) reported average releases of 7–20 $\times 10^7$ Bq (2–5 $\times 10^{-3}$ Ci)/MW(e)y for BWR and 2–20 $\times 10^6$ Bq (5–50 $\times 10^{-5}$ Ci)/MW(e)y from PWR. There were, however, wide variations in release rates between various individual installations.

Important experience of the environmental effects of radiodiodine released accidentally from a reactor resulted from the "Windscale Accident" in the UK [44]. Irradiated fuel elements became overheated, the cladding ruptured and volatile fission products were released through the stack. It has been estimated that 7×10^{14} Bq (20,000 Ci) of ^{131}I were discharged to the environment ([26], p. 129).

WASH-1400 ([33], Table VI-2, p. 9) gives estimated releases of negligible to 60%, with the corresponding probabilities, for a number of different PWR and BWR.

MRC in 1975 ([39], Table 4, p. 16) estimated percentage releases of 0.2–100% of the postulated iodine content depending on the type of cladding and other reactor variables.

c) Deposition. The local collective dose commitment for the release of ^{131}I from reactors is given by UNSCEAR ([6], Sect. 118, pp. 195–196) as 6×10^{-12} man-Gy per Bq (22 man-rad per Ci), which, on the basis of operating experience gives, for

BWR	1×10^{-3} man-Gy (0.1 man-rad)/MW(e)y
PWR, GCR, HWR	1×10^{-5} man-Gy (1×10^{-3} man-rad)/MW(e)y

ICRP Committee 4 has calculated the food chain contamination resulting at 1000 m downwind from an assumed continuous atmospheric release of ^{131}I to terrestrial environment. The following values are given in [43]:

Assumed ground-level release rate	3.7×10^{10} Bq (1 Ci)/yr
Effects at 1 km:	
– deposition on grass and vegetables	11 Bq (300 pCi)/m^2
– concentration in cows' milk	4 Bq (120 pCi)/L
– rate of intake by infant drinking 0.7 L milk/day	3 Bq (84 pCi)/d.

From this can be calculated the annual intake and the annual dose to the thyroid (from the ALI).

d) Pathways to Man. In the "Windscale Accident" of 1957 which released 7×10^{14} Bq (20,000 Ci) of ^{131}I, the maximum concentration in milk was 5×10^4 Bq (1.4 μCi)/L ([26], pp. 129, 132).

For a single ground-level release of ^{131}I which was assumed to deposit on pasture 3.7×10^4 Bq (1 μCi)/m^2 the MRC ([39], p. 22) calculated an integrated concentration in fresh cows' milk of 5×10^4 Bq-days (1.4 μCi-days)/L and thus an intake by an infant of 3.7×10^4 Bq (1 μCi) of ^{131}I. The resulting dose to the thyroid was estimated to be 0.2 Sv (16 rems) for a child.

ICRP Committee 4 ([43], Table 1.7, p. 65) assumed an acute ground-level release of 3.7×10^{10} Bq (1 Ci) of ^{131}I to the atmosphere; the calculated results at a distance of 1,000 m downwind were:

Assumed ground-level release	3.7×10^{10} Bq (1 Ci)
Effects at 1 km:	
– deposition on grass and vegetables	1.1×10^{10} tr (3.5 μCi·d)/m^2
– beef	1.1×10^{9} tr (0.34 μCi·d)/kg
– milk	4.1×10^{9} tr (1.3 μCi·d)/L

This would give a dose to the infant thyroid of 0.15 Sv (15 rems).

3. Exposure Due to Fuel Reprocessing

a) Production and Release. As mentioned in Sect. 2a the amount of ^{131}I in the irradiated fuel depends very much on the cooling time. Experience in the UK ([6], Table 24, p. 201; Sect. 161, p. 205) in the 1970's has shown a normalized release of ^{131}I of 3×10^8 Bq (9×10^{-3} Ci)/MW(e)y and in the USA of 3×10^5 Bq (8×10^{-6} Ci)/MW(e)y.

b) Deposition and Pathways to Man. Because irradiated reactor fuel is allowed to "cool" for some months before reprocessing, most of the ^{131}I will have disappeared by radioactive decay. The environmental effects are therefore due to the longer-lived ^{129}I (half-life 1.6×10^7 yr). Local collective dose commitments from the ^{129}I released in fuel reprocessing have been published by UNSCEAR ([6], Sect. 163, p. 205).

4. Annual Limits on Intake

ICRP Committee 2 has calculated that the following intakes of ^{131}I give an effective dose equivalent of 50 mSv [7]:

Ingestion	4×10^6 Bq (1×10^2 μCi)
Inhalation (Class D)	6×10^6 Bq (2×10^2 μCi)

The ALI calculated by ICRP Committee 2 to give a dose of 0.5 Sv to the thyroid gland of Reference Man are:

Ingestion	1×10^6 Bq (30 μCi)
Inhalation (Class D)	2×10^6 Bq (55 μCi)

For infants these ALI's should be reduced by a factor of 10.

Caesium-137

Exposure Due to Man-Made Sources

1. Exposure Due to Nuclear Bombs

a) Production. About six atoms of ^{137}Cs are produced per hundred fissions. Assuming that 1.45×10^{23} fissions yield 1 kt of energy [36] it may be calculated that an explosion of 1 megaton produces 6.3×10^{15} Bq (0.17 MCi) of ^{137}Cs.

b) Release and Deposition. The calculation above shows that the activity yield of ^{137}Cs from a nuclear explosion is 1.7 times that of ^{90}Sr (see p. 251). In addition, UNSCEAR has reported ([6], Sect. 97, p. 141; [11], Vol. 1, Sect. 222, p. 51) that the ratio of $^{137}Cs/^{90}Sr$ is fairly constant at about 1.6 in fallout deposited at many different times and sites. Thus the more extensive data on ^{90}Sr levels can frequently be used to compute the ^{137}Cs levels.

Since production and deposition ratios are nearly the same, one may conclude that ^{137}Cs and ^{90}Sr have equal deposition velocities, when they are averaged over long periods and many different atmospheric conditions.

c) Pathways to Man. ^{137}Cs deposited on the earth from the air finds its way into human diets mainly through grain, meat and milk ([6], Table 17, p. 143). The transfer factor from deposition to total diet is taken by UNSCEAR ([6], Sect. 105, p. 143) to be 0.1 Bq (4 pCi) per g of potassium in food per 3.7×10^7 Bq (1 mCi) deposited per km^2 ($P_{23} = 4 \times 10^{19}$ Bq (gK^{-1}) per Bq km^{-2} or 4 pCi (gK^{-1}) per mCi km^{-2}). The concentration in milk is representative of that in total diet.

For residents of Chicago, Gustafson et al. [45] reported that the dietary contributions to ^{137}Cs intake were approximately as follows: milk, 30%; grain, 25%; meat, 20%; fruits, 10%; vegetables, 10%; other, 5%.

The following facts about the potassium metabolism of Reference Man ([30], pp. 327, 403) permit calculation of daily intakes and equilibrium body contents:

– Body content of K at all ages $\cong$ 2 g/kg		
– Daily intake in food,	infant	0.5 g K/day
	10-year-old	3 g K/day
	adult	3.3 g K/day

Similar quantitative conclusions were arrived at by the NCRP [46] who reported that at a continued deposition rate of 3.7×10^7 Bq (1 mCi)/km^2 per year the dietary level would reach 0.1 Bq (3 pCi) ^{137}Cs/g K which would lead to a constant body content of 0.3 Bq (9 pCi)/g K which is a total of 48 Bq (1,300 pCi) for a 70 kg Reference Man.

^{137}Cs that finds its way into fresh water may find its way to man's food through the fish that live in those waters. These fish may have ^{137}Cs concentrations several thousand times higher than the water [47]. In [43] it was calculated that a constant release of ^{137}Cs into surface water sufficient to maintain a constant concentration of 0.04 Bq (1 pCi)/L would lead to a concentration in fish of 111 Bq (3,000 pCi)/kg and this would be the dominant route of intake by two orders of magnitude.

A special pathway for the transfer of fallout ^{137}Cs to the diet of sub-polar peoples is by way of lichens and reindeer or caribou meat. It is well known that lichens and mosses trap airborne pollutants [48] and that the deposition of bomb-produced fission products is higher in northern latitudes. Since reindeer and caribou graze lichens in winter their intake of ^{137}Cs may be high with consequent elevation of the concentration in meat. This meat is an important item of diet for native peoples in the Arctic and sub-Arctic who had body contents of ^{137}Cs, and the resulting dose commitments, in the 1960's ([11], Vol. 1, Sect. 230, p. 52; Sect. 233 and Fig. XXI, p. 53), 50–100 times higher than other inhabitants of the northern hemisphere.

2. *Exposure Due to Nuclear Reactor Operations*

a) Production. Some estimates of the amount of ^{137}Cs in irradiated reactor fuel are:

- Fuel irradiated to 20,000 MWd per tonne contains 2.5×10^{15} Bq (6.67×10^4 Ci)/tonne of ^{137}Cs ([39], Table 3, p. 15).

- In the "reference" irradiated fuel of WASH-1400, 0.15% of the total activity was due to ^{137}Cs ([33], Table VI-1, p. 6).
- The fuel of a light water reactor irradiated to 33,000 MWd per tonne contained 3.7×10^{15} Bq (0.1 MCi)/t of ^{137}Cs [40].

b) Release. Releases of ^{137}Cs to the environment from nuclear power production may arise from three different sources, (i) reactor operations, (ii) fuel reprocessing, or (iii) accidents.

i) For both PWR and BWR, UNSCEAR has given ([6], pp. 188–189) the rate of airborne release of ^{137}Cs as 7.4×10^5 Bq (20 μCi)/MW(e)y.
The same reference gives waterborne releases of about 7.4×10^4 Bq (2 μCi)/MW(e)y for PWR and about 9.3×10^5 Bq (25 μCi)/MW(e)y for BWR.

ii) In fuel reprocessing UNSCEAR ([6], Sect. 150–151, p. 203) reports airborne releases of 10^{-7} to 10^{-10} of the ^{137}Cs in the fuel, and waterborne releases to the sea of 10^{-2} to 10^{-3} of the activity in the fuel.

iii) For accidental fuel melt-down the MRC ([39], Table 3, p. 15) has postulated a release of 100% of the ^{137}Cs in irradiated fuel (2.5×10^{11} (6.7 Ci)/t in fuel irradiated to 20,000 MWd/t). Cladding of the fuel may reduce this to a few percent ([39], Table 4, p. 16). In WASH-1400 fractional releases of Cs varying from a few tens of a percent to negligible (along with their probabilities) were postulated for a number of PWR and BWR ([33], Table VI-1, p. 6).

c) Deposition. In a short-term release such as that occurring in an accident the deposition velocity could vary from 0.1 to 30 cm/s according to ICRP [43].

d) Pathways to Man. i) Reactor operations. UNSCEAR ([6], Table 21, p. 196) estimates a local collective dose equivalent commitment of 1.4×10^{-5} man-Gy (1.4×10^{-3} man-rad) per MW(e)y from airborne effluents of PWR and 1.1×10^{-5} man-Gy (1.1×10^{-3} man-rad) per MW(e)y from BWR. For waterborne releases ([6],Table 23, p. 200) the dose commitments are:

PWR	1.1×10^{-7} man-Gy (1.1×10^{-5} man-rad)/MW(e)y
BWR	1.4×10^{-6} man-Gy (1.4×10^{-4} man-rad)/MW(e)y
GCR	3.8×10^{-6} man-Gy (3.8×10^{-4} man-rad)/MW(e)y

ii) Fuel reprocessing. For effluents from fuel reprocessing plants UNSCEAR ([6], Tables 26–27, p. 206) estimates the following whole-body dose commitments from ^{137}Cs:

Airborne	6.6×10^{-7} man-Gy (6.6×10^{-5} man-rad)/MW(e)y
Salt waterborne	8×10^{-4} man-Gy (0.08 man-rad)/MW(e)y
Fresh waterborne	3×10^{-4} man-Gy (0.03 man-rad)/MW(e)y

WASH-1400 ([33], Table VI-17, p. 60) estimates that a ground deposition of ^{137}Cs of 3.7×10^7 Bq (1 mCi)/km^2 gives a whole body dose equivalent rate of 7×10^{-7} Sv (7×10^{-5} rem)/yr.

iii) Accidents. The MRC concluded ([39], pp. 26–30) that when ^{137}Cs was released to the air the major source of internal contamination is by the pasture → cow → milk food chain. It was estimated that deposition of 3.7×10^4 Bq (1 μCi)/m^2 on pasture would give a total integrated concentration in milk of 3.8×10^{10} tr (12 μCi-days) per litre, 3.2×10^{10} [10] of these being received in the first year. If these integrated concentrations are multiplied by the daily intake of milk (0.7 L/d for a child and 0.5 L/d for an adult) the number of Bq ingested will be obtained.

ICRP Committee 4 ([43], p. 32; Figs. 1-12–1-14, pp. 90–92) calculated the consequences of an acute ground-level release of 3.7×10^{10} Bq (1 Ci) of ^{137}Cs. At a distance of 1000 m downwind, computations of the results for the 1,000 days after the release gave deposition rates for soil and foliage (3.7×10^4 Bq (μCi)/m^2), concentration in milk (3.7×10^4 Bq (μCi)/L) and in beef (3.7×10^4 Bq (μCi)/kg). From the results it was concluded that eating leafy vegetables would be the most critical food path for both infants and adults; otherwise drinking milk was the most critical path for infants and eating beef for adults.

The whole-body dose for adults was calculated to be 0.09 Sv (9 rems) in the first year, but because of the large uncertainty in deposition velocity (0.1–30 cm/sec) the uncertainty in dose gave a similar range (5×10^{-3}–1.8 Sv (0.5–176 rem)/yr).

3. *Annual Limits on Intake*

ICRP has published the following ALI for ^{137}Cs [7]:

Ingestion	4×10^6 Bq (1×10^2 μCi)
Inhalation (Class D)	6×10^6 Bq (2×10^2 μCi)

Radium-226

Exposure Due to Natural Sources

a) Production. Radium-226 is an intermediate member of the radioactive decay chain of uranium-238 found in nature. Most samples of "radium" will contain several short-lived daughters; the principal radioactive emissions will include the α-particles from ^{226}Ra, ^{222}Rn, ^{218}Po, and ^{214}Po; the β-particles from ^{214}Pb, ^{214}Bi, and ^{210}Tl along with a mixture of γ-rays mainly from ^{226}Ra and ^{214}Bi ([42], p. 246).

Since the uranium-238 radioactive family occurs in nature, ^{226}Ra and its parents and daughters are normal constituents of the earth's crust. They occur in higher concentrations in uranium ores.

^{226}Ra is important as an environmental contaminant not only because of its ubiquity, leading to daily intakes by inhalation and ingestion, but also because of its first radioactive daughter, radon-222. ^{222}Rn, being a noble gas, is transported in air to be breathed by man or to contaminate the environment more widely by deposition of its radioactive daughters. Environmental contamination by ^{222}Rn is in itself a large and important subject which will not be dealt with in this section which is restricted to ^{226}Ra alone.

b) Pathways to Man. Small amounts of radium are found in the air due to resuspension of soil particles. In most regions of the earth this is responsible for the daily inhalation of about 3.7×10^{-5} Bq (10^{-15} Ci) ([6], Sect. 112, p. 59).

Uncontaminated surface waters usually contain such small amounts of radium that drinking water is a minor source of intake. Some wells and hot springs, however, may contain 0.04–0.4 Bq (1–10 pCi)/L of radium ([6], Sect. 114, p. 59).

For the population in general food is the main source of radium intake which, for an average diet, may be about 0.04 Bq (1 pCi)/day ([6], Sect. 113, p. 59).

Larger dietary intakes may be due to eating some items of food containing higher concentrations of radium (Brazil nuts and Pacific salmon) or to living in areas (found in India and Brazil) with high concentrations of natural uranium and thorium in the soil ([6], Sect. 113, 115, p. 59).

For purposes of dosimetry, UNSCEAR ([6], pp. 60–61) gives data on the radium content of various human tissues, especially bone, as a function of dietary content and daily intake. The average activity concentrations in four important tissues are shown in the table below. The resulting α-particle doses per year were calculated assuming that two-thirds of the ^{222}Rn daughter escaped from the tissue.

Organ or tissue	^{226}Ra concentration in Bq/kg (pCi/kg)	Yearly α dose in Gy (mrad)
Lung	4.8×10^{-3} (0.13)	1×10^{-7} (0.01)
Gonads	4.8×10^{-3} (0.13)	1×10^{-7} (0.01)
Bone	0.3 (8)	
Red bone marrow	4.8×10^{-3} (0.13)	3×10^{-7} (0.03)
Bone lining cells	4.8×10^{-3} (0.13)	2.7×10^{-6} (0.27)

The effective dose equivalent (H_E) is about 4 μSv (4×10^{-4} rem)/yr.

Exposure Due to Man-Made-Sources

1. Exposure Due to Coal-Fired Power Plants ([6], pp. 86–88)

a) Production and Release. Coal contains all the radioactive elements found naturally in the earth and when the coal is burned these radionuclides are emitted through the stack in the fly ash. If the activity concentration of ^{226}Ra in fly ash is 0.04 Bq (1 pCi)/g and the flow of fly ash through the stack is 0.7 to 30 t (representative mean 10 tonne) per megawatt-year of electrical-energy, the activity of ^{226}Ra discharged would be 3.7×10^5 Bq (10^{-5} Ci)/MW(e)y.

b) Pathways to Man. From assumptions about diffusion of the emitted fly ash and population density, the following collective a dose commitments to various tissues have been calculated.

Radionuclide	Collective α dose commitment in man-Gy per MW(e)y [man-rad per MW(e)y]		
	Lung	Red bone marrow	Bone lining cells
^{226}Ra	2×10^{-6} (2×10^{-4})	4×10^{-8} (4×10^{-6})	3×10^{-7} (3×10^{-5})
Total (^{238}U, ^{226}Ra, ^{210}Pb, ^{228}Ra, ^{228}Th, ^{232}Th)	1×10^{-4} (10^{-2})	2×10^{-5} (2×10^{-3})	2×10^{-4} (2×10^{-2})

The effective dose equivalent (H_E) is about 5×10^{-6} man-Sv (5×10^{-4} man-rem)/MW(e)y for ^{226}Ra and 6×10^{-4} man-Sv (6×10^{-2} man-rem)/MW(e)y for all nuclides emitted. The collective effective dose commitments from these releases are about 5 man-Sv (5×10^2 man-rem) and 600 man-Sv (6×10^4 man-rem), respectively.

2. Exposure Due to Inhaled Phosphates ([6], pp. 89–91)

a) Production and Release. Phosphate-containing rock is mined in appreciable quantities (130 million tonnes in USA in 1973) to provide industrial phosphates, one-half for fertilizer, the other half for chemicals and gypsum building materials. The most important natural radionuclide in this rock is ^{226}Ra which occurs in concentrations from 0.04–5 Bq (1–130 pCi)/g.

b) Pathways to Man. The largest collective dose commitments resulting from the various uses are those from the use of phosphogypsum as a building material, viz., 0.01 man-Gy (1 man-rad) of whole body dose from γ-rays per tonne of rock marketed (38×10^6 t/yr in USA), giving a collective dose commitment of 4×10^5 man-Sv/yr. Lesser doses, shown in the following table, result from the use of the phosphate rock for fertilizers (2×10^7 t/yr in USA).

Radionuclide	Collective α dose commitment in man-Gy per tonne (man-rad per tonne)			
	Lung	Gonads	Red bone marrow	Bone lining cells
^{226}Ra	4×10^{-8} (4×10^{-6})	4×10^{-8} (4×10^{-6})	1×10^{-7} (1×10^{-5})	1×10^{-6} (1×10^{-4})
Total (^{238}U, ^{226}Ra, ^{210}Pb)	6×10^{-7} (6×10^{-5})	1×10^{-6} (1×10^{-4})	1×10^{-6} (1×10^{-4})	7×10^{-6} (7×10^{-4})

The resulting collective dose commitments are 30 man-Sv/yr for ^{226}Ra and 270 man-Sv/yr for all nuclides emitted.

3. Exposure Due to Luminous Timepieces ([6], pp. 96, 97)

a) Production and Release. Although largely replaced as a luminizer for watch dials, ^{226}Ra is still widely used for clocks.

b) Pathways to Man. UNSCEAR has concluded that the population doses resulting from this application arise mainly from external exposure to γ-rays. They have also calculated that the annual dose to the gonads could be 2×10^8 Bq (6 mrad) from a wristwatch and 3.7×10^6 Bq (0.1 mrad) from an alarm clock.

4. *Exposure Due to Uranium Milling* ([6], pp. 167–170)

a) Production and Release. ^{226}Ra is the most important radionuclide in uranium mill wastes; the concentration in liquid effluent may vary from 9.3–18.5 Bq (250–500 pCi)/L. In dry tailings the concentration may be 20.7 Bq (560 pCi)/g.

A mill processing 6×10^5 t of uranium ore per year in the USA released, airborne, about 3.7×10^8 Bq (10 mCi) of ^{226}Ra per year, which is equivalent to 3.7×10^4 Bq (1 µCi) per MW(e)y.

b) Pathways to Man. UNSCEAR calculated the collective dose commitments from this release as:

	Collective dose commitments in man-Gy/MW(e)y [man-rad/MW(e)y]			
Route of exposure	Body	Lung	Bone marrow	Bone lining cells
External	6×10^{-7} (6×10^{-5})	-	-	-
Ingestion	-	-	5×10^{-8} (5×10^{-6})	6×10^{-8} (6×10^{-6})
Inhalation	-	1×10^{-7} (1×10^{-5})	5×10^{-8} (5×10^{-6})	4×10^{-8} (4×10^{-6})

5. *Exposure Due to Uranium Fuel Fabrication* ([6], p. 171)

In fuel fabrication residual amounts of ^{226}Ra are removed from the uranium compounds produced in milling. In the USA it was estimated that fuel fabrication operations released about 3.7×10^9 Bq (0.1 Ci)/yr of ^{226}Ra in liquid effluents, equivalent to 1.3×10^5 Bq (3.4×10^{-6} Ci)/MW(e)y.

6. *Annual Limits on Intake* [7]

ICRP Committee 2 has calculated that the following intakes will give an effective dose equivalent of 50 mSv:

Ingestion	5×10^4 Bq (1 µCi)
Inhalation (Class W)	8×10^3 Bq (0.2 µCi)

This ALI for ingestion could be used to calculate a maximum permissible concentration of ^{226}Ra in soil, using Canada as an example, from the following facts:

- ICRP maximum permissible annual intake of ^{226}Ra for individuals in the population (critical group) $= 0.1 \times ALI = 4 \times 10^3$ Bq (0.1 μCi) ([3], Sect. 119, p. 23);
- Annual consumption of vegetables, other than potatoes = 86 kg [49];
- Vegetable fresh weight/dry weight ≅ 11 [50];

$$\frac{^{226}\text{Ra per g dry weight vegetable}}{^{226}\text{Ra per g dry soil}} = 0.25 \text{ [51]};$$

- Permissible concentration of ^{226}Ra in soil to give a daily ingestion of 3.7×10^3 Bq (0.1 μCi) = 185 Bq (5 nCi)/kg.

Plutonium-239

Exposure Due to Man-Made Sources

1. Exposure Due to Nuclear Bombs

a) Production. ^{239}Pu is produced from neutron capture in ^{228}U according to the following scheme:

$$^{238}\text{U}_{92} + \text{n} \rightarrow {}^{239}\text{U}_{92} \xrightarrow[\text{rapid}]{\beta^-} {}^{239}\text{Np}_{93} \xrightarrow[\text{rapid}]{\beta^-} {}^{239}\text{Pu}_{94}$$

Capture of neutrons by ^{239}Pu leads to isotopes of plutonium with higher atomic weights and other transuranic daughters. A diagram illustrating these relations has been published by UNSCEAR ([6], Fig. 1, p. 204).

The plutonium that occurs in the environment is usually a mixture of ^{239}Pu and ^{240}Pu which are difficult to distinguish; therefore, hereafter, "Pu" will be used to indicate the mixture of $^{239\&240}Pu$.

b) Release and Deposition. It has been estimated that 1.5×10^{16} Bq (400 kCi) of Pu have been released in weapons testing and that 1.2×10^{16} Bq (320 kCi) have been dispersed around the world [52], 9.3×10^{15} Bq (250 kCi) in the northern hemisphere and 2.6×10^{15} Bq (70 kCi) in the southern hemisphere ([6], Sect. 127, p. 148). Isotopic analyses have indicated that the ratio of activities $^{239}Pu/^{240}Pu = 60/40$ [52]. Most of this came from tests conducted before 1963.

Many measurements over several years have shown that in the stratosphere and in surface air the activity ratio Pu/^{90}Sr has remained fairly constant at 0.017 ([6], Sect. 127, p. 148). By assuming the same deposition velocity as for ^{90}Sr (1–4 cm/s) the deposition can be calculated when monitoring data are not available. Bennett has published the results of fallout monitoring and computation for New York for the twenty years 1954–1974 ([52], Table I, p. 368).

Plutonium deposited on soil moves slowly downward and displays the same depth profile as ^{137}Cs ([52], pp. 375–376). Ninety-five percent of plutonium entering the sea and freshwater lakes is quickly deposited in sediments where its behaviour is similar to that of ^{137}Cs [27, 53].

In January 1968 a B-52 aeroplane loaded with a nuclear bomb crashed at Thule, Greenland, dispersing about 9.3×10^{11} Bq (25 Ci) of plutonium into the sea. Environmental monitoring carried out between 1968 and 1974 discovered the presence of some plutonium in bottom sediments, molluscs and worms but none in higher vertebrates such as fish, seabirds and marine mammals [54].

c) Pathways to Man. UNSCEAR has concluded ([6], Sects. 131–136, p. 148) that the most important route to man is by inhalation of contaminated air. Bennett [52] has calculated that residents of New York inhaled a total of about 1.5 Bq (40 pCi) of Pu in the two decades from 1954–1974. Thus, of the 9.3×10^{15} Bq (250 kCi) deposited in the northern hemisphere during the same period, about 10^{-16} was inhaled by an average individual. This fraction should be kept in mind for assessing some of the absurd estimates of the consequences of having plutonium fuel in reactors [55].

The calculated body content at the end of the 20-year inhalation intake by New York residents was 0.09 Bq (2.5 pCi) ([6], Sects. 131–136, p. 148). The contents of various organs and tissues were calculated; the results agreed reasonably well for all tissues, except kidneys, with those found by analysis of members of the population in the USA [52, 56].

UNSCEAR estimated that the population-weighted dose, up to 2000 A.D., from bomb plutonium was 1×10^{-5} Gy (1 mrad) in the northern hemisphere and 3×10^{-6} Gy (0.3 mrad) in the southern ([6], par. 131–136, p. 148).

Ingestion of environmental plutonium by man may result from its deposition on land or its entry into surface waters.

Bennett ([52], Table IV, p. 374) has reported measured values of the ratio

$$\frac{\text{pCi per g fresh weight of vegetables}}{\text{pCi per g of soil}}$$

for a number of plants including vegetables in the human diet; most of the values ranged from 1×10^{-3} to 1×10^{-4}.

Miettinen [22] measured Pu in two food chains in Finland. In the terrestrial one the Pu content in lichens and reindeer liver, respectively, were 8.1 Bq (220 pCi)/kg and 0.7 Bq (20 pCi)/kg in 1963 and 0.7 Bq (20 pCi)/kg and 0.07 Bq (2 pCi)/kg in 1973. In a typical marine food chain from the Gulf of Finland the following concentrations of Pu were found:

Sediment	7.4 Bq (200 pCi)/kg
Brown algae (fresh wt)	0.2 Bq (5 pCi)/kg
Blue mussel (fresh wt, whole animal)	0.02 Bq (0.6 pCi)/kg
Fish (fresh wt)	1.5×10^{-3} to 5.2×10^{-3} Bq (0.04–0.14 pCi)/kg

Measurements were made of ^{239}Pu in several marine invertebrates, including mussels, clams, oysters and scallops, from Cape Cod [57]. Mean body concentrations ranged from 4×10^{-3} to 1.8×10^{-2} Bq (0.11 to 0.49 pCi)/kg fresh wt (body), 100–500 times greater than concentrations in the environment.

In a limnological study of Lake Michigan and other Great Lakes, Edgington et al. [27] measured the concentration of Pu in sediments, mixed plankton, zooplankton, planktivorous fish, piscivorous fish and water. The concentration declined quite regularly through each stage from 3.7 Bq (100 pCi)/kg to about 3.7×10^{-5} Bq (10^{-3} pCi)/kg by about one order of magnitude per stage.

From the data available it seems clear that the concentration of plutonium declines as one proceeds along food chains from soil to man.

Bennett [52] reported the results of a dietary analysis in New York in 1972 indicating that the annual ingestion of Pu was 0.06 Bq (1.6 pCi). From this the transfer coefficient P_{23} was calculated by UNSCEAR ([6], Sect. 138, p. 150) as

$$\frac{0.06\ \mathrm{Bq\ y^{-1}}}{6.3 \times 10^5\ \mathrm{Bq\ km^{-2}\ y^{-1}}} = 9.5 \times 10^{-8}\ \mathrm{Bq/Bq\ km^{-2}}$$

$$\text{or}\ \frac{1.6\ \mathrm{pCi\ y^{-1}}}{0.017\ \mathrm{mCi\ km^{-2}\ y^{-1}}} = 94\ \mathrm{pCi/mCi\ km^{-2}}.$$

If this coefficient is multiplied by the estimated deposition for each year from 1954–1974 a total ingestion of 9.25 Bq (250 pCi) is derived. From this UNSCEAR estimates the mean population dose commitment to be 1.2×10^{-7} Gy (1.2×10^{-2} mrad) and the collective dose commitment from all test explosions to be 3×10^{-12} man-Gy/Bq (10 man-rad/Ci) of Pu released, to the bone lining cells and to the lungs ([6], Sect. 140, 142, p. 150).

A somewhat different estimate of individual dose commitment is obtained using the dosimetry calculations of ICRP Committee 2 [7] as follows:

- The ingestion of 7.4×10^5 Bq (0.02 mCi) gives an effective dose equivalent of 0.5 Sv to bone surfaces.
- The ingestion of 7.4 Bq (200 pCi) would give 5 μSv to bone surfaces.
- Since ICRP uses Q = 20 for α-particles, 5 μSv = 2.5×10^{-7} Gy (2.5×10^{-2} mrad)
- This is to be compared with the 1.2×10^{-7} Gy (1.2×10^{-2} mrad) estimated by UNSCEAR (above).

2. *Exposure Due to Nuclear Reactor Operations*

a) Production and Release. At present the greatest production of ^{239}Pu is for nuclear weapons, but understandably very little information is available concerning the amounts produced or released.

The environmental statement for the LMFBR of the USA, quoted in [58], postulates a release to the atmosphere of 3.7×10^6 Bq (0.1 mCi) of Pu per 1,000 MW(e)y. It is also given as 10^{-9} of the Pu made and burned in the fuel cycle [59].

Because of its low volatility and because it gives rather low dose commitments when released to the environment, plutonium is not often considered in assessing the consequences of reactor accidents. For example, in the "reference" fuel of WASH-1400 ([33], Table VI-1, p. 6; Table VI-2, p. 9) 2.5×10^{-4}%

of the radioactivity is due to Pu and of this, less than 10^{-3} is released in the postulated accident.

b) Pathways to Man. Estimates of the dose commitments to a population in the USA as a result of operating the LMFBR fuel cycle have been referred to in [58] and [59]. In [59] it is assumed that 10^{-5} of the Pu released [1.5×10^3 Bq (4×10^{-5} mCi)/MW(e)y [58]] is inhaled by man. Thus the intake by inhalation $= 1.5 \times 10^{-2}$ man-Bq (0.4 man-pCi)/MW(e)y. According to ICRP Committee 2 the inhalation of 200 Bq (5 nCi) gives an effective dose equivalent to the skeleton of 0.5 Sv [7]. Therefore the inhalation of 1.5×10^{-2} man-Bq (0.4 man-pCi) gives an effective population dose equivalent of 4×10^{-5} man-Sv/MW(e)y.

3. Exposure Due to Fuel Reprocessing

a) Production and Release. Pu is one of the nuclides of greatest concern for internal contamination of workers in fuel processing and reprocessing [60]; it is not considered one of the major nuclides for environmental contamination. UNSCEAR ([6], Table 25, p. 202) reports a release to liquid effluents of 2.6×10^5 Bq (7×10^{-6} Ci)/MW(e)y from the NFS plant in the USA and a normalized release rate, for all plants, into liquid effluents of 2.2×10^{10} Bq (0.6 Ci)/MW(e)y to salt water and 2.6×10^5 Bq (7×10^{-6} Ci)/MW(e)y to fresh water ([6], Table 27, p. 206).

b) Pathways to Man. UNSCEAR ([6], Table 25, p. 202) has calculated a collective dose from the 2.2×10^{10} Bq (0.6 Ci)/MW(e)y of Pu, released in fuel processing, to be 1×10^{-5} man-Gy (1×10^{-3} man-rad)/MW(e)y.

4. Annual Limits on Intake

ICRP Committee 2 [7] has calculated that the following intakes of Pu give an effective dose equivalent of 50 mSv:

Ingestion		1×10^6 Bq (30 μCi)
Inhalation	(Class W)	4×10^2 Bq (10 nCi)
	(Class Y)	6×10^2 Bq (20 nCi)

and that the following intakes give a dose equivalent of 500 mSv to bone lining cells:

Ingestion		8×10^5 Bq (20 μCi)
Inhalation	(Class W)	2×10^2 Bq (5 nCi)
	(Class Y)	5×10^2 Bq (15 nCi).

Larsen and Oldham [61] found that 1 ppm of chlorine in drinking water that contained approximately 0.2 Bq (5 pCi)/mL of ^{239}Pu (IV) resulted in a 75% oxidation to ^{239}Pu (VI). The significance of this for the maximum permissible concentration of plutonium in drinking water was nullified when Sullivan et al. [62] found that, in contradiction to Weeks et al. [63], there was no appreciable difference in the absorption of intragastrically injected ^{238}Pu (IV) and ^{238}Pu (VI) by rats or guinea pigs allowed food *ad libitum*.

References

1. Internat. Comm. Rad. Units Measurements: Radiation Quantities and Units. ICRU Rep. 19; ICRU Publications: Washington, D.C. 1971
2. Internat. Comm. Rad. Units Measurements: Phys. Med. Biol. *20,* 1029 (1975)
3. Recommendations of the Internat. Comm. on Radiological Protection. ICRP Publication 26, Ann. ICRP *1,* 1 (1977)
4. Mayneord, W.V., Clarke, R.H.: Brit. J. Radiol., Suppl. No. *12,* 15 (1975)
5. Brown, J.M.: Health Phys. *31,* 231 (1976)
6. UN Sci. Comm. Effects of Atomic Radiation: Sources and Effects of Ionizing Radiation. Offic. Rec. of the General Assembly, 32nd Session, Suppl. No. 40 (A/32/40); United Nations: New York, N.Y. 1977
7. Recommendations Internat. Comm. Radiological Protection, Rep. Committee 2: Limits for Intakes of Radionuclides by Workers. ICRP Publication 30, Part 1, Ann. ICRP *2* (3/4), 1 (1979)
8. Recommendations Internat. Comm. Radiological Protection, Rep. Committee 4: The Assessment of Intern. Contamination Resulting from Recurrent or Prolonged Uptakes. ICRP Publication 10A; Pergamon Press: Oxford 1971
9. Pochin, E.E.: Estimated Population Exposure from Nuclear Power Production and Other Radiation Sources; Organization for Economic Cooperation and Development: Paris 1976
10. Rep. UN Sci. Comm. Effects of Atomic Radiation, Offic. Rec. General Assembly, 13th Session, Suppl. No. 17 (A/3838); United Nations: New York, N.Y. 1958. *Ibid.,* 17th Session, Suppl. No. 16 (A/5216); United Nations: New York, N.Y. 1962. *Ibid.,* 19th Session, Suppl. No. 14 (A/5814); United Nations: New York, N.Y. 1964. *Ibid.,* 21st Session, Suppl. No. 14 (A/6314); United Nations: New York, N.Y. 1966. *Ibid.,* 24th Session, Suppl. No. 13 (A/7613); United Nations: New York, N.Y. 1969
11. UN Sci. Comm. Effects of Atomic Radiation: Ionizing Radiation: Levels and Effects. Offic. Rec. General Assembly, 27th Session, Suppl. No. 25 (A/8725); United Nations: New York, N.Y. 1972
12. Recommendations Internat. Comm. Radiological Protection, Report for Committee 1: The Evaluation of Risks from Radiation. ICRP Publication 8; Pergamon Press: Oxford 1966
13. Recommendations Internat. Comm. Radiological Protection. Rep. Committee 1: Radiosensitivity and Spatial Distribution of Dose. ICRP Publication 14; Pergamon Press: Oxford 1969
14. Advisory Committee on the Biological Effects of Ionizing Radiations (BEIR Report): The Effects on Populations of Exposure to Low Levels of Ionizing Radiation; US Nat. Acad. Sci., Nat. Res. Council: Washington, D.C. 1972
15. New Primary Sites of Malignant Neoplasms in Canada, 1975. Statistics Canada: Ottawa 1978
16. Jacobi, W. in: Radiation Protection Measurement – Philosophy and Implementation. EUR-5397; Comm. Europ. Communities: Luxembourg 1975; p. 63
17. Butler, G.C. (ed.): Principles of Ecotoxicology. SCOPE Rep. 12; John Wiley: Chichester-New York-Brisbane-Toronto 1978
18. Internat. Atomic Energy Agency: Principles for Establishing Limits for the Release of Radioactive Materials into the Environment. Safety Series No. 45, STI/PUB/477; IAEA: Vienna 1978
19. Recommendations Internat. Comm. Radiological Protection, Rep. Committee 4: Principles of Environmental Monitoring Related to the Handling of Radioactive Materials. ICRP Publication 7; Pergamon Press: Oxford 1965
20. Internat. Atomic Energy Agency: Effects of Ionizing Radiation on Aquatic Organisms and Ecosystems. Techn. Rep. Ser. No. 172, STI/DOC/10/172; IAEA: Vienna 1976
21. Rogers, L.E., Rickard, W.H. (eds.): Ecology of the 200 Area Plateau Waste Management Environs: A Status Report. PNL-2253; Battelle Pacific Northwest Laboratories: Richland, Washington 1977
22. Miettinen, J.K. in: Miller, M.W., Stannard, J.N. (eds.): Environmental Toxicity of Aquatic Radionuclides: Models and Mechanisms; Ann Arbor Science Publishers Inc.: Ann Arbor, Michigan 1976; pp. 29–43

23. Dahlman, R.C., Bondietti, E.A., Eyman, L.D. in: Friedman, A.M. (ed.): Actinides in the Environment; Amer. Chem. Soc., Washington, D.C. 1976; pp. 47–80
24. Singh, H., Marshall, J.S.: Health Phys. *32,* 195 (1977)
25. Nat. Res. Council, Rep. by the Panel on Radioactivity in the Marine Environment of the Committee on Oceanography: Radioactivity in the Marine Environment; US Nat. Acad. Sci., Washington, D.C. 1971
26. Medical Res. Council: The Hazards to Man of Nuclear and Allied Radiations. A Second Rep. to the Medical Research Council. Command 1225; Her Majesty's Stationery Office: London 1960
27. Edgington, D.N., Wahlgren, M.A., Marshall, J.S. in: Miller, M.W., Stannard, J.N. (eds.): Environmental Toxicity of Aquatic Radionuclides: Models and Mechanisms; Ann Arbor Science Publishers Inc.: Ann Arbor, Michigan 1976; pp. 45–79
28. Bacq, Z.M., Alexander, P.: Fundamentals of Radiobiology, 2nd ed.; Pergamon Press: Oxford 1961
29. Templeton, W.L. in: Miller, M.W., Stannard, J.N. (eds.): Environmental Toxicity of Aquatic Radionuclides: Models and Mechanisms; Ann Arbor Science Publishers Inc.: Ann Arbor, Michigan 1976; pp. 287–301
30. Internat. Comm. Radiological Protection: Report of the Task Group on Reference Man. ICRP Publication 23; Pergamon Press: Oxford 1975
31. Gorman, D.J., Wong, K.Y.: Environmental Aspects of Tritium from CANDU Station Releases. H.P.D.-78-2; Ontario Hydro: Toronto, Ont. 1978
32. Nat. Council on Radiation Protection and Measurements: Krypton-85 in the Atmosphere – Accumulation, Biological Significance, and Control Technology. NCRP Report No. 44; NCRP: Washington, D.C. 1975
33. US Atomic Energy Commission: An Assessment of Accident Risks in U.S. Commercial Nuclear Power Plants. USAEC Reactor Safety Study, WASH-1400 (draft); USAEC: Washington, D.C. Aug. 1974; Appendix VI
34. Zuker, M., Ridgeway, J.M., Miller, D.R. in: Biological Implications of Radionuclides Released from Nuclear Industries. IAEA-SM-237/26; Internat. Atomic Energy Agency: Vienna 1979
35. Crouch, E.A.C.: Atomic Data and Nuclear Data Tables *19* (5), 417 (1977)
36. Halden, N.A. et al. in Health and Safety Laboratory Fallout Program Quarterly Summary Report. HASL-117; HASL: New York, N.Y. 1961; pp. 194–199
37. Hardy, E.P., Jr.: Environmental Measurements Laboratory Environmental Quarterly and Appendix, Sept. 1-Dec. 1, 1977. EML-334; Dept. of Energy, Environmental Measurements Lab.: New York, N.Y. 1978
38. Bennett, B.G.: Private communication 1978
39. Medical Res. Council: Criteria for Controlling Radiation Doses to the Public after Accidental Escape of Radioactive Material. Her Majesty's Stationery Office: London 1975
40. Oak Ridge Nat. Laboratory: Siting of Fuel Reprocessing Plants and Waste Management Facilities. ORNL-4451; Oak Ridge Nat. Laboratory: Oak Ridge, Tennessee 1970
41. Bryant, P.M., Jones, J.A. in: Environmental Behavior of Radionuclides Released in the Nuclear Industry. STI/PUB/345; Internat. Atomic Energy Agency: Vienna 1973
42. Wilson, B.J. (ed.): The Radiochemical Manual, 2nd ed.; The Radiochemical Centre: Amersham 1966
43. Recommendations Internat. Comm. Radiological Protection, Rep. Committee 4: Radionuclide Release into the Environment: Assessment of Doses to Man. ICRP Publication 29. Ann. ICRP *2* (2), 1 (1979)
44. Atomic Energy Office: Accident at Windscale No. 1 Pile on October 10, 1957, Command 302; Her Majesty's Stationery Office: London 1957
45. Gustafson, P.F., Miller, J.E.: Health Phys. *16,* 167 (1969)
46. Recommendations Nat. Council on Radiation Protection and Measurements: Cesium-137 from the Environment to Man: Metabolism and Dose. NCRP Rep. No. 52; NCRP: Washington, D.C. 1977; p. 13
47. Häsänen, E., Miettinen, J.K.: Nature *200,* 1018 (1963)
48. Goodman, G.T., Roberts, T.M.: Nature *231,* 287 (1971)

49. Apparent per Capita Disappearance of Food in Canada. Catalogue 32226; Statistics Canada: Ottawa, Ontario 1975
50. Spector, W.S. (ed.): Handbook of Biological Data; W.B. Saunders Co.: Philadelphia, Pennsylvania 1956
51. Kirchmann, R. et al.: Etude du cycle biologique parcouru par la radioactivité. BLG-477; Centre d'Etude de l'Energie Nucléaire: Mol, Belgique 1973
52. Bennett, B.G. in: Transuranium Nuclides in the Environment. STI/PUB/410; Internat. Atomic Energy Agency: Vienna 1976; pp. 367–383
53. Hetherington, J.A. in: Miller, M.W., Stannard, J.N. (eds.): Environmental Toxicity of Aquatic Radionuclides: Models and Mechanisms; Ann Arbor Science Publishers Inc.: Ann Arbor, Michigan 1976; pp. 81–106
54. Aarkrog, A.: Health Phys. *32,* 271 (1977)
55. Gofman, J.W.: J. Am. Med. Assoc. *236,* 284 (1976)
56. Campbell, E.E. et al. in: Annual Report of the Biomedical and Environmental Research Program of the LASL Health Division, Jan.–Dec. 1973. LA-5633-PR; Los Alamos Scientific Laboratory: Los Alamos, New Mexico 1974; pp. 27–33
57. Noshkin, V.E.: Health Phys. *22,* 537 (1972)
58. Cuddihy, R.G. et al. in: Transuranium Nuclides in the Environment. STI/PUB/410; Internat. Atomic Energy Agency: Vienna 1976; pp. 657–669
59. Barr, N.F. in: Transuranium Nuclides in the Environment. STI/PUB/410; Internat. Atomic Energy Agency: Vienna 1976; pp. 649–656
60. Cohen, B.L.: Health Phys. *32,* 359 (1977)
61. Larsen, R.P., Oldham, R.D.: Health Phys. *35,* 918 (1978)
62. Sullivan, M.F. et al.: Radiat. Res. *80,* 116 (1979)
63. Weeks, M.H. et al.: Radiat. Res. *4,* 339 (1956)

Subject Index

GPSR Compliance

The European Union's (EU) General Product Safety Regulation (GPSR) is a set of rules that requires consumer products to be safe and our obligations to ensure this.

If you have any concerns about our products, you can contact us on ProductSafety@springernature.com

In case Publisher is established outside the EU, the EU authorized representative is:

Springer Nature Customer Service Center GmbH
Europaplatz 3
69115 Heidelberg, Germany

Batch number: 10620881

Printed by Printforce, the Netherlands